ANALYSIS

ANALYSIS

Editor

Krista Lenz

scitus
academics

Analysis

Edited by **Krista Lenz**

This Book is Printed on Acid-Free Paper. ♾

ISBN: 9781681179568

Library of Congress Control Number: 2017949090

SCITUS Academics LLC,

616, Corporate Way, Suite 2, 4766,

Valley Cottage, NY 10989

www.scitusacademics.com

Notice

Printed in United States of America.

Preface

Analysis is the process of breaking a complex topic or substance into smaller parts in order to gain a better understanding of it. The technique has been applied in the study of mathematics and logic since before Aristotle (384–322 B.C.), though analysis as a formal concept is a relatively recent development. Mathematical analysis includes a very large part of mathematics. The part of mathematics in which functions and their generalizations are studied by the method of limits. The concept of limit is closely connected with that of an infinitesimal quantity, therefore it could be said that mathematical analysis studies functions and their generalizations by infinitesimal methods. These theories are usually studied in the context of real and complex numbers and functions. Analysis evolved from calculus, which involves the elementary concepts and techniques of analysis. Analysis may be distinguished from geometry; however, it can be applied to any space of mathematical objects that has a definition of nearness (a topological space) or specific distances between objects.

This book presents studies that discuss several mathematical analysis methods and their respective applications. It deals with recent developments in and the latest research on mathematics, statistics and their applications. It covers a broad spectrum of modern harmonic analysis, functional differential equations, complex analysis, special functions, function spaces, summability theory, Fourier and wavelet analysis, and numerical analysis of great interest to the research community. All chapters are contributed by eminent academics, scientists, researchers and scholars in their respective fields, hailing from around the world. This book will be of valuable to young researchers and professionals whose work involves the use of mathematical analysis.

Table of Contents

CHAPTER 1

Mathematical Analysis of Financial Model on Market Price with Stochastic Volatility

Mitun Kumar Mondal[1], Md. Abdul Alim[2], Md. Faizur Rahman[3], Md. Haider Ali Biswas[1]

[1]*Mathematics Discipline, Science Engineering and Technology School, Khulna University, Khulna, Bangladesh*

[2]*Department of Mathematics, University of Chittagong, Chittagong, Bangladesh*

[3]*South East Engineering College, University of Rajshahi, Rajshahi, Bangladesh*

ABSTRACT

The Heston model is one of the most popular stochastic volatility models for option pricing to measure the volatility of different parameters in the financial market. In this work, we study the statistical analysis of Heston Model by partial differential equations. The model proposed by Heston takes into account non-lognormal distribution of the assets returns, leverage effect and the important mean reverting property of volatility. We have assayed on the return distribution on the basis of different values of correlation parameter and volatility, then we measure the effects of parameters ρ (correlation coefficient) and σ (standard deviation) for different situation such as $\rho > 0$, $\sigma > 0$, $\rho = 0$, $\sigma = 0$, $\rho < 0$, $\sigma < 0$ etc. On return distribution of Heston Model which indicates market condition for buyers and sellers to buy and sell options. All solvers used in this analysis are implemented using MATLAB codes and the simulation results are presented graphically.

Keywords: Stochastic Volatility, Black Scholes Biases, Heston Model, Black-Scholes Equation, Calibration, Characteristic Functions

1. INTRODUCTION

In 1970, Fischer Black, Myron Scholes and Robert Merton derived the "Black- Scholes model" (sometimes known as "Black-Scholes-Merton") which changed the way and impact the world of pricing derivatives using stocks as the underlying asset [1] [2]. For this great contribution to study and analyze the financial market, Scholes and Merton were awarded the Nobel Prize in economics in 1997 (Fischer Black died in 1995). It is then easy to understand the significant impact and necessity of this formula in studying the market price in the economics and finance. Since then many researchers have been aptly used to describe Black and Scholes' model to option pricing theory. Despite subsequent development of option theory, the original Black-Scholes (BS) formula for a European call option remains the most successful and widely used application. This formula is particularly useful because it relates the distribution of spot returns to the cross-sectional properties of option prices.

Another great obtainment in the financial market is Heston's stochastic volatility model, which helps to resolve a shortcoming of the BS model. More precisely, we can say that models based on BS assume that the underlying volatility is constant over the life of the derivative and unaffected by the changes in the price level of the underlying security. Howsoever, these models cannot explain long-observed features of the implied volatility surface like as volatility smile and skew, which indicate that implied volatility does tend to vary with respect to strike price and expiry. By assuming that the volatility of the underlying price is a stochastic process rather than a constant, it becomes possible to model derivatives more accurately.

Now-a-days, Heston model is considered as one of the most popular stochastic volatility option pricing models, which is motivated by the widespread evidence that volatility is stochastic and that the distribution of risky asset returns has tail (s) longer than that of a normal distribution [3]. A Stochastic volatility model which has correlation of price and volatility innovations can address both empirical stylized facts. The Stochastic volatility option pricing model was developed by the contributions of many researchers. Among them, Johnson and Shanno [4] showed the option pricing for the changing of variance, Wiggins [5] presented a numerical solution of the call option valuation problem given a fairly general continuous stochastic process for return volatility, Hull and White gave an analysis of the bias in option pricing caused by a stochastic volatility in [6] and the pricing of options on assets with stochastic volatilities in [7]. Scott [8] developed an option pricing model when the variance changes randomly. Stein and Stein studied the stock price distributions that arise when prices follow a diffusion process with a stochastically varying volatility parameter in [9] and Heston showed a closed solution for the options with stochastic volatility in [3]. Also a semi-closed form solution was derived based on characteristic function of the price distribution [3]. All of the authors used a mathematical model to explain the option pricing model because of mathematical modeling can play a significant role to discuss different types of real phenomena which lead to design better prediction. As a result many authors use mathematical modeling in different spheres. For example, Biswas et al. [10] studied the potential impacts of Global Climate Change in Bangladesh. Mondal and Biswas [11] developed a mathematical model to describe the transmission of Nipah virus between bats and human. Neilan and Lenhart [12] showed the application optimal control strategy in disease modelling. Biswas and Haque [13] discussed the necessity of nonlinear dynamical system to control the infectious disease. We also refer readers to [1] [14] [15] [16] and the references within for the study of more applications of mathematical modeling in real life problems.

The real issue was whether the ideal behavior was defensible in the derivation of this formula, Heston made several simplifying assumptions [3]. The theoretical models may tell us, ultimately the prices of financial instruments such as options are determined by the market. One of the key pricing parameters in the Heston model is the volatility or the expected standard deviation of the underlying assets assuming that asset price follows a geometric Brownian motion with constant volatility consequently the options on the same asset, but with different strike prices and maturity dates. Since this number varies across different options with the same underlying, and when plotted on a graph against the strike price of these options formed smile characteristic. But this does not appear acceptable as the volatility should only depend on the underlying asset. This event suggests that the original Black-Scholes model is not structurally able to accurately describe option prices and their dynamics. Prior solution can be categorized in two situations where the assumption made by Black-Scholes that volatility be constant had been relaxed in both cases. Heston models add an additional source of randomness to the model. On the other hand, local volatility models make use of a surface which describes the volatility over different maturities and underlying asset prices. Only key solution of local volatility is to determine local volatility function by calibration observed market prices. The famous Heston model has been intensively studied and used as the foundation for almost any option pricing formula in today's financial markets. But we introduced an interpolators approach called finite difference. Finite difference is a function of the distance of the point to the origins which are particularly useful in interpolating data and are widely used in both pattern recognition and surface investigation. This investigation helps investor to know price options more accurately. It focuses on the reconstruction of unknown functions from known data by scientific way. Generally finite difference method is multivariate and they may be solutions of partial differential equations satisfying certain additional conditions. However, the reconstruction of multivariate functions from data can only be done if the space furnishing the trial functions and makes function sets as interpolators. Finite difference methods also provide interpolation of function values given at irregularly positioned points for any value of observation index.

The main goal of this paper is to measure true volatility and experimental volatility as well as compare them to examine the present situation of the share market. We want to measure the effect of different parameters of Heston model on return distribution and effect on implied volatility surface with strike and maturity as well as discuss the pricing through the Partial Differential Equation.

This work is actually a statistical analysis of option price Heston model and a calculation of different error measurement to test the consistency of different functions.

2. BLACK-SCHOLES EQUATION

Black and Scholes [17] first proposed this equation bringing a huge change in the financial market, and it was the first time when people knew how to make a price for an option. The model was developed under the following assumptions:

A1. The stock price follows the stochastic process *ds*=□*sdt*+□*sdw* with fixed and;

A2. Unrestricted short-selling of stock, with full use of short-sale proceeds;

A3. No transactions costs and taxes;

A4. No dividends are paid during the life of the option;

A5. There are no riskless arbitrage opportunities:

A6. It is based on European options;

A7. The risk-free rate of interest r is constant and same for all maturities;

A8. Continuous trading;

In order to make a price for a call option on a non-dividend paying stock with the BS Equation, we need to know current stock price, strike price, risk-free interest rate, volatility and time to maturity. It is easy to get all above inputs variables in the market except the volatility. For the price of a non-dividend paying call option, the BS equation is described as:

$$C(S,t) = SN(d_1) - Ke^{-r(T-t)}N(d_2)$$

where, $d_1 = \frac{\ln\left(\frac{S}{k}\right) + \left(r + \frac{\sigma^2}{2}\right)(T-t)}{\sigma\sqrt{(T-t)}}$ and $d_2 = d_1 - \sigma\sqrt{(T-t)}$

Here, S is the stock price at time t, T is the maturity date, K is the strike price, $N(d_2)$is the cumulative normal distribution, σ;is the volatility. Although Black-Scholes equation is still widespread used in the market, much evidence has shown that the assumption of fixed volatility is not suitable for actual data. Consequently, in this dissertation, we consider the volatility following a stochastic process rather than a constant during the life of a call option.

3. THE HESTON STOCHASTIC VOLATILITY MODEL

The crude assumption of constant volatility in the Black-Scholes formula causes problem. One model where the volatility is a stochastic process is the Heston Stochastic Volatility Model [3] , which is an extended version of the Black- Scholes stochastic differential equations (SDE) with a volatility. The Heston Model takes the non-log normal distribution of the assets returns and the leverage effect into account, the correlation between the two Wiener processes. The Heston Model defined by following stochastic processes:

$$dS_t = \mu S_t dt + \sqrt{V_t} S_t dW_t^1 \qquad (1)$$

$$dV_t = \kappa(\theta - V_t)dt + \sigma\sqrt{V_t}dW_t^2 \qquad (2)$$

and where W_t^1 and W_t^2 are correlated Wiener processes with ρ, i.e.

$$dW_t^1 dW_t^2 = \rho dt \tag{3}$$

where, μ is the drift coefficient of the stock price, θ is the long term mean of variance, k is the rate of mean reversion, σ;is the volatility of volatility, s_t and v_t are the price and volatility process respectively, where $\{v_t\}_{t\geq 0}$ is a square root mean reverting process, first used by with long run mean, θ and rate of reversion k. To take into account the leverage effect, stock returns and implied volatility are negatively correlated, W_t^1 and W_t^2 are correlated wiener process and the correlation coefficients is ρ. All the parameters μ, k, θ, σ, ρ are the time and state homogeneous.

4. ANALYSIS OF HESTON MODEL

4.1. Partial Differential Equation Approach of Heston Model

For some stochastic volatility models, one can find a partial differential equation (PDE), the value of any option must be satisfied by such a PDE. For Heston's Stochastic Volatility model, a PDE exists, but calculation is quite complicated due to the difficult estimation of the market price of volatility risk. In order to price options in the SV model, we can apply no-arbitrage arguments, or use the risk-neutral valuation method. First we discuss the no-arbitrage method. The riskless portfolio is constructed as in the Black-Scholes model. But the construction method is different..In the SV option pricing model, there is only one traded risky asset S but two random sources dW_t^1 and dW_t^2. So the market is incomplete. We cannot perfectly replicate the option solely with the underlying stock. No-arbitrage arguments are not enough to give the option price. We need additional assumptions. In the following derivation, equilibrium arguments are also employed. We know that the market can be completed by adding any option written on stock S. Simply, the market is complete when we have two traded assets, the underlying asset S and a benchmark option V_1. Then all other options can be replicated by these two traded assets.

To proceed, consider a self-financing/risk-less portfolio with value ⍰ consisting of an option with value $V(S,v\ t\ )$which we want to price, - Δ units of the underlying asset S and, in order to hedge the risk associated with the random volatility, - $\Delta_1$units of benchmark option with value$V_1(S,V,t)$.

Hence,

$$\Pi = V - \Delta S - \Delta_1 V_1 \tag{4}$$

The portfolio is self-financing i.e. for risk-less portfolio, so that

$$d\prod = dV - \Delta dS - \Delta_1 dV_1 \tag{5}$$

By applying two dimensional form of Ito's formula, we have

$$\begin{aligned} d\prod = &\left\{\frac{\partial V}{\partial t} + \frac{1}{2}vS^2\frac{\partial^2 V}{\partial S^2} + \rho\sigma Sv\frac{\partial^2 V}{\partial S\partial v} + \frac{1}{2}v\sigma^2\frac{\partial^2 V}{\partial v^2}\right\}dt + \frac{\partial V}{\partial S}dS + \frac{\partial V}{\partial v}dv \\ &-\Delta_1\left\{\frac{\partial V_1}{\partial t} + \frac{1}{2}vS^2\frac{\partial^2 V_1}{\partial S^2} + \rho\sigma Sv\frac{\partial^2 V_1}{\partial S\partial v} + \frac{1}{2}v\sigma^2\frac{\partial^2 V_1}{\partial v^2}\right\} - \Delta_1\frac{\partial V_1}{\partial S}dS \\ &-\Delta_1\frac{\partial V_1}{\partial v}dv - \Delta dS \end{aligned} \tag{6}$$

Now, we can rewrite it by collecting the terms of *dS*, *dt* and *dv*

$$\begin{aligned} d\prod = &\left\{\frac{\partial V}{\partial t} + \frac{1}{2}vS^2\frac{\partial^2 V}{\partial S^2} + \rho\sigma Sv\frac{\partial^2 V}{\partial S\partial v} + \frac{1}{2}v\sigma^2\frac{\partial^2 V}{\partial v^2}\right\}dt + \left(\frac{\partial V}{\partial v} - \Delta_1\frac{\partial V_1}{\partial v}\right)dv \\ &-\Delta_1\left\{\frac{\partial V_1}{\partial t} + \frac{1}{2}vS^2\frac{\partial^2 V_1}{\partial S^2} + \rho\sigma Sv\frac{\partial^2 V_1}{\partial S\partial v} + \frac{1}{2}v\sigma^2\frac{\partial^2 V_1}{\partial v^2}\right\}dt \\ &+\left(\frac{\partial V}{\partial S} - \Delta_1\frac{\partial V_1}{\partial S} - \Delta\right)dS \end{aligned} \tag{7}$$

To make the portfolio riskless, we choose

$$\frac{\partial V}{\partial v} - \Delta_1\frac{\partial V_1}{\partial v} = 0 \tag{8}$$

$$\frac{\partial V}{\partial S} - \Delta_1\frac{\partial V_1}{\partial S} - \Delta = 0 \tag{9}$$

To eliminate the terms *dS* and *dv*, we solve the Equation (8) and (9) as

$$\Delta_1 = \frac{\frac{\partial V}{\partial v}}{\frac{\partial V_1}{\partial v}} \tag{10}$$

$$\Delta = \frac{\partial V}{\partial S} - \Delta_1 \frac{\partial V_1}{\partial S} = \frac{\partial V}{\partial S} - \frac{\frac{\partial V}{\partial v}}{\frac{\partial V_1}{\partial v}} \frac{\partial V_1}{\partial S} \tag{11}$$

The portfolio is risk free if we rebalance the Equation (7) according to (10) and (11). On the other hand, the riskless portfolio must earn a risk free rate, i.e. the return of this risk-free portfolio must equal the (deterministic) risk-free rate of return. Otherwise, there would be an arbitrage opportunity.

$$d\prod = \left\{ \frac{\partial V}{\partial t} + \frac{1}{2} vS^2 \frac{\partial^2 V}{\partial S^2} + \rho\sigma Sv \frac{\partial^2 V}{\partial S \partial v} + \frac{1}{2} v\sigma^2 \frac{\partial^2 V}{\partial v^2} \right\} dt$$

$$-\Delta_1 \left\{ \frac{\partial V_1}{\partial t} + \frac{1}{2} vS^2 \frac{\partial^2 V_1}{\partial S^2} + \rho\sigma Sv \frac{\partial^2 V_1}{\partial S \partial v} + \frac{1}{2} v\sigma^2 \frac{\partial^2 V_1}{\partial v^2} \right\} dt$$

$= r\prod dt = r\left(V - \Delta S - \Delta_1 V_1\right) dt$ (by using Equation (7))

$$\therefore\ d\prod = r\left(V - \Delta S - \Delta_1 V_1\right) dt \tag{12}$$

Using above two equations we have,

$$\frac{\frac{\partial V}{\partial t} + \frac{1}{2} vS^2 \frac{\partial^2 V}{\partial S^2} + \rho\sigma Sv \frac{\partial^2 V}{\partial S \partial v} + \frac{1}{2} v\sigma^2 \frac{\partial^2 V}{\partial v^2} + rS \frac{\partial V}{\partial S} - rV}{\frac{\partial V}{\partial v}}$$

$$= \frac{\frac{\partial V_1}{\partial t} + \frac{1}{2} vS^2 \frac{\partial^2 V_1}{\partial S^2} + \rho\sigma Sv \frac{\partial^2 V_1}{\partial S \partial v} + \frac{1}{2} v\sigma^2 \frac{\partial^2 V_1}{\partial v^2} + rS \frac{\partial V_1}{\partial S} - rV_1}{\frac{\partial V_1}{\partial v}} \tag{13}$$

Notice that the left-hand side is a function of V only and the right-hand side is a function of V_1 only. The only way that this equation holds is that both sides are equal to some function, i.e. g only depends on the independent variables *s,v* and *t*. Setting $g = \left(\kappa(\theta - V) - \Lambda(S,v,t)\sigma\sqrt{V}\right)$ yields a special case of a so-called affine diffusion process. For this class of processes, the pricing PDE is tractable analytically. In this case we have,

$$\frac{\partial V}{\partial t}+\frac{1}{2}vS^2\frac{\partial^2 V}{\partial S^2}+\rho\sigma Sv\frac{\partial^2 V}{\partial S\partial v}+\frac{1}{2}v\sigma^2\frac{\partial^2 V}{\partial v^2}+rS\frac{\partial V}{\partial S}-rV$$
$$=\left(\kappa(\theta-V)-\Lambda(S,v,t)\sigma\sqrt{V}\right)\frac{\partial V}{\partial v} \tag{14}$$

Now replacing *V* by *U(S,V,t,T)* and *v* by *V*, we have

$$\frac{1}{2}VS^2\frac{\partial^2 U}{\partial S^2}+\rho\sigma SV\frac{\partial^2 U}{\partial S\partial V}+\frac{1}{2}V\sigma^2\frac{\partial^2 U}{\partial V^2}+rS\frac{\partial U}{\partial S}-rU$$
$$+\frac{\partial U}{\partial t}+\left(\kappa(\theta-V)-\Lambda(S,v,t)\sigma\sqrt{V}\right)\frac{\partial U}{\partial V}=0 \tag{15}$$

$\Lambda(s,v\ t)$ is called the market price of volatility risk. According to Heston's assumption, the market price of volatility risk is proportional to volatility i.e. to the square root of the variance.

$$\Lambda(S,v,t)\ \alpha\ \sqrt{V} \Rightarrow \Lambda(S,v,t)=k\sqrt{V}\ , \tag{16}$$

where *k* is the proportional constant.

Multiplying both sides of Equation (16) by $\sigma\sqrt{V}$, then we have,

$$\Lambda(S,v,t)\sigma\sqrt{V}=k\sigma V$$

Thus Equation (16) becomes,

$$\frac{1}{2}VS^2\frac{\partial^2 U}{\partial S^2}+\rho\sigma SV\frac{\partial^2 U}{\partial S\partial V}+\frac{1}{2}V\sigma^2\frac{\partial^2 U}{\partial V^2}+rS\frac{\partial U}{\partial S}-rU$$
$$+\frac{\partial U}{\partial t}+\left(\kappa(\theta-V)-\lambda V\right)\frac{\partial U}{\partial V}=0 \tag{17}$$

Therefore $\lambda(s,\ v\ t)$ represents the market price of volatility risk. The price of volatility risk is independent of particular asset. It can be obtained theoretically from any asset depending on volatility risk. $\lambda(s,\ v\ t)$ Assume the strike price to be *K* and expiring time *T*. The price is considered in rectangular area of $[0,\infty]\times[\ 0,\infty]$ and on horizontal time$[0,T]$. Then For European call option the option price obeys Equation (4.14) with boundary,

$$\left.\begin{aligned}
&U(S,v,t)=\max(0,S-k)\\
&U(0,v,t)=0\\
&\frac{\partial U}{\partial S}(\infty,v,t)=1\\
&rS\frac{\partial U}{\partial S}(S,0,t)+\kappa\theta\frac{\partial U}{\partial V}(S,0,t)-rU(S,0,t)+\frac{\partial U}{\partial t}(S,0,t)=0\\
&U(S,\infty,t)=S
\end{aligned}\right\} \tag{18}$$

This choice of market price of volatility risk gives us analytical advantages. The drift term of the specified process (4) is an affine function of the state variable itself. The affinity makes the model easier to solve. Since the diffusion of the variance process is also proportional to the square root of the variance, the product of the market price of risk and the diffusion is proportional to variance itself. As a result, the drift term will remain affine under the Equivalent Martingale Measure (EMM). This particular market price of volatility risk helps the model to have a closed-form solution. We can also apply the risk-neutral valuation method to the SV model. The market is incomplete. But it is still free of arbitrage. The equivalent martingale measure is not unique. We have to choose one of all these measures to price the options. So the price of the option is also not unique. It will depend on which equivalent martingale measure we use.

4.2. The Effect of Parameters on Return Distribution

There are many economic, empirical, and mathematical reasons for choosing a model with such a form for a detailed statistical/ empirical analysis).Empirical studies have shown that an asset's log-return distribution is non-Gaussian. It is characterized by heavy tails and high peaks (leptokurtic). There is also empirical evidence and economic arguments that suggest that equity returns and implied volatility are negatively correlated (also termed 'the leverage effect'). This departure from normality plagues the Black-Scholes-Merton model with many problems. In this work, we will show effect of effects of ρ and σ on return distribution.

4.3. Effects of ρ on Return Distribution

Consider that ρ denotes the correlating factor between the sources of randomness for the underlying and the volatility. ρ can be interpreted as the correlation between the returns and the volatility of the asset.

Therefore it captures the leverage effect, affecting the heaviness of the tails, thus the skewness of the return distribution. Intuitively, if $\rho < 0$, then volatility will increase as the asset price return decreases, this will spread the left tail and squeeze the right tail of the distribution creating a fat left-tailed distribution. Conversely, if $\rho > 0$, then volatility will increase as the asset price/return increases. This will spread the right tail and squeeze the left tail of the distribution creating a fat right-tailed distribution and if $\rho = 0$ the skewness is close to zero. As a result ρ, affects the skewness of the distribution. Figures 1-5 show the effect ρ for different values. However, Figures 4-12 are similar to those presented in [18] , but in this study we have investigated the behaviors of the volatility by changing the values of the parameter ρ and σ. The simulation results have been shown in Figures 4-12 respectively.

4.4. Effects of σ on Return Distribution

The σ affects to the kurtosis (peak) of the distribution. When σ is zero the volatility is deterministic, because the diffusion process in dV_t will be dropped and hence the returns will be normally distributed as in the BSM-model. Increasing σ will increase the peak (kurtosis), creating heavy tails on both sides, i.e. the increase in σ represents the market volatility is more volatile and higher σ shows higher peaks than less one.

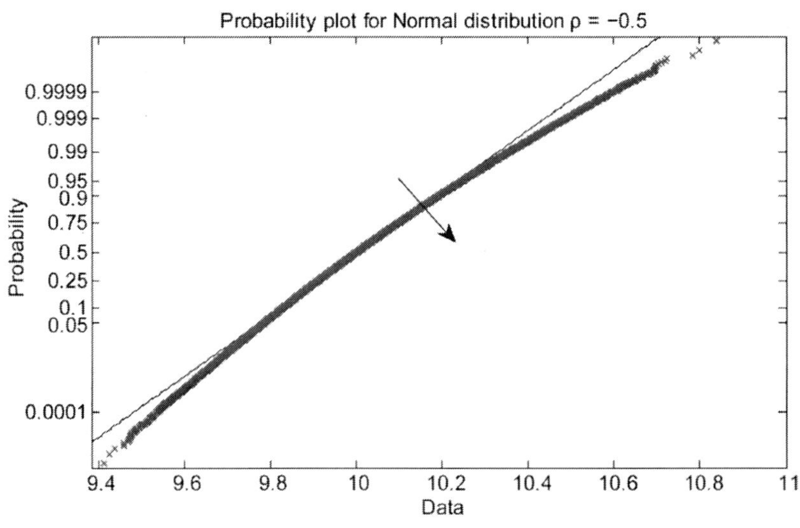

Figure 1. Probability plot, when ρ is smaller than zero asset returns is left skewed.

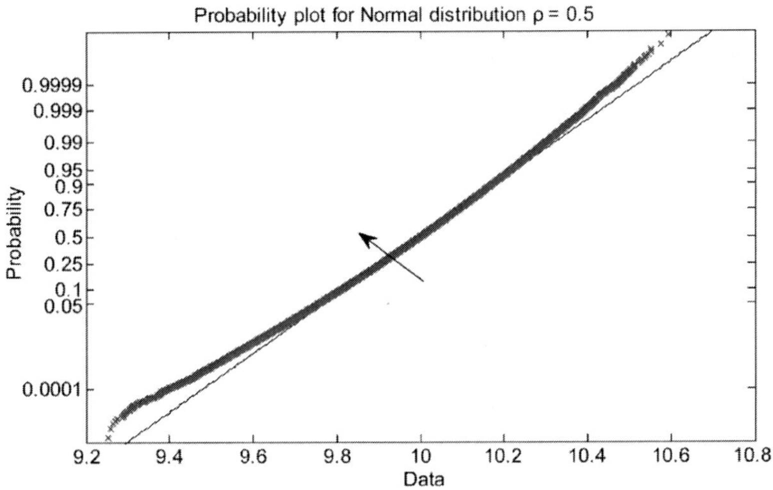

Figure 2. Probability plot, when ρ is greater than zero asset returns is right skewed.

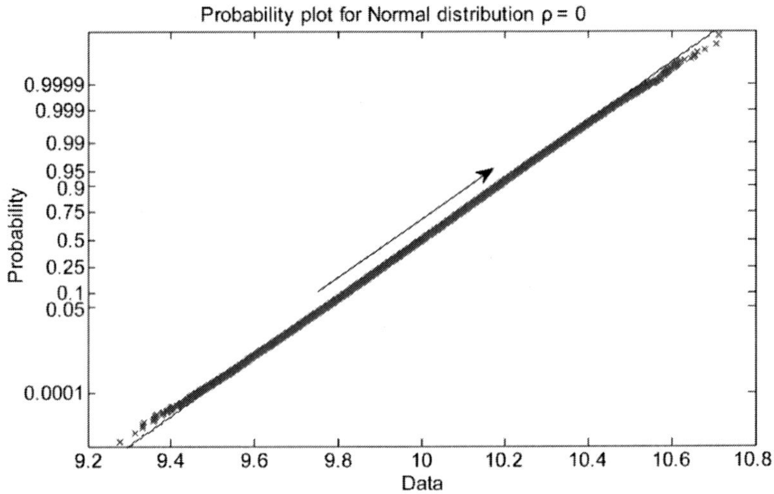

Figure 3. Probability plot, when ρ is equal to zero asset returns is close to zero.

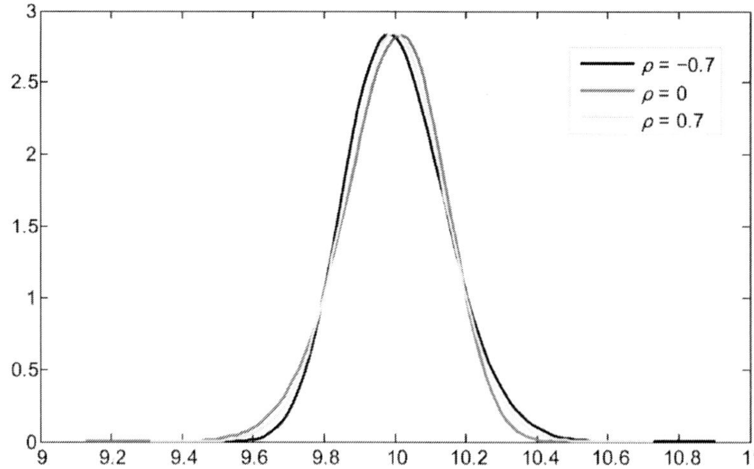

Figure 4. The effect of ρ on the skewness of the density function.

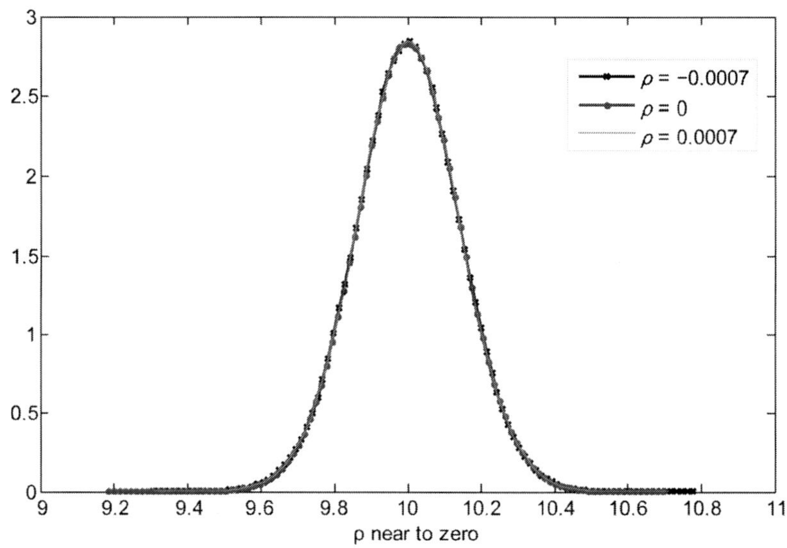

Figure 5. The effect of ρ on the skewness of the density function near to zero.

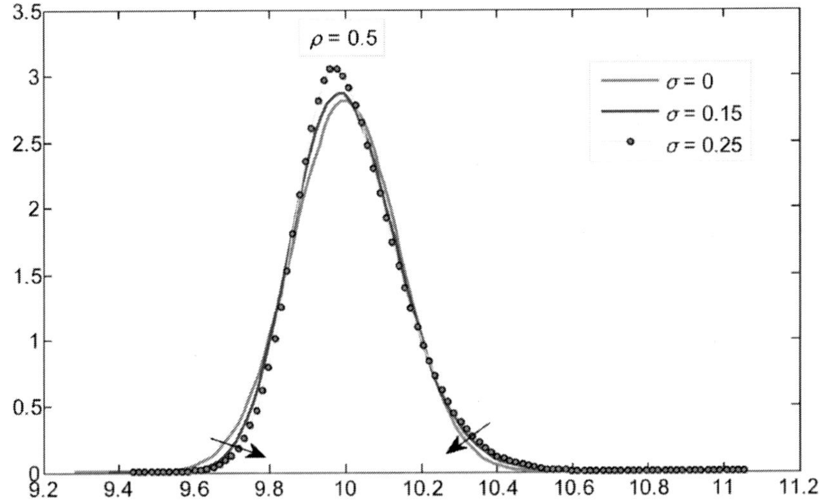

Figure 6. The effect of σ on the kurtosis of the density function.

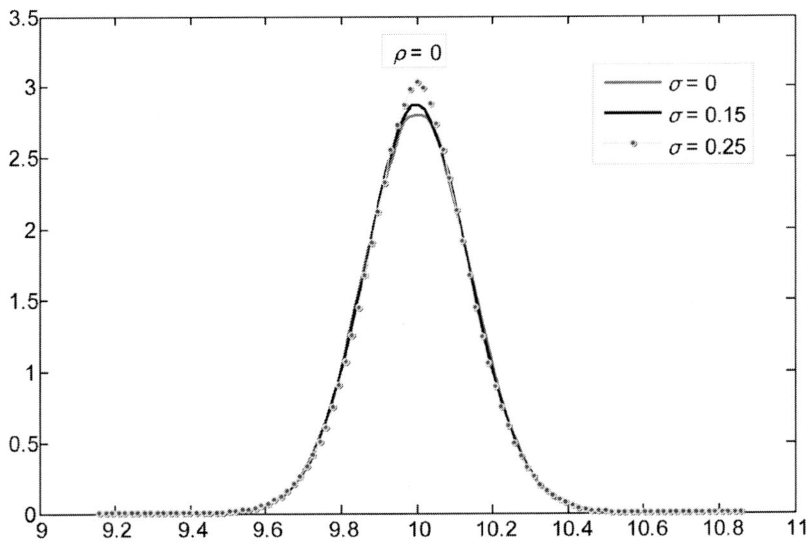

Figure 7. The effect of σ on the kurtosis of the density function to zero.

In Figures 8-10, we investigate the effect of ρ on the implied volatility surface generated under Heston's model. Here Maturity (years), Strike and Implied Volatility are denoted by x-label, y-label and z-label respectively. In Figure 8 and Figure 9, ρ = 0.007 and ρ = 0. Away from the-money options have higher implied volatilities than near-the-money options. This is consistent with the 'smile' shape of implied volatilities in some financial market, e.g. currency options markets. This observation can be explained by the fat-tailed distribution of returns. We can also find that the "smile" flattens. When the time to maturity increases. This is also consistent with the real financial markets. The BS model tends to work well for options with long maturities as a result of the corresponding flattened smile.

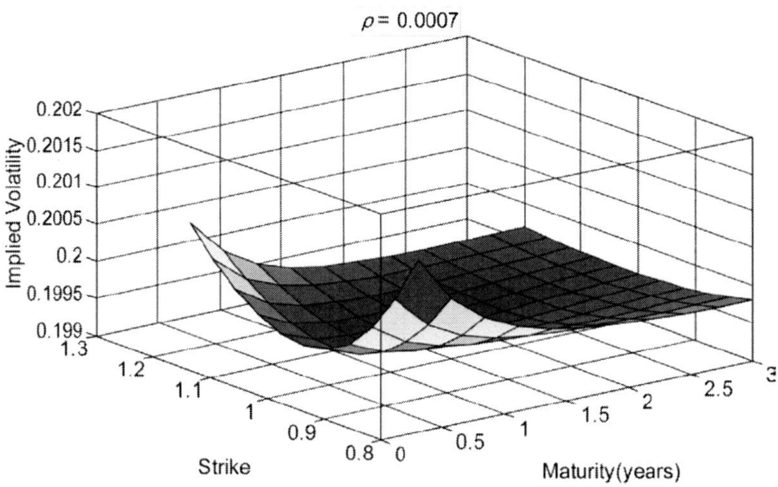

Figure 8. Implied Volatility; ρ = 0.0007; κ = 2; σ = 0.01; θ = 0.04; v_0 = 0.04; r = 1%; s = 1 strikes 0.06 - 1.2; maturities: 0.4 - 3.0 years.

In Figure 11, ρ is negative. We can find that in-the-money calls have higher implied volatilities, whereas out-of-the-money calls have lower implied volatilities. This is consistent with the phenomenon of 'volatility skew' in some financial markets, especially the equity options markets.

In Figure 12, ρ is positive which can generate an opposite skew shape compared with Figure 11such kind of volatility skew may appear in energy options markets.

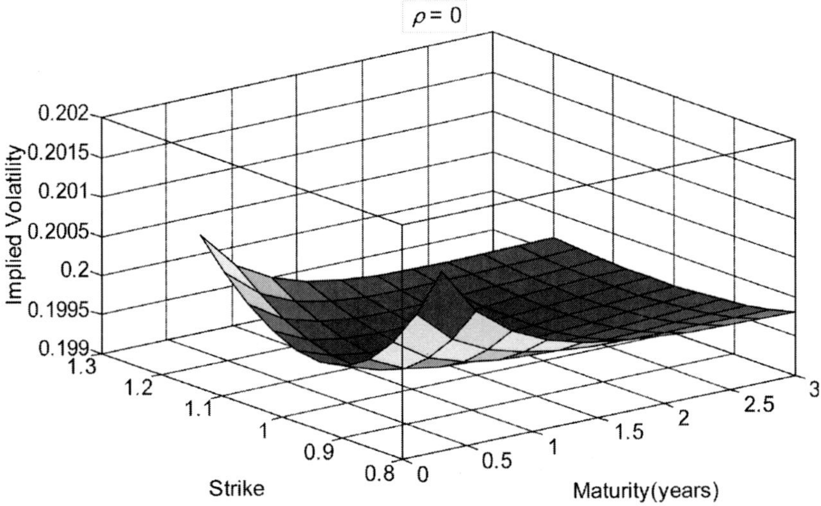

Figure 9. Implied Volatility; $\rho = 0$; $\kappa = 2$; $\sigma = 0.01$; $\theta = 0.04$; $v_0 = 0.04$; $r = 1\%$; $s = 1$ strikes 0.06 - 1.2; maturities: 0.4 - 3.0 years.

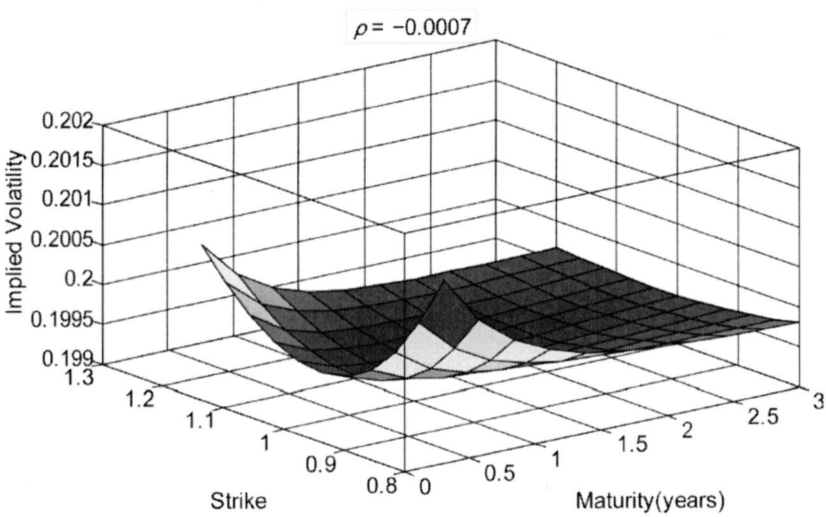

Figure 10. Implied Volatility; $\rho = -0.0007$; $\kappa = 2$; $\sigma = 0.01$; $\theta = 0.04$; $v_0 = 0.04$; $r = 1\%$; $s = 1$ strikes 0.06 - 1.2; maturities: 0.4 - 3.0 years.

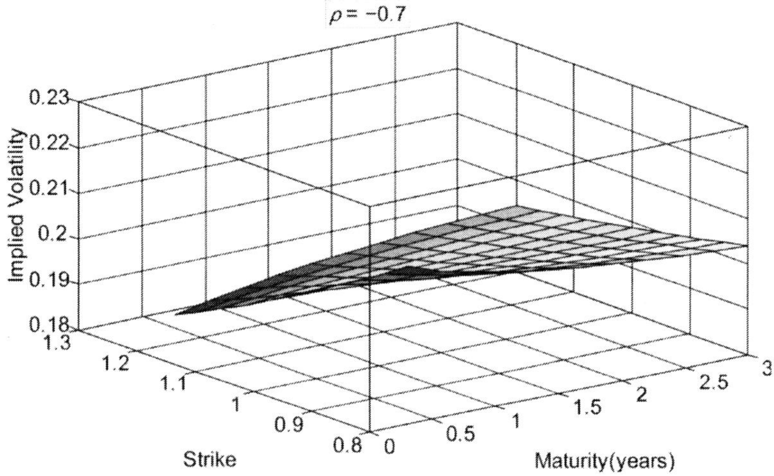

Figure 11. Implied Volatility; $\rho = -0.7$; $\kappa = 2$; $\sigma = 0.01$; $\theta = 0.04$; $v_0 = 0.04$; $r =$ 1%; $s = 1$ strikes 0.06 - 1.2; maturities: 0.4 - 3.0 years.

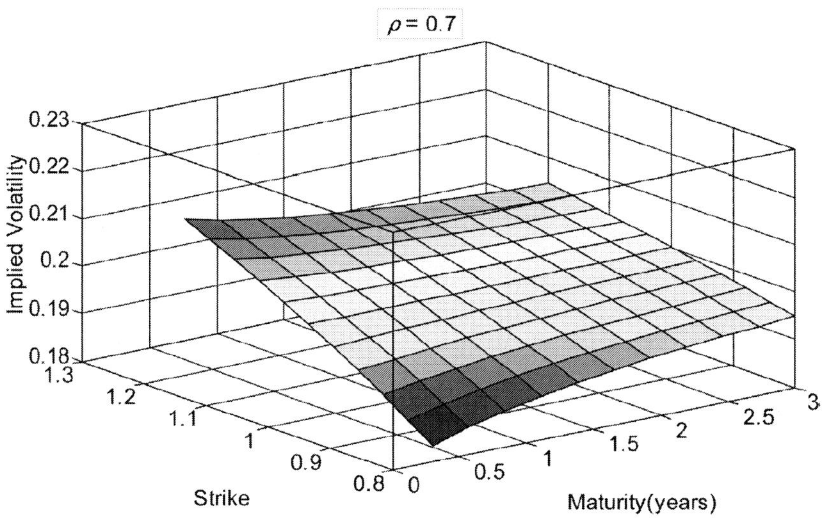

Figure 12. Implied Volatility; $\rho = 0.7$; $\kappa = 2$; $\sigma = 0.01$; $\theta = 0.04$; $v_0 = 0.04$; $r =$ 1%; $s = 1$ strikes 0.06 - 1.2; maturities: 0.4 - 3.0 years.

5. CONCLUSIONS

Stochastic volatility models tackle one of the most restrictive hypotheses of the Black-Scholes model framework, which assumes that volatility remains constant during the option's life. However, by observing financial markets it becomes apparent that volatility may change dramatically in short time periods and its behavior is clearly not deterministic. Among stochastic volatility models, the Heston model presents two main advantages. First, it models an evolution of the underlying asset which can take into account the asymmetry and excess kurtosis that are typically observed (and expected) in financial asset returns. Second, it provides closed form solutions for the pricing of European options.

The study made in this paper demonstrated a technique for constructing smile and skew consistent prices by violating one of the crude assumptions in the BS model, constant volatility. The result shows that the Heston approximation works really well and only face big problems when options with high time to maturity are to be priced. Another problem is that the approximation gives us incorrect prices when the moneyless is below one. To reduce this problem further studies of the volatility smile could be done and were the skew of options that are not in the money could be compare to options that are in the money and trying to repair this. As one could observe from the results above is that the Heston approximation loses its accuracy as the time to maturity increases, but Black and Scholes is also facing the same type of problem. Since the Heston model was not built on the assumption on non-constant volatility, it showed an improvement of modeling stocks and receiving smile consistent option prices.

ACKNOWLEDGEMENTS

The authors would like to express sincere thanks to anonymous reviewers for helpful comments on an earlier version of this paper, and they are also grateful to an author assistant of Scientific Research Publishing and an editorial assistant of Journal of Mathematical Finance.

REFERENCES

1. Biswas, M.H.A. (2014) On the Evolution of AIDS/HIV Treatment: An Optimal Control Approach. Current HIV Research, 12, 1-12. https://doi.org/10.2174/ 1570162 X1201140716094638
2. Hull, J. and John, C. (2008) Options Futures and Other Derivatives. 7th Edition, Pearson Prentice Hall, Upper Saddle River

3. Heston, S.L. (1993) A Closed Solution for Options with Stochastic Volatility, with Application to Bond and Currency Options. Review of Financial Studies, 6, 327-343. https://doi.org/10.1093/rfs/6.2.327
4. Johnson, H. and Shanno, D. (1987) Option Pricing When the Variance Is Changing. Journal of Financial and Quantitative Analysis, 22, 143-151. https://doi.org/10.2307/2330709
5. Wiggins, J.B. (1987) Option Values under Stochastic Volatility: Theory and Empirical Estimates. Journal of Financial Economics, 19, 351-372
6. Hull, J. and White, A. (1988) An Analysis of the Bias in Option Pricing Caused by a Stochastic Volatility. Advances in Futures and Options Research, 3, 27-61
7. Hull, J. and White, A. (1987) The Pricing of Options on Assets with Stochastic Volatilities. Journal of Finance, 42, 281-300. https://doi.org/10.1111/j.1540-6261.1987.tb02568.x
8. Scott, L.O. (1987) Option Pricing When the Variance Changes Randomly: Theory, Estimation and an Application. Journal of Financial and Quantitative Analysis, 22, 419-438. https://doi.org/10.2307/2330793
9. Stein, E. and Stein, C.J. (1991) Stock Priced Distributions with Stochastic Volatility: An Analytical Approach. Review of Financial Studies, 4, 727-752. https://doi.org/10.1093/rfs/4.4.727
10. Biswas, M.H.A., Rahman, T. and Haque, N. (2016) Modeling the Potential Impacts of Global Climate Change in Bangladesh: An Optimal Control Approach. Journal of Fundamental Applied Science, 8, 1-19. https://doi.org/10.4314/jfas.v8i1.1
11. Mondal, M.K. and Biswas, M.H.A. (2015) Modeling the Transmission Dynamics of the Nipah Virus Infection. Proceedings of 1st International Conference on Mathematics and It's Applications, Khulna, 23 December 2015, 22-26
12. Neilan, R.M. and Lenhart, S. (2010) An Introduction to Optimal Control with an Application in Disease Modeling. DIMACS Series in Discrete Mathematics, 75, 67-81. https://doi.org/10.1090/dimacs/075/03
13. Biswas, M.H.A. and Haque, M.M. (2016) Nonlinear Dynamical Systems in Modeling and Control of Infectious Diseases. In: Pinelas, S., Doslá, Z., Dosly, O. and Kloeden, P., Eds., Differential and Difference Equations with Applications, Springer Proceedings in Mathematics & Statistics, Vol. 164, Springer, Cham., 149-158. https://doi.org/10.1007/978-3-319-32857-7_15
14. Berezansky, L., Idels, L. and Kipnis, M. (2011) Mathematical Model of Marine Protected Areas. IMA Journal of Applied Mathematics, 76, 312-325. https://doi.org/ 10.1093/imamat/hxq043
15. Biswas, M.H.A. (2014) Optimal Control of Nipah Virus (NiV) Infections: A Bangladesh Scenario. Journal of Pure and Applied Mathematics: Advances and Applications, 12, 77-104
16. Murray, J.D. (1989) Mathematical Biology: Biomathematics. 2nd Edition, Springer, New York. https://doi.org/10.1007/978-3-662-08539-4
17. Black, F. and Scholes, M. (1973) Valuation of Options and Corporate Liabilities. Journal of Political Economy, 81, 637-654. https://doi.org/10.1086/260062
18. Moodley, N. (2005) The Heston Model: A Practical Approach with Matlab Code. B.Sc. Thesis, University of the Witwatersrand, Johannesburg

CHAPTER 2

The Analysis of Real Data Using a Stochastic Dynamical System Able to Model Spiky Prices*#

Lorella Fatone[1], Francesca Mariani[2], Maria Cristina Recchioni[3], Francesco Zirilli[4]

[1]*Dipartimento di Matematica e Informatica, Università di Camerino, Camerino, Italy*

[2]*CERI—Centro di Ricerca "Previsione, Prevenzione e Controllo dei Rischi Geologici", Università di Roma "La Sapienza", Roma, Italy*

[3]*Dipartimento di Scienze Sociali "D. Serrani", Università Politecnica delle Marche, Ancona, Italy*

[4]*Dipartimento di Matematica "G. Castelnuovo", Università di Roma "La Sapienza", Roma, Italy*

ABSTRACT

In this paper we use filtering and maximum likelihood methods to solve a calibration problem for a stochastic dynamical system used to model spiky asset prices. The data used in the calibration problem are the observations at discrete times of the asset price. The model considered has been introduced by V. A. Kholodnyi in [1,2] and describes spiky asset prices as the product of two independent stochastic processes: the spike process and the process that represents the asset prices in absence of spikes. A Markov chain is used to regulate the transitions between presence and absence of spikes. As suggested in [3] in a different context the calibration problem for this model is translated in a maximum likelihood problem with the likelihood function defined through the solution of a filtering problem. The estimated values of the model parameters are the coordinates of a constrained maximizer of the likelihood function. Furthermore, given the calibrated model, we develop a sort of tracking procedure able to forecast forward asset prices. Numerical examples using synthetic and real data of the solution of the calibration problem and of the performance of the tracking procedure are presented. The real data studied are electric power price data taken from the UK electricity market in the years 2004-2009. After calibrating the model using the spot prices, the forward prices forecasted with the tracking procedure and the observed forward prices are compared. This comparison can be seen as a way to validate the model, the formulation and the solution of the calibration problem and the forecasting procedure. The result of the comparison is satisfactory. In thewebsite: http://www.econ.univpm.it/recchioni/finance/w10 some auxiliary material including animations that helps the understanding of this paper is shown. A more general reference to the work of the authors and of their coauthors in mathematical finance is the website: http://www. econ.univpm.it/recchioni/finance.

Keywords: stochastic dynamical system; inverse problem; spiky prices; data time series analysis

1. INTRODUCTION

A spike in an asset price is an abrupt movement of the price followed by an abrupt movement of approximately the same magnitude in the opposite direction. The modeling of spikes in asset prices is a key problem in finance. In fact spiky prices are encountered in several contexts such as, for example, in electric power prices and, more in general, in commodity prices.

In this paper the model introduced by V. A. Kholodnyi in [1,2] to describe spiky prices is combined with some ideas introduced by the authors and some coauthors to study calibration problems in mathematical finance, see [3-9]. That is we introduce a stochastic dynamical system to model spikes in asset prices and we study a calibration problem for the dynamical system introduced. The method proposed to solve the calibration problem is tested doing the analysis of data time series. We consider synthetic and real data. The real data studied are electric power price data taken from the UK electricity market in the years 2004-2009.

Following V. A. Kholodnyi (see [1,2]) we model spiky asset prices as a stochastic process that can be represented as the product of two independent stochastic processes: the spike process and the process that describes the asset prices in absence of spikes. The spike process models spikes in asset prices and it is either equal to the multiplicative amplitude of the spike during the spike periods or equal to one during the regular periods, that is during the periods between spikes. The second stochastic process of the Kholodnyi model describes prices in absence of spikes. This last process has been chosen as a diffusion process. Finally we use a two-state Markov chain in continuous time to determine whether asset prices are in the spike state, that is during a spike, or in the regular state, that is between spikes.

The model for spiky asset prices studied depends on five real parameters. Two of them come from the process that describes the asset prices in absence of spikes, one of them comes from the spike process and the last two parameters come from the two-state Markov chain used to model the transitions between spike and regular states.

The calibration problem studied consists in estimating these five parameters from the knowledge at discrete times of the asset prices (observations of the spiky prices). That is the calibration problem is a parameter identification problem or, more in general, is an inverse problem for the stochastic dynamical system that models the asset prices. This calibration problem is translated in a constrained optimization problem for a likelihood function (maximum likelihood problem) with the likelihood function defined through the solution of a filtering problem. The likelihood function is defined using the probability density function associated with the diffusion process modeling asset prices in absence of spikes. This formulation of the calibration problem is inspired to the one introduced in [3] in the study of the Heston stochastic volatility model that has been later extended to the study of several other calibration problems in mathematical finance (see [4,5,8, 9]).

The filtering and the maximum likelihood problems mentioned above are solved numerically. The resulting numerical solution of the calibration problem determines the values of the (unknown) parameters that make most likely the observations actually made. Note that in the processing of numerical data to improve the robustness and the quality of the solution of the calibration problem some preliminary steps are introduced in the optimization procedure used to solve the calibration problem and the results obtained in these preliminary steps are used to penalize the likelihood function obtained from the filtering problem. That is the maximum likelihood problem originally formulated in analogy to [3] is reformulated adding penalization terms to the likelihood function and choosing an ad hoc initial guess for the optimization procedure to improve the robustness and the quality of its solution. This reformulated optimization problem is solved numerically using a method based on a variable metric steepest ascent method.

Furthermore, as a byproduct of the solution of the filtering problem, we develop a tracking procedure that, given the calibrated model, is able to forecast forward asset prices.

The method used to solve the calibration problem and the tracking procedure are used to analyze data time series. Numerical examples of the solution of the calibration problem and of the performance of the tracking procedure using synthetic and real data are presented. The synthetic data are obtained computing one trajectory of the stochastic dynamical system that models spiky asset prices. We generate daily synthetic data for a period of two years. The first year of data is generated with one choice of the model parameters, the second year of data is generated with a different choice of the model parameters. The second year of data is generated using as initial point of the trajectory the last point of the first year of data. In the solution of the calibration problem we choose as observation period a period of one year, that is we use as data the daily observations corresponding to a time period of one year, and we move one day at the time this observation period through the two years of data. The calibration problem is solved for each choice of the observation period. The two choices of the model parameters used to generate the data and the time when the model parameters change value are reconstructed satisfactorily by the calibration procedure. The real data studied are daily electric power price data taken from the UK. electricity market. These electric power price data are spiky data. We choose the data of the calibration problems considered as done above in the study of synthetic data extracting the observation periods from a time series of five years (i.e. the years 2004-2009) of daily electric power (spot) price data taken from the UK market. The results obtained show that the model is able to establish a stable relationship between the data time series and the estimated model parameter values. Note that in the real data time series for each observation day we have the electric power spot price and the associated forward prices observed that day for a variety of delivery periods. That is for each spot price there is a set of forward prices associated to it corresponding to different delivery periods. Moreover in the calibration problem only spot prices are used as data. To exploit this fact we proceed as follows. After calibrating the model using as data the spot prices observed in the first three years of the data time series, we use the calibrated model, the tracking procedure and the spot prices not used in the calibration to forecast the forward prices associated to these last spot prices. We compare the forward prices forecasted with the tracking procedure with the observed forward prices. The comparison is satisfactory and establishes the effectiveness of the model, the validity of the proposed formulation and solution of the calibration problem and the quality of the forecasted prices.

We note that the model studied is too simplistic to be of practical value in electricity markets. In fact our model is able to capture only one property of the electric power prices: the presence of spikes. It does not consider, for example, the mean-reverting property and the presence of weekly and season cycles in electricity prices. This study aims to be a first attempt to solve, with the strategy presented in Section 3, calibration problems involving stochastic dynamical systems that can be used to describe electric power prices. That is the methodology discussed in this paper can be applied in the calibration of more sophisticated stochastic dynamical models that can be used in electricity markets (see for example [10]).

The website: http://www.econ.univpm.it/recchioni/finance/w10 contains some auxiliary material including some animations that helps the understanding of this paper. A general reference to the work of the authors and of their coauthors in mathematical finance is the website: http://www.econ.univpm.it/recchioni/finance.

The paper is organized as follows. In Section 2 we describe the model for spiky asset prices. In Section 3 we formulate and we solve the filtering and the calibration problems for the model presented in Section 2 and we introduce the tracking procedure used to forecast forward prices. In Section 4 some numerical examples of the solution of the calibration problem and of the performance of the tracking procedure introduced in Section 3 using synthetic and real data are presented.

2. THE STOCHASTIC DYNAMICAL SYSTEM FOR SPIKY ASSET PRICES

Let us introduce a stochastic dynamical system used to model spiky asset prices, see [1,2]. In this model the spiky prices are defined as the product of two stochastic processes: the spike process and the process that describes asset prices in absence of spikes. As said in the Introduction the transitions between spike and regular states are regulated through a two state Markov chain in continuous time.

Let t be a real variable that denotes time, $t \geq 0$, M_t be the (continuous time) two-state (i.e. regular state, spike state) Markov chain and let

$$P(T,t) = \begin{pmatrix} P_{ss}(T,t) & P_{sr}(T,t) \\ P_{rs}(T,t) & P_{rr}(T,t) \end{pmatrix}, \quad 0 < t \leq T \tag{1}$$

be its 2 × 2 transition probability matrix. Note that in (1) $P_{ss}(T, t)$ and $P_{rs}(T, t)$ denote the transition probabilities of going from the spike state at time t respectively to the spike or to the regular state at time T and similarly $P_{sr}(T, t)$ and $P_{rr}(T, t)$ denote the transition probabilities of going from the regular state at time t respectively to the spike or to the regular state at time T.

Note that the Chapman-Kolmogorov equation for the Markov chain M_t, $t \geq 0$ can be written as follows:

$$P(T,t) = P(T,\tau)P(\tau,t),$$
$$\text{for every } t,\, T,\, \tau \text{ such that } 0 < t \leq \tau \leq T \tag{2}$$

together with the condition that $P(T,t)$ is the 2 × 2 identity matrix when $t = T$

Let 0 < t ≤ T and $p_r(t)$ $p_s(t)$, $p_r(T)$, $p_s(T)$ be respectively the probabilities of being in the regular state and in the spike state at time t and at time T. We have:

$$\begin{pmatrix} p_s(T) \\ p_r(T) \end{pmatrix} = P(T,t)\begin{pmatrix} p_s(t) \\ p_r(t) \end{pmatrix}, \quad 0 < t \leq T \tag{3}$$

In the time-homogeneous case it is possible to write an explicit formula for the matrix $P(T,t)$, t ≤ T as a function of the parameters that define the generator of $P(T,t)$, t ≤ T These parameters control the duration and the frequency of the spikes, that is they control the expected lifetime of the spike periods and the expected lifetime of the periods between spikes. It is easy to see that in the case of a time-homogeneous Markov chain M_t, $t \geq 0$, the transition probability matrix $P(T,t)$, $0 < t \leq T$ is a function only of the difference $T - t$ and that we have [1]:

$$P(T,t) = \begin{pmatrix} \dfrac{b + ae^{-(T-t)(a+b)}}{a+b} & \dfrac{b - be^{-(T-t)(a+b)}}{a+b} \\ \dfrac{a - ae^{-(T-t)(a+b)}}{a+b} & \dfrac{a + be^{-(T-t)(a+b)}}{a+b} \end{pmatrix}, \quad 0 < t \leq T \tag{4}$$

where a and b are non negative real parameters. Moreover we assume that a, b are not both zero.

From now on in our model for spiky prices we always assume that the two-state Markov chain M_t, $t \geq 0$ is time-homogeneous and, as a consequence, we assume that its transition probability matrix $P(T,t)$, $0 < t \leq T$ is given by (4).

The process that describes the asset prices in absence of spikes is a diffusion process defined through the following stochastic differential equation:

$$d\hat{S}_t = \mu\hat{S}_t dt + \sigma\hat{S}_t dW_t, \quad t > 0 \tag{5}$$

with the initial condition:

$$\hat{S}_0 = \hat{S}^* \tag{6}$$

where $\tilde{S}_t > 0$ denotes the random variable that describes the asset price in absence of spikes at time t, $t > 0$, μ, σ are constants, $t > 0, W_t$ is the standard Wiener process, $t > 0$, dW_t, $W_0 = 0$ is its stochastic differential and the random variable $\hat{S}^* > 0$ is a given initial condition. For simplicity we assume that the random variable $\hat{S}^*$ is concentrated in a point with probability one and, with abuse of notation, we denote this point with $\hat{S}^*$. Recall that μ is the drift coefficient and $\sigma > 0$ is the volatility coefficient.

Equation (5) defines the asset price dynamics of the celebrated Black-Scholes model. Note that the solution of (5), (6) is a Markov process known as geometric Brownian motion. Note that several other Markov processes different from the one defined in (5), (6) can be used to model asset price dynamics in absence of spikes. For example we can use one of the so called stochastic volatility models that have been introduced recently in mathematical finance to refine the Black-Scholes asset dynamics model.

Let us define the spike process λt, $t \geq 0$ that models the amplitude of the spikes in the asset prices. Let ξ_t, $t \geq 0$ be a stochastic process made of independent positive random variables with given probability density functions $\sum(t, \xi)$, $\xi > 0$, $t \geq 0$. We assume that if the Markov chain M_t, $t \geq 0$ is in the regular state then the spike process λ_t, $t \geq 0$ is equal to one. If the Markov chain M_t, $t \geq 0$ transits from the regular state to the spike state at time τ, $\tau > 0$ then the spike process λ_t, $t \geq 0$ is equal to a value sampled from the random variable ξ_r during the entire time period beginning with $t = \tau$ that the Markov chain M_t, $t \geq 0$ remains in the spike state. We assume that M_t, $t \geq 0$ is in the regular state at time $t = 0$ with probability one, so that the spike process λ_t starts with $\lambda_0 = 1$ with probability one. Let us observe that in the case of spikes with constant amplitude $\lambda > 1$ the probability density function $\Sigma(t,\xi) = \Sigma(\xi)$, $\xi > 0$, $t \geq 0$ is the Dirac delta function $\delta(\xi - \lambda)$ with support in $\xi = \lambda$ that is we have:

$$\Sigma(t,\xi)=\delta(\xi-\lambda),\ \xi>0,\ t\geq 0 \tag{7}$$

Finally we say that the spike process λ_t, $t \geq 0$ is in the spike state or in the regular state if the Markov chain M_t, $t \geq 0$ is in the spike state or in the regular state respectively. In our model for spiky prices we assume that the spikes have constant amplitude $\lambda > 1$ that is we assume (7).

Let us define now the process that describes the asset prices with spikes. Let us denote with $S_t > 0$ the random variable that models the price (eventually) with spikes of the asset at time $t \geq 0$. We assume that the spike process λt, $t \geq 0$ and the process $\hat{S}_t$, $t \geq 0$ that describes asset prices in absence of spikes, are independent. Following [1,2] we define the process S_t, $t \geq 0$ as the product of the spike process λ_t, $t \geq 0$ and of the process $\hat{S}_t$, $t \geq 0$ that is:

$$S_t=\lambda_t\hat{S}_t,\ t\geq 0 \tag{8}$$

Note that the process S_t, $t \geq 0$ for spiky asset prices is in the spike state or in the regular state depending from the fact that the spike process λ_t, $t \geq 0$ is in the spike state or in the regular state respectively, or equivalently, depending from the fact that the Markov chain M_t, $t \geq 0$ is in the spike state or in the regular state respectively.

In [1] it has been shown that the process S_t, $t > 0$ is able to reproduce spikes in asset prices. This can be seen in **Figure 1** where synthetic data sampled from S_t, $t > 0$ are shown. These synthetic data are those used in the numerical examples presented in Section 4. Note that depending on the values of the parameters defining the model, the trajectories of S_t, $t > 0$ can exhibit a variety of behaviours. In particular, when M_t, $t \geq 0$ is time-homogeneous and $a \neq 0$, $b \neq 0$ the expected times $\bar{t}_s$ and $\bar{t}_s$ spent by the process S_t, $t > 0$ in "spike mode" and in "regular (i.e. inter-spike) mode", respectively, are given by (see [1]):

$$\overline{t_s}=\frac{1}{a},\ \overline{t_r}=\frac{1}{b} \tag{9}$$

If $\bar{t}_s$ is small in comparison with the characteristic time of change of the process $\hat{S}_t$, $t > 0$ then the process $\hat{S}_t$, $t > 0$ exhibits spikes. For example, if $\hat{S}_t$, $t > 0$ is the diffusion process defined in (5) then the previous condition can be stated as follows:

$$\sigma^2 \bar{t}_s = \frac{\sigma^2}{a} \ll 1 \text{ and } \mu \bar{t}_s = \frac{\mu}{a} \ll 1 \tag{10}$$

We interpret $\bar{t}_s$ as the expected lifetime of a spike, and $\bar{t}_r$ as the expected time elapsed between two consecutive spikes. That is the parameters *a* and *b* control the duration and the frequency of the spikes through (9). For example, the relations (9) suggest that in order to model short-lived spikes the parameter *a* must be large, while to model rare spikes the parameter *b* must be small.

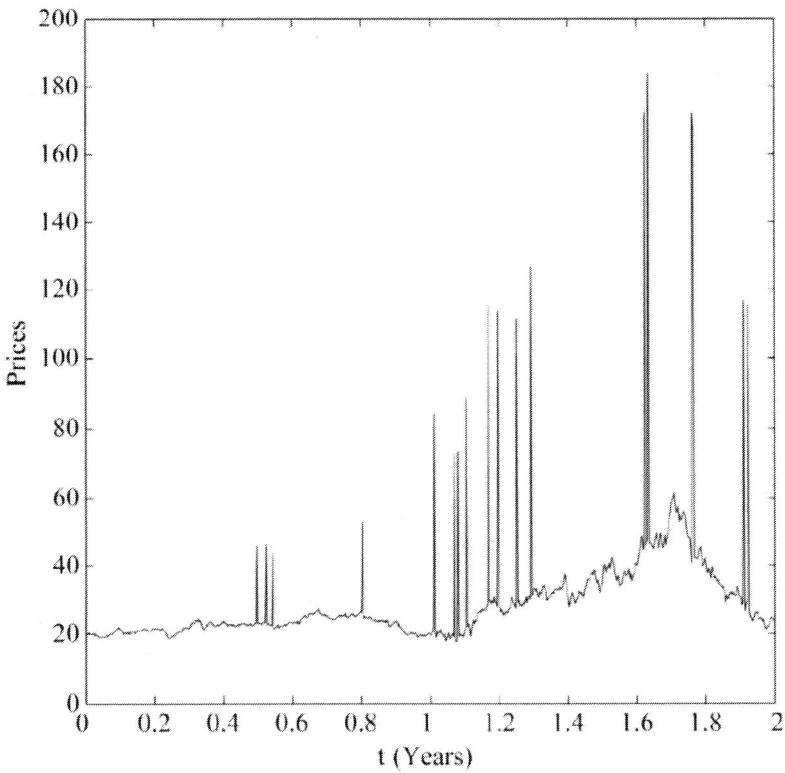

Figure 1. The synthetic data.

Using some heuristic rules to recognize spikes on the data time series considered thank to their relation with the spike duration and frequency a rough estimate of the parameters *a* and *b* can be obtained directly from observed spiky prices (market data). For example, the relations (9) suggest that initial estimates a^0 and b^0 of a and b respectively, can be obtained as follows:

$$a^0 = \frac{1}{\overline{t}_s^0}, \quad b^0 = \frac{1}{\overline{t}_r^0} \tag{11}$$

where $\overline{t}_s^0$ and $\overline{t}_r^0$ are computed (using some heuristic rules) directly from the data available as the observed average lifetime of the spikes and the observed average time between two consecutive spikes, respectively.

Finally it is easy to see that the asymptotic probabilities of the time-homogeneous Markov chain M_t, $t \geq 0$ of being in the spike and in the regular states are respectively:

$$p_s(\infty) = \pi_s = \frac{b}{a+b}, \quad p_r(\infty) = \pi_r = \frac{a}{a+b} \tag{12}$$

With the previous choices the model for spiky asset prices introduced in [1,2] depends on five parameters, that is: μ, σ, a, b, λ. The parameters μ and σ are those appearing in (5) and are relative to the asset price model in absence of spikes, the remaining parameters are relative to the spike process (i.e. λ) and to the Markov chain that regulates the transitions between spike and regular states (i.e. a, b).

3. THE FILTERING AND THE CALIBRATION PROBLEMS

The parameters μ, σ, a, b, λ are grouped in the following vector:

$$\underline{\Theta} = (\mu, \sigma, a, b, \lambda)^T \tag{13}$$

where the superscript T denotes the transpose operator. The vector $\underline{\Theta}$ is the unknown of the calibration problem that must be determined from the data available. The vectors $\underline{\Theta}$, describing admissible sets of parameter values, must satisfy some constraints, that is we must have $\underline{\Theta} \in M$ where:

$$\mathcal{M} = \left\{ \underline{\Theta} = (\mu, \sigma, a, b, \lambda)^T \in \mathbb{R}^5 \middle| \sigma \geq 0, a \geq 0, b \geq 0, \lambda \geq 1 \right\} \tag{14}$$

and $\mathbb{R}^5$ is the five dimensional real Euclidean space.

The constraints contained in M express some elementary properties of the model introduced in Section 2. Note that, when necessary to take care of special circumstances, more constraints can be added to **M**.

The data of the calibration problem are the observation times: $0 = t_0 < t_1 < t_2 < \cdots < t_n < +\infty$, and the spiky asset price $\tilde{S}_{t_i}$ observed at time t_i, i = 0,1,...,n, where $\tilde{S}_0 = \hat{S}_0 = \hat{S}^*$ The observed price $\tilde{S}_i$ is a positive real number that is understood as being sampled from the random variable $S_{t_i}, \ i = 0,1,\cdots,n$ Starting from these data we want to solve the following problems:

Calibration Problem: Find an estimate of the vector $\underline{\Theta} = (\mu,\sigma,a,b,\lambda)^T$.

Filtering Problem (Forecasting Problem): Given the value of the vector $\underline{\Theta} = (\mu,\sigma,a,b,\lambda)^T$ estimated forecast the forward asset prices.

Note that the prices $\tilde{S}_j, \ j = 0,1,\cdots,n$ are called spot asset prices and that with forward asset prices we mean prices "in the future", that is, in our setting, we mean the forward prices at time t_i associated to the spot price $\tilde{S}_i, \ i = n$ These are the prices at time t_i of the asset with delivery time $t > t_i$, $i = n$ Let us point out that in correspondence of a spot price we can forecast several forward prices associated to different delivery periods. The delivery periods of interest are those actually used in real markets, in fact for these delivery periods observed forward prices are available and a comparison between observed forward prices and forecasted forward prices is possible. In Section 4 in the analysis of real data we consider (some of) the delivery periods used in the UK electricity market in the years 2004-2009 and we make this comparison.

Let us mention that the forecasted forward asset prices associated to the spiky price model introduced in Section 2 do not exhibit spikes while, of course, the corresponding (spot) asset prices exhibit spikes (see [1,2] for more details).

The calibration problem consists in estimating the value of the vector $\underline{\Theta}$ that makes "most likely" the observations. The observations available at time $t > 0$ are denoted with:

$$\mathcal{F}_t = \left\{\tilde{S}_i : t_i \le t\right\}, \quad t > 0 \tag{15}$$

Note that for simplicity we assume that the transitions from regular state to spike state or viceversa happen at the observation times.

Let $p(\hat{S},t|\mathcal{F}_t,\underline{\Theta}),\ \hat{S}>0,\ t>0$ be the probability density function of the stochastic process $\hat{S}_t$ at time $t > 0$ conditioned to the observations F_t and let $p_i(\hat{S},t|\underline{\Theta}) = p(\hat{S},t|\mathcal{F}_{t_i},\underline{\Theta}),\ \hat{S}>0,\ t_i<t\le t_{i+1}$ be the probability density function of the stochastic process $\hat{S}_t$ conditioned to the observations made up to time $t = t_i$, when $t_i<t\le t_{i+1},\ i=0,1,\cdots,n$ where for convenience we define $t_{n+1}=+\infty$.

The probability density functions $p_i(\hat{S},t|\underline{\Theta}),\ \hat{S}\ge 0\ t_i<t\le t_{i+1},\ \underline{\Theta}\in\mathcal{M},\ i=0,1,\cdots,n$ are the solutions of the following initial value problems for the Fokker-Planck equation associated to the Black-Scholes model (5):

for $i = 0,1,\ldots,n$,

$$\frac{\partial p_i}{\partial t}=\frac{1}{2}\sigma^2\hat{S}^2\frac{\partial^2 p_i}{\partial\hat{S}^2}+\mu\hat{S}\frac{\partial p_i}{\partial\hat{S}} \tag{16}$$

$$\hat{S}\ge 0,\quad t_i<t\le t_{i+1},\quad \underline{\Theta}\in\mathcal{M},$$

$$p_i(\hat{S},t_i|\underline{\Theta})=f_i(\hat{S};\underline{\Theta}),\quad \hat{S}\ge 0,\quad \underline{\Theta}\in\mathcal{M} \tag{17}$$

where

$$f_0(\hat{S};\underline{\Theta})=\delta(\hat{S}-\hat{S}^*),\quad \hat{S}>0,\quad \underline{\Theta}\in\mathcal{M} \tag{18}$$

$$f_i(\hat{S};\underline{\Theta})=p_r(t_i)\delta(\hat{S}-\tilde{S}_i)+p_s(t_i)\delta\left(\hat{S}-\frac{\tilde{S}_i}{\lambda}\right)$$

$$\hat{S}>0,\quad i=1,2,\cdots,n,\quad \underline{\Theta}\in\mathcal{M}, \tag{19}$$

where $p_r(t_i)$ and $p_s(t_i)$ are, respectively, the probabilities defined through the time-homogeneous Markov chain M_t, $t \ge 0$, of being in the regular and in the spike state at time $t=t_i,\ i=0,1,\cdots,n$ (see (3)).

The probability density functions p_i, $i = 0,1,\ldots,n$, solutions of the initial value problems for the FokkerPlanck equations (16)-(19), can be written as follows:

$$p_i(\hat{S},t|\underline{\Theta})=\int_0^{+\infty}d\hat{S}'p_f(\hat{S},t,\hat{S}',t_i)f_i(\hat{S}';\underline{\Theta})$$

$$\hat{S}\ge 0,\ t_i<t\le t_{i+1},\ i=0,1,\cdots,n,\ \underline{\Theta}\in\mathcal{M}, \tag{20}$$

where p_f is the fundamental solution of the FokkerPlanck equation (16) associated to the Black-Scholes model (5), that is:

$$p_f\left(\hat{S},t,\hat{S}',t'\right)=\frac{1}{\sigma\hat{S}'\sqrt{2\pi\left(t-t'\right)}}$$
$$\times e^{-\frac{\left[\log\left(\frac{\hat{S}}{\hat{S}'}\right)+\left(\mu-\frac{1}{2}\sigma^2\right)\left(t-t'\right)\right]^2}{2\sigma^2\left(t-t'\right)}}$$
$$\hat{S}\geq 0,\ \hat{S}'\geq 0,\ \ t,\ t'\geq 0,\ \ t-t'>0. \tag{21}$$

In order to measure the likelihood of the vector $\underline{\Theta}$ we introduce the following (log-)likelihood function:

$$F\left(\underline{\Theta}\right)=\sum_{i=0}^{n-1}\log p_i\left(\tilde{S}_{i+1},t_{i+1}\middle|\underline{\Theta}\right),\quad \underline{\Theta}\in\mathcal{M} \tag{22}$$

It is worthwhile to note that definition (22) contains an important simplification. In fact a more correct definition of the (log-)likelihood function should be:

$$\tilde{F}\left(\underline{\Theta}\right)=\sum_{i=0}^{n-1}\log p_i\left(\tilde{\tilde{S}}_{i+1},t_{i+1}\middle|\underline{\Theta}\right),\quad \underline{\Theta}\in\mathcal{M} \tag{23}$$

where $\tilde{\tilde{S}}_{i+1}=\tilde{S}_{i+1}$ or $\tilde{\tilde{S}}_{i+1}=\frac{\tilde{S}_{i+1}}{\lambda}$ depending on whether at time t_{i+1} the asset price is in the regular state or in the spike state respectively, $i=0,1,\cdots,n-1$. However, since when dealing with real financial data the decision about the character of the state (regular or spike state) of the observed prices is dubious, we prefer to adopt the definition (22) for the (log-)likelihood function. In fact the choice made in (22), at the price of introducing some inaccuracy, avoids the necessity of defining a (dubious) criterion to recognize regular and spike states in order to evaluate the (log-)likelihood function. The validity of this choice is supported by the fact that in the numerical experience shown in Section 4 the simplification introduced in (22) using $\tilde{S}_{i+1}$ instead of $\tilde{\tilde{S}}_{i+1}$, $i=0,1,\cdots,n-1$, is sufficient to obtain satisfactory results.

The solution of the calibration problem is given by the vector $\underline{\Theta}$ that solves the following optimization problem:

$$\max_{\underline{\Theta}\in\mathcal{M}} F\left(\underline{\Theta}\right) \tag{24}$$

Problem (24) is called maximum likelihood problem. In fact the vector $\underline{\Theta}^{*}\in\mathcal{M}$ solution of (24) is the vector that makes "most likely" the observations actually made. Problem (24) is an optimization problem with nonlinear objective function and linear inequality constraints.

Note that (22), (24) is one possible way of formulating the calibration problem considered using the maximum likelihood method. Many other formulations of the calibration problem are possible and legitimate. Moreover the formulation of the calibration problem (22), (24) can be easily extended to handle situations where we consider calibration problems associated to data set different from the one considered here, such as, for example, data set containing asset and option prices or only option prices.

The optimization algorithm used to solve the maximum likelihood problem (22), (24) is based on a variable metric steepest ascent method. The variable metric technique is used to handle the constraints.

Let $\|\cdot\|$ be the absolute value of ·, and $\|\cdot\|$ be the Euclidean norm of the vector ·, Beginning from an initial guess:

$$\underline{\Theta}^{0}=\left(\mu^{0},\ \sigma^{0},\ a^{0},\ b^{0},\ \lambda^{0}\right)^{T}\in\mathcal{M} \tag{25}$$

we update at every iteration the current approximation of the solution of the optimization problem (24) with a step in the direction of the gradient with respect to $\underline{\Theta}$ computed in a suitable variable metric of the (log-)likelihood function (22).

In particular let us fix a tolerance value *tol* > 0 and a maximum number of iterations *iter* > 0. Given $\underline{\Theta}=\underline{\Theta}^{0}\in\mathcal{M}$ the optimization procedure can be summarized in the following steps:

1) Set $k = 0$ and initialize $\underline{\Theta}=\underline{\Theta}^{0}$

2) Evaluate $F\left(\underline{\Theta}^{k}\right)$ If $k > 0$ and $\left|F\left(\underline{\Theta}^{k}\right)-F\left(\underline{\Theta}^{k-1}\right)\right|$ go to item 7;

3) Evaluate the gradient of the (log-)likelihood function in $\underline{\Theta}=\underline{\Theta}^{k}$ that is:

$$\nabla F\left(\underline{\Theta}^k\right)=\left(\frac{\partial F}{\partial \mu},\frac{\partial F}{\partial \sigma},\frac{\partial F}{\partial a},\frac{\partial F}{\partial b},\frac{\partial F}{\partial \lambda}\right)^T\left(\underline{\Theta}^k\right) \tag{26}$$

If If $\left\|\nabla F\left(\underline{\Theta}^k\right)\right\| < tol$ go to item 7;

4) Perform the steepest ascent step, evaluating $\underline{\Theta}^{k+1}=\underline{\Theta}^k+\eta_k \nabla F\left(\underline{\Theta}^k\right)$ where η_k is a quantity that is used to define the the step made. The quantity η_k can be chosen as a scalar or, more in general, as a matrix of suitable dimensions. The choice of η_k involves the use of the "variable metric". When η_k is chosen to be a scalar we have a "classical" steepest ascent method;

5) If $\left\|\underline{\Theta}^{k+1}-\underline{\Theta}^k\right\| < tol$ go to item 7;

6) Set $k = k + 1$ if $k < iter$ go to item 2;

7) Set $\underline{\Theta}^* = \underline{\Theta}^k$ and stop.

The vector $\underline{\Theta}^*$ obtained in step 7 is the (numerically computed) approximation of the maximizer of the (log-) likelihood function.

Numerical experiments have shown that the (log-) likelihood function $F(\underline{\Theta})$ defined in (22) is a flat function. That is there are many different vectors $\underline{\Theta}$ in **M** that make likely the data. In the optimization of objective functions with flat regions when local methods, such as the steepest ascent method, are used to solve the corresponding optimization problems (i.e.: (24)), special attention must be paid to the choice of the initial guess of the optimization procedure. That is in actual computations a "good" initial guess $\underline{\Theta}^0 \in \mathcal{M}$ is important to find a "good" solution of the optimization problem (24). More specifically in the problem considered here the use of good initial guesses of the volatility σ and of the drift μ improves substantially the quality of the estimates of all the parameters contained in the vector $\underline{\Theta}$. That is in the solution of problem (24) the parameters μ and σ are the most "sensitive" parameters of the vector $\underline{\Theta}$.

For these reasons in order to improve the robustness of the solution of problem (24) that is obtained with the optimization method described above it is useful to introduce some ad hoc steps that lead to a simple reformulation of the (log-)likelihood function (22), of the calibration problem (24) and of the method used to solve problem (24). First of all an ad hoc preliminary simplified optimization problem is solved to produce a high quality initial guess for the steepest ascent method summarized in steps 1) - 7). We proceed as follows. First we estimate two initial values of μ and σ, let us say $\bar{\mu}$ and $\bar{\sigma}$, directly from the data available, that is we compute the historical volatility and the historical drift of the data available (see, for example, [11]). These estimates are used as initial guesses to maximize the objective function (22) as a two variables function of the parameters μ and σ, with the constraint $\sigma > 0$ Note that in this preliminary step we consider as data used to define the (log-)likelihood function (22) only the "most regular portion of the input data". The most regular portion of the input data is chosen using elementary empirical rules that try to find one or more regular periods in the data considered. In this preliminary optimization problem the remaining components of the vector $\underline{\Theta}$ are initialized as follows $a = b = \bar{a}^0 = \bar{b}^0 = 0$ and $\lambda = \bar{\lambda}^0 = 1$ and these initial values are kept constants in the optimization procedure with respect to μ and σ. Let $(\bar{\mu}^*, \bar{\sigma}^*)$ be the maximizer found at the end of this preliminary step. In order to keep memory in the maximum likelihood problem (24) of the values $\bar{\mu}^*$ and $\bar{\sigma}^*$ found in the preliminary step we add to the objective function (22) the following penalization term:

$$k_1 \left|\mu - \bar{\mu}^*\right|^2 + k_2 \left|\sigma - \bar{\sigma}^*\right|^2, \; k_1 > 0, k_2 > 0 \tag{27}$$

where k_1, k_2 are positive penalization constants. That is we consider the following modified (log-)likelihood function:

$$\begin{aligned} \bar{F}(\underline{\Theta}) &= F(\underline{\Theta}) + k_1 \left|\mu - \bar{\mu}^*\right|^2 + k_2 \left|\sigma - \bar{\sigma}^*\right|^2 \\ &= \sum_{i=0}^{n-1} \log p_i \left(\tilde{S}_{i+1}, t_{i+1} \middle| \underline{\Theta}\right) \\ &\quad + k_1 \left|\mu - \bar{\mu}^*\right|^2 + k_2 \left|\sigma - \bar{\sigma}^*\right|^2, \\ k_1 > 0, \;\; & k_2 > 0, \;\; \underline{\Theta} \in \mathcal{M}. \end{aligned} \tag{28}$$

The variable metric steepest ascent method 1) - 7) is now applied to solve problem (24) when the (five variables) (log-)likelihood function $\bar{F}$ defined in (28) replaces the (log-)likelihood function F defined in (22) starting from the initial guess:

$$\underline{\Theta}^0 = \left(\bar{\mu}^*, \bar{\sigma}^*, a^0, b^0, \lambda^0\right)^T \in \mathcal{M} \tag{29}$$

where the initial values a^0 and b^0 are estimated directly from the input data through (11) and, finally, the initial value λ^0 is chosen as a rough estimate of the average amplitude of the spikes appearing in the input data. That is problem (24) is solved replacing F with $\bar{F}$ starting from an initial guess that uses the results obtained in the preliminary step of the optimization procedure and tries to exploit the data using the "physical" meaning of the model parameters.

These ad hoc steps used to reformulate problem (24), when applied to the analysis of synthetic data, lead to a substantial improvement in the accuracy of the estimates of the model parameters obtained when compared to the estimates obtained solving directly (22), (24) and to a great saving in the computational cost of solving the calibration problem. This reformulation of problem (24) is used to analyze the data considered in Section 4.

As a byproduct of the solution of the calibration problem we obtain a technique to forecast forward asset prices. Let us consider a filtering problem. We assume that the vector $\underline{\Theta}$ solution of the calibration problem associated to the data $\mathcal{F}_t = \{\tilde{S}_i : t_i \leq t\}, \quad t = t_n > 0$ is known. From the knowledge of $\underline{\Theta} = (\mu, \sigma, a, b, \lambda)^T$ at time $t = t_n$ we can forecast the forward asset prices with delivery period $\Delta t > 0$ deep in the future and delivery time $\tau_n = t_n + \Delta_t$ as follows:

$$\mathbb{E}\left(S_{\tau_i} \middle| \mathcal{F}_{t_n}, \underline{\Theta}\right) = \mathbb{E}\left(\lambda_{\tau_i}\right)\mathbb{E}\left(\hat{S}_{\tau_i}\right)$$
$$= \left(1 \cdot p_r(\tau_i) + \lambda \cdot p_s(\tau_i)\right)\mathbb{E}\left(\hat{S}_{t_i}\right)e^{\mu(\tau_i - t_i)}$$
$$\tau_i = t_i + \Delta t, \quad i = n, \tag{30}$$

where E(·) denotes the expected value of ·.

In the numerical experiments presented in Section 4 we use the following approximation:

$$\mathbb{E}\left(\hat{S}_{t_i}\right) \simeq \frac{1}{10}\sum_{k=0}^{9}\tilde{S}_{i-k}, \quad i = n \tag{31}$$

Note that using formula (31) we have implicitly assumed n > 9. In fact in the data time series considered in Section 4 the average in time of the (spiky) observations appearing in (31) gives a better approximation of the "spatial" average $\mathbb{E}\left(\hat{S}_{t_i}\right)$ of the price without spikes than the individual observation $\tilde{S}_i$ made at time $t = t_i$ of the spiky price. However the average in time of the observations approximates the "spatial" average only if short time periods are used. This is the reason why we limit the mean contained in (31) to the data corresponding to ten consecutive observation times, that is corresponding to a period of ten days when we process daily data as done in Section 4. Note that in (31) an average of spiky prices is used to approximate the expected value of non spiky prices.

4. NUMERICAL RESULTS

The first numerical experiment presented consists in solving the calibration problem discussed in Section 3 using synthetic data. This experiment does the analysis of a time series of daily data of the spiky asset price during a period of two years. The time series studied is made of 730 daily data of the spiky asset prices, that is the prices $\tilde{S}_i$ at time t_i, t_i, $i=0,1,\cdots,729$These data have been obtained computing one trajectory of the stochastic dynamical system used to model spiky asset prices defined in Section 2 looking at the computed trajectory at time $t=t_i$, $i=0,1,\cdots,729$ with $t_i-t_{i-1}=1/365$ where we have chosen t_0 = 0 and $\tilde{S}_0=\hat{S}_0=\hat{S}^*=20$. We choose the vector $\underline{\Theta}$ that specifies the model used to generate the data equal to $\underline{\Theta}=(\mu,\ \sigma,\ a,\ b,\ \lambda)^T=\underline{\Theta}_1=(0.03,\ 0.3,\ 70,\ 1,\ 2)^T$ in the first year of data (i.e. when $t=t_i$, $i=0,\ 1,\ \cdots,\ 364$), and equal to $\underline{\Theta}=(\mu,\ \sigma,\ a,\ b,\ \lambda)^T=\underline{\Theta}_2=(0.1,\ 0.8,\ 150,\ 5,\ 4)^T$ in the second year of data (i.e. when $t=t_i i=365,\ 366,\ \cdots,\ 729$. The data are generated using as initial value of the second year the last datum of the first year. The synthetic data generated in this way are shown in **Figure 1**. These data are spiky data and the fact that the first year of data is generated using a different choice of $\underline{\Theta}$ than the choice made in the second year of data can be seen simply looking at **Figure 1**.

We solve problem (24) with the ad hoc procedure described in Section 3 using the data associated to a time window made of 365 consecutive observation times, that is 365 days (one year), and we move this window across the two years of data discarding the datum corresponding to the first observation time of the window and inserting the datum corresponding to the next observation time after the window. Note that numerical experiments suggest that it is necessary to take a large window of observations to obtain a good estimate the parameters a, b and λ. The calibration problem is solved for each choice of the data time window applying the procedure described in Section 3, that is it is solved 365 times. The 365 vectors $\underline{\Theta}$ constructed in this way are compared to the two choices of the model parameters $\underline{\Theta}$ used to generate the data. Moreover the time when $\underline{\Theta}$ changes from being $\underline{\Theta}$ to being $\underline{\Theta}_2$ is reconstructed.

To represent the 365 vectors obtained solving the calibration problem in correspondence to the 365 data time windows considered, we associate to each reconstructed vector $\underline{\Theta}\in R^5$ a point on a straight line (see **Figure 2**). Let us explain how this correspondence is established. We first represent the vectors $\underline{\Theta}_1$ and $\underline{\Theta}_2$ that generate the data as two points on the straight line mentioned above having a distance proportional to $\|\underline{\Theta}_1-\underline{\Theta}_2\|^*$ measured in $\|\underline{\Theta}_1\|^*$ units, where $\|\underline{\Theta}\|^*=\frac{1}{5}\sum_{i=1}^{5}|\underline{\Theta}(i)|$ and $\underline{\Theta}=(\Theta(1),\Theta(2),\cdots,\Theta(5))^T\in R^5$. We choose the origin of the straight line to be the mid point of the segment that joins $\underline{\Theta}_1$ and $\underline{\Theta}_2$ In **Figure 2** the diamond represents the vector $\underline{\Theta}_1$ and the square represents the vector $\underline{\Theta}_2$The unit length is $\|\underline{\Theta}_1\|^*$. The vectors (points) solution of the 365 calibration problems are represented as (green or red) stars. A point $P=\underline{\Theta}$ is plotted around $\underline{\Theta}_1$ when the quantity $\frac{\|\underline{\Theta}-\underline{\Theta}_1\|^*}{\|\underline{\Theta}_1\|^*}$ is smaller than the quantity $\frac{\|\underline{\Theta}-\underline{\Theta}_2\|^*}{\|\underline{\Theta}_2\|^*}$, otherwise the point *p* is plotted around $\underline{\Theta}_2$.

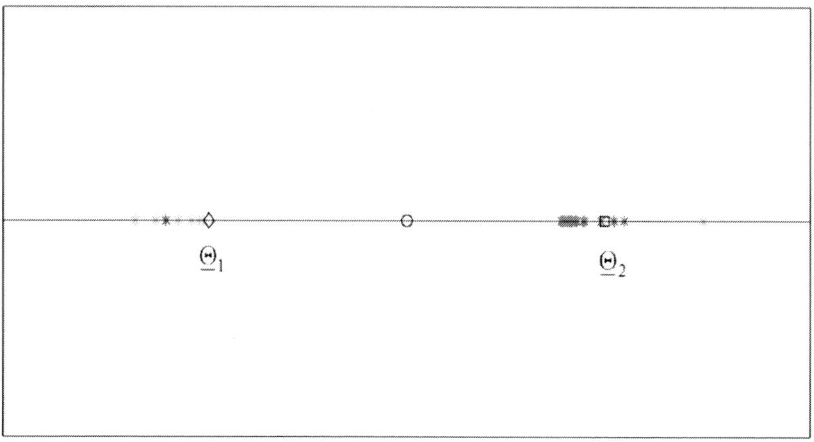

Figure 2. Reconstruction of the parameter vectors $\underline{\Theta}_1$ and $\underline{\Theta}_2$.

The distance of the point *P* from $\underline{\Theta}_1$ when *P* is plotted around $\underline{\Theta}_1$ (or from $\underline{\Theta}_2$ when *P* is plotted around is $\|\underline{\Theta}_1-\underline{\Theta}\|^*$ measured in $\|\underline{\Theta}_1\|^*$ units (or $\|\underline{\Theta}_2-\underline{\Theta}\|^*$ measured in $\|\underline{\Theta}_1\|^*$ units). The point $P=\underline{\Theta}$ plotted around $\underline{\Theta}_1$ is plotted to the right or to the left of $\underline{\Theta}_1$ according to the sign of the second component of $\underline{\Theta}-\underline{\Theta}_1$ (negative second component of $\underline{\Theta}-\underline{\Theta}_1$ is plotted to the left of $\underline{\Theta}_1$). Remind that the second component of $\underline{\Theta}$ is the volatility coefficient σ. A similar statement holds for the points plotted around $\underline{\Theta}_2$. The results obtained in this experiment are shown in **Figure 2**. In this figure the green stars represent the solutions of the calibration problems associated to the first 183 data time windows, that is the first "half" of data time windows, while the red stars represent those associated to the second "half" of the data time windows (that is the second 182 time windows). **Figure 2** shows that the points (vectors) obtained solving the calibration problems are concentrated on two disjoint segments one to the left and one to the right of the origin and that they form two disjoint clusters around $\underline{\Theta}_1$ and $\underline{\Theta}_2$. That is, the solution of the 365 calibration problems corresponding to the 365 time windows described previously shows that two sets of parameters seems to generate the data studied. This is really the case. Moreover, as expected, the points of the cluster around $\underline{\Theta}_2$ are in majority red stars, that is they are in majority the points obtained by the analysis of data time windows containing a majority of observations made in the second year (the second "half" of the data time windows), and a similar statement holds for the points of the cluster around $\underline{\Theta}_1$ (in majority green stars).

The second numerical experiment is performed using real data. The real data studied are electric power price data taken from the UK electricity market. These data are "spiky" asset prices. The data time series studied is made of the observation times $0 = t_0 < t_1 < t_2 < \cdots < t_n = 1395 < +\infty$ (days), and of the spiky asset price $\tilde{S}_i$ observed at time t_i, where $\tilde{S}_i$ is the daily electric power spot price (GBP/MWh), named Day-Ahead price, $i = 0, 1, \cdots, n$ Excluding week-end days and holidays this data time series corresponds to more than 5 years of daily observations going from January 5, 2004 to July 10, 2009. Remind that GBP means Great Britain Pound and that MWh means Mega-Watt/hour.

Moreover the data time series studied in correspondence of each spot price contains a series of forward prices associated to it for a variety of delivery periods. These prices include: forward price 1 month deep in the future (Month-Ahead price), forward price 3 months deep in the future (Quarter-Ahead price), forward price 4 months deep in the future (Season-Ahead price), forward price 1 year deep in the future (One Year-Ahead price). These forward prices are observed each day t_i and the forward prices observed at time t_i are associated to the spot price $\tilde{S}_i$, $i = 0, 1, \cdots, 1395$ The spot and the forward prices contained in the data time series mentioned above are shown in **Figure 3**.

The observed electric power prices generate data time series with a complicated structure. The stochastic dynamical system studied in this paper does not pretend to fully describe the properties of the electric power prices. Indeed it is able to model only one property of these prices: the presence of spikes. In addition the electricity prices have many other properties, for example, they are mean reverting and have well defined periodicity, that is they have diurnal, weekly and seasonal cycles. A lot of specific models incorporating (some of) these features are discussed in the literature, see for example [10].

Let us begin performing the analysis of the data using the model introduced in Section 2. The first question to answer is: the model for spiky prices presented in Section 2 is an adequate model to analyze the time series of the spot prices? We answer to this question analyzing the relationship between the data and the model parameters established through the solution of the calibration problem. Let us begin showing that the relation between the data and the model parameters established through the calibration problem is a stable relationship. We proceed as follows. We have more than 5 years of daily observations. We apply the calibration procedure to the data corresponding to a window of 257 consecutive observation times. Note that 257 is approximately the number of working days contained in a year and remind that we have data only in the working days.

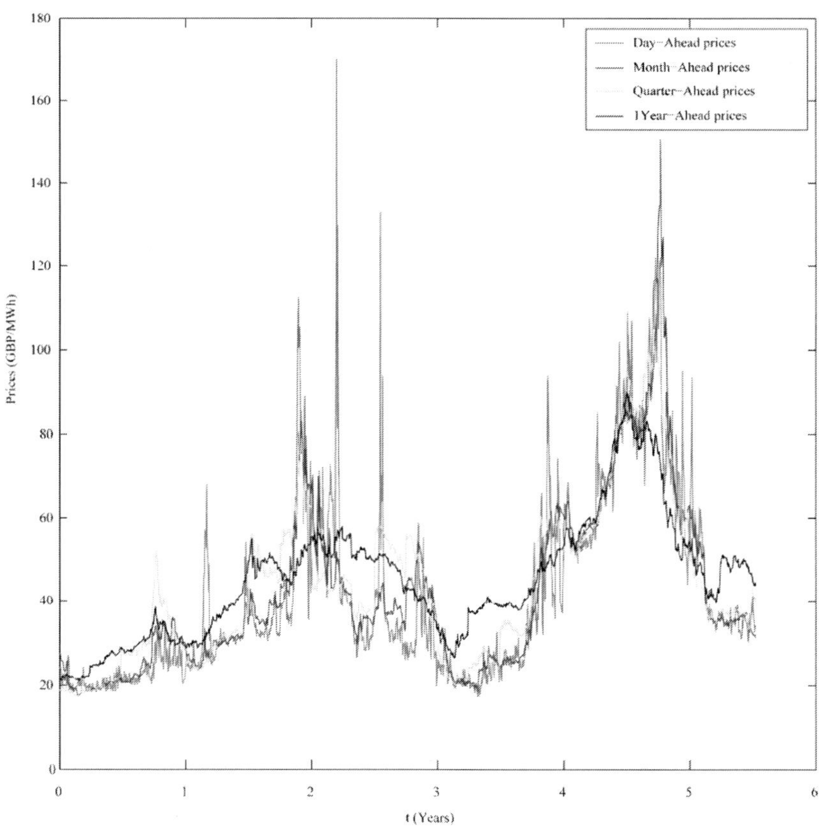

Figure 3. The UK electric power price data.

We move this window through the data time series discarding the datum corresponding to the first observation time of the window and adding the datum corresponding to the next observation time after the window. In this way we have 1396 − 257 = 1139 data windows and for each one of these data windows we solve the corresponding calibration problem. The calibration problems are solved using the optimization procedure (including the ad hoc steps) described in Section 3. The reconstructions of the parameters obtained moving the window along the data are shown in Figures 4 and 5. In Figures 4 and 5 the abscissa represents the data windows used to reconstruct the model parameters numbered in ascending order according to the order in time of the first day of the window. Figures 4 and 5 show that the model parameters, with the exception of λ, are approximately constant functions of the data window. The parameter λ reconstructed shown in **Figure 5** is a piecewise constant function. These findings support the idea that the model and the formulation of the calibration problem presented respectively in Sections 2 and 3 are adequate to interpret the data. In fact they establish a stable relationship between the data and the model parameters as shown in Figures 4 and 5.

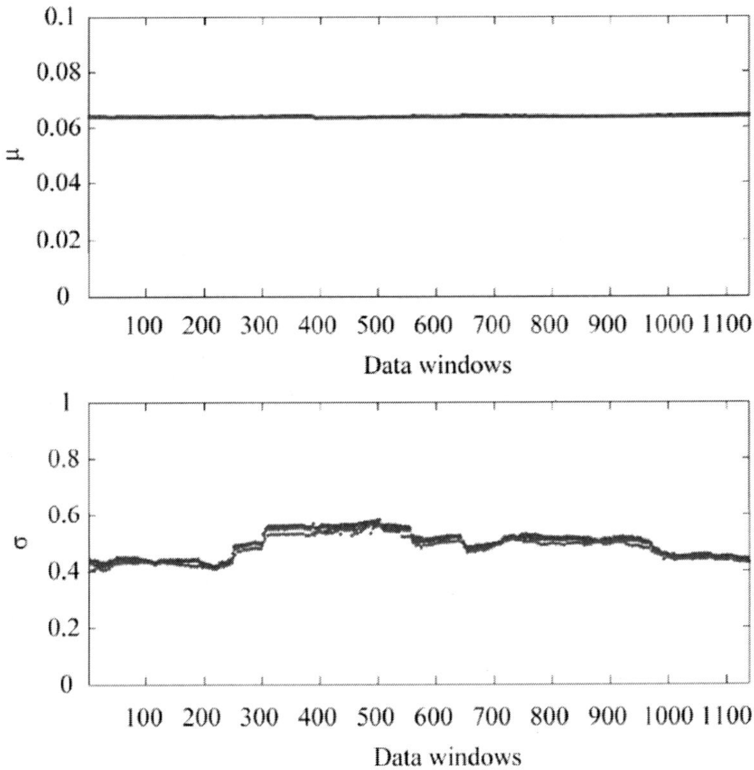

Figure 4. Reconstruction of the parameters of the BlackScholes model: μ, σ.

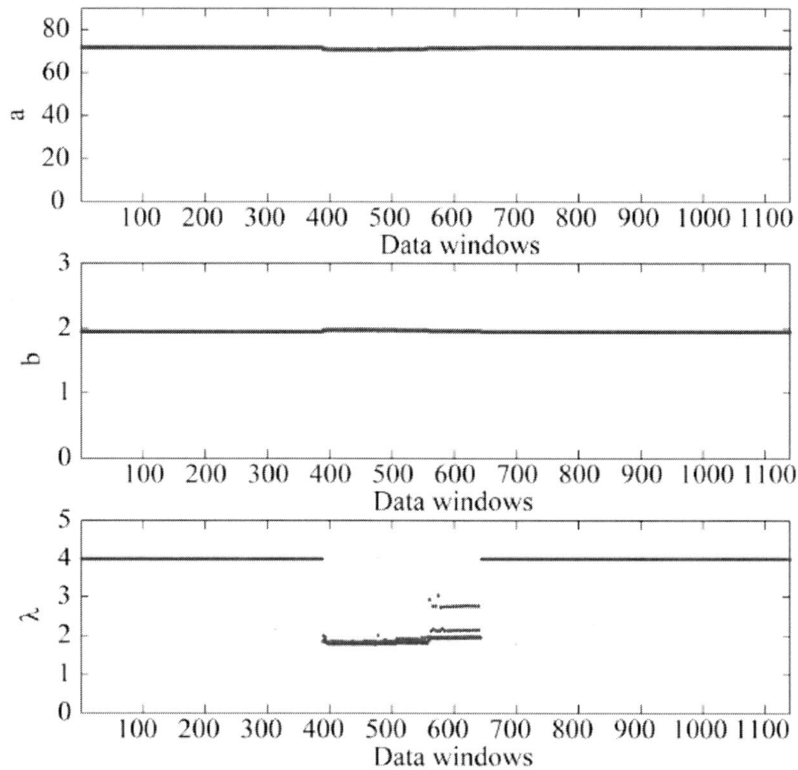

Figure 5. Reconstruction of the parameters used to model the spikes: a, b, λ.

In the analysis of the real data time series the second question to answer is: the solution of the calibration problem and the tracking procedure introduced in Section 3 can be used to forecast forward prices? To answer this question we compare the observed and the forecasted forward prices. We apply the calibration procedure to a data window made of the first three years of consecutive observations of the spot price taken from the data time series shown in **Figure 3** and we use the solution of the calibration problem found and formulae (30), (31) to calculate the forecasted forward prices associated to the spot prices of the data time series not included in the data window mentioned above used in the calibration problem. In Figures 6-8 the forward electric power prices forecasted are shown and compared to the observed forward electric power prices. In Figures 6-8 the abscissa is the day of the spot price associated to the forward prices computed. The abscissa of Figures 6-8 is coherent with the abscissa of **Figure 3**.

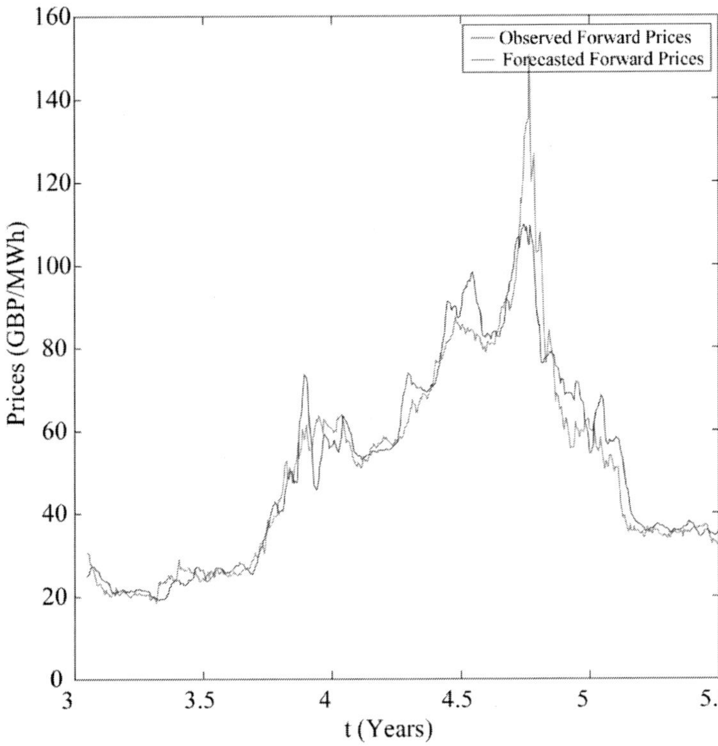

Figure 6. Month-ahead prices.

Table 1 summarizes quantitatively the results shown in Figures 6-8 and gives the average relative error $e_{\text{forward prices}}$ committed using the forecasted forward prices, that is the average relative error committed approximating the observed forward prices with the forecasted forward prices.

Table 1 and Figures 6-8 show the high quality of the forecasted forward prices answering the second question posed about the analysis of the data time series in the affirmative.

We can conclude that the data analysis presented shows that the model introduced to describe spiky prices, the formulation of the calibration problem and its numerical Lorella Fatone, Francesca Mariani, Maria Cristina Recchioni, Francesco Zirilli solution presented in this paper have the potential of being tools of practical value in the analysis of data time series of spiky prices.

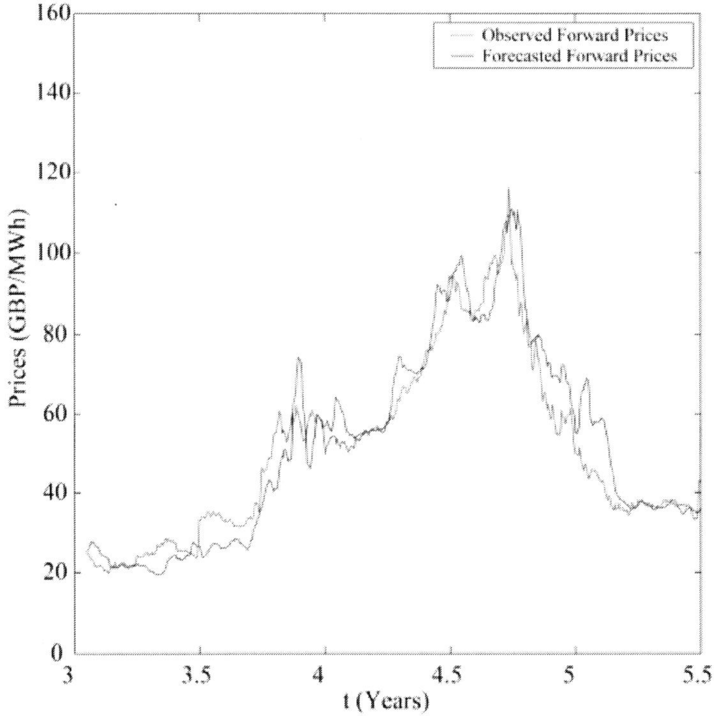

Figure 7. Quarter-ahead prices.

Table 1. Average relative errors of the forecasted forward electric power prices when compared to the observed forward prices.

Number of days in the future	$e_{\text{forward prices}}$
30 (Month-Ahead prices)	0.0796
90 (Quarter-Ahead prices)	0.1160
365 (One Year-Ahead prices)	0.2183

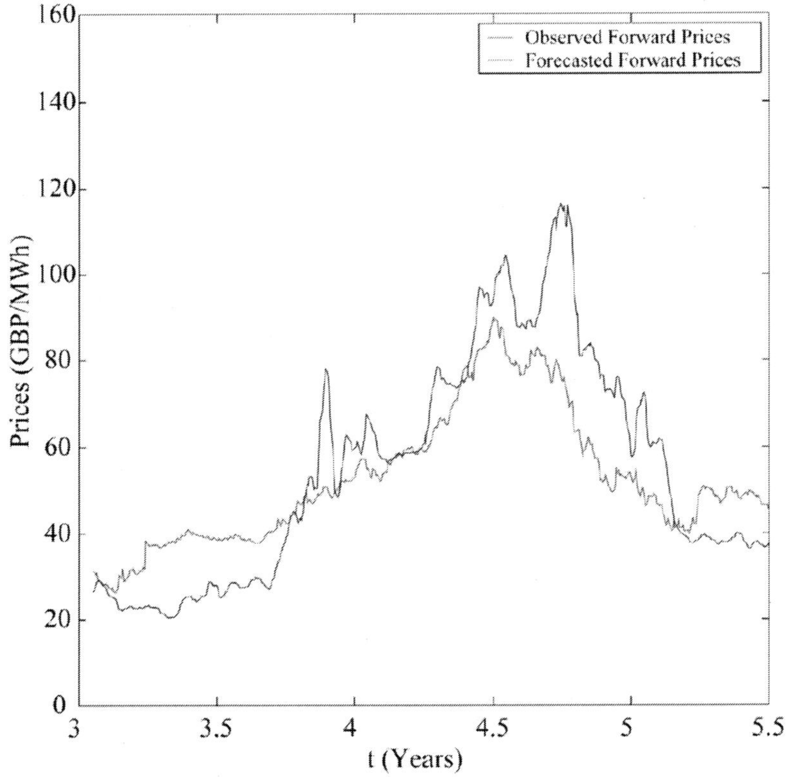

Figure 8. One year-ahead prices.

REFERENCES

1. V. A. Kholodnyi, "Valuation and Hedging of European Contingent Claims on Power with Spikes: A Non-Markovian Approach," Journal of Engineering Mathematics, Vol. 49, No. 3, 2004, pp. 233-252. doi:10.1023/B:ENGI.0000031203.43548.b6
2. V. A. Kholodnyi, "The Non-Markovian Approach to the Valuation and Hedging of European Contingent Claims on Power with Scaling Spikes," Nonlinear Analysis: Hybrid Systems, Vol. 2, No. 2, 2008, pp. 285-304. doi:10.1016/j.nahs.2006.05.002
3. F. Mariani, G. Pacelli and F. Zirilli, "Maximum Likelihood Estimation of the Heston Stochastic Volatility Model Using Asset and Option Prices: An Application of Nonlinear Filtering Theory," Optimization Letters, Vol. 2, No. 2, 2008, pp. 177-222.doi: 10.1007/s11590-007-0052-7
4. L. Fatone, F. Mariani, M. C. Recchioni and F. Zirilli, "Maximum Likelihood Estimation of the Parameters of a System of Stochastic Differential Equations that Models the Returns of the Index of Some Classes of Hedge Funds," Journal of Inverse and Ill Posed Problems, Vol. 15, No. 5, 2007, pp. 493-526. doi:10.1515/jiip.2007.028
5. L. Fatone, F. Mariani, M. C. Recchioni and F. Zirilli, "The Calibration of the Heston Stochastic Volatility Model Using Filtering and Maximum Likelihood Methods," G. S.

Ladde, N. G. Medhin, C. Peng and M. Sambandham, Eds., Proceedings of Dynamic Systems and Applications, Dynamic Publishers, Atlanta, Vol. 5, 2008, pp. 170-181.
6. L. Fatone, F. Mariani, M. C. Recchioni and F. Zirilli, "Calibration of a Multiscale Stochastic Volatility Model Using European Option Prices," Mathematical Methods in Ecomics and Finance, Vol. 3, No. 1, 2008, pp. 49-61.
7. L. Fatone, F. Mariani, M. C. Recchioni and F. Zirilli, "An Explicitly Solvable Multi-Scale Stochastic Volatility Model: Option Pricing and Calibration," Journal of Futures Markets, Vol. 29, No. 9, 2009, pp. 862-893. doi:10.1002/fut.20390
8. L. Fatone, F. Mariani, M. C. Recchioni and F. Zirilli, "The Analysis of Real Data Using a Multiscale Stochastic Volatility Model," European Financial Management, 2012, in press.
9. P. Capelli, F. Mariani, M. C. Recchioni, F. Spinelli and F. Zirilli, "Determining a Stable Relationship between Hedge Fund Index HFRI-Equity and S&P 500 Behaviour, Using Filtering and Maximum Likelihood," Inverse Problems in Science and Engineering, Vol. 18, No. 1, 2010, pp. 93-109.
10. C. R. Knittel and M. R. Roberts, "An Empirical Examination of Restructured Electricity Prices," Energy Economics, Vol. 27, No. 5, 2005, pp. 791-817.doi:10.1016/ j.eneco. 2004.11.005
11. J. H. Hull, "Options, Futures, and Other Derivatives," 7th Edition, Pearson Prentice Hall, London, 2008.

NOTES

*The research reported in this paper is partially supported by MUR—Ministero Università e Ricerca (Roma, Italy), 40%, 2007, under grant: "The impact of population ageing on financial markets, intermediaries and financial stability". The support and sponsorship of MUR are gratefully acknowledged.

#The numerical experience reported in this paper has been obtained using the computing grid of ENEA (Roma, Italy). The support and sponsorship of ENEA are gratefully acknowledged.

CHAPTER 3

Numerical Simulation Analysis of a Mathematical Model of Circadian Pacemaker Neurons

Takaaki Shirahata

Kagawa School of Pharmaceutical Sciences, Tokushima Bunri University, Sanuki, Japan

ABSTRACT

Sim and Forger have proposed a mathematical model of circadian pacemaker neurons in the suprachiasmatic nucleus (SCN). This model, which has been formulated on the Hodgkin-Huxley mo- del, is described by a system of nonlinear ordinary differential equations. An important feature of the SCN neurons observed in electrophysiological recording is spontaneous repetitive spiking, which is reproduced using this model. In the present study, numerical simulation analysis of this model was performed to evaluate variations in two system parameters of this model: the maximal conductance of calcium current (g_{Ca}) and the maximal conductance of sodium current (g_{Na}). Simulation results revealed the spontaneous repetitive spiking states of the model in the (g_{Ca}, g_{Na})-parameter space.

Keywords: Mathematical Model, Numerical Simulation, Neuron, Spiking

1. INTRODUCTION

Repetitive spiking activity is an important feature of excitable cells such as neurons and muscle cells. In the field of applied mathematics, this activity can be extensively analyzed using various mathematical models, which are described by a system of nonlinear ordinary differential equations (ODEs). One example is the Hodgkin-Huxley model, which can generate repetitive spiking because of an interaction between the sodium conductance and the potassium conductance (page 25 in [1]). In this model, the sodium conductance plays an excitatory role, whereas the potassium conductance plays an inhibitory role. The other example is the Morris-Lecar model, which can also generate repetitive spiking because of an interaction between the calcium conductance and the potassium conductance (page 49 in [1]). In this model, the calcium conductance and not the sodium conductance plays an excitatory role.

Another type of mathematical model of the repetitive spiking activity is the one in which the repetitive spiking is generated by an interaction among sodium, calcium, and potassium conductances. One example is a mathematical model of circadian pacemaker neurons in the suprachiasmatic nucleus (SCN) [2], which is described by a system of nonlinear ODEs (see the Materials and Methods in [2] as well as the Materials and Methods of this manuscript). Considering the Hodgkin-Huxley and Morris-Lecar models described above, it is assumed that both the sodium and calcium conductances play excitatory roles in the circadian pacemaker neuron model. How- ever, the difference in the contribution to the repetitive spiking between the sodium and calcium conductances was not clarified in detail in a previous study [2]. In particular, the influence of variations in these two conductance values on the repetitive spiking was not studied in detail in that study [2]. Taking into consideration that it is not only interesting but also very important to analyze the ionic conductance in detail (page 26 in [3]), it is necessary to investigate the influence of variations in these two conductance values and reveal the difference between these two conductances. Therefore, in the present study, we focused on two system parameters of the circadian pacemaker neuron model, namely the maximal conductance of calcium current (g_{Ca}) and the maximal conductance of sodium current (g_{Na}), and we performed numerical simulation analysis.

2. MATERIALS AND METHODS

A mathematical model of the circadian pacemaker neuron, which has been used in the present study, was a model developed by Sim and Forger [2].

This model is described by a system of nonlinear ODEs and contains six state variables: a membrane potential of the circadian pacemaker neuron [V(mV)] and five gating variables of ionic currents (m, h, n, r, and f). The time evolution of the six state variables is described as follows:

$$C\frac{dV}{dt} = I_{app} - I_{Na}(V,m,h) - I_K(V,n) - I_L(V) - I_{Ca}(V,r,f) \tag{1}$$

$$\frac{dX}{dt} = \frac{1}{\tau_X(V)}\left(X_\infty(V) - X\right) \quad (X = m,h,n,r,f) \tag{2}$$

where C is membrane capacitance (5.7 pF); $I_{Na}(V, m, h)$, $I_K(V, n)$, $I_L(V)$, and $I_{Ca}(V, r, f)$ are a sodium current, a potassium current, a leak current, and a calcium current, respectively, which are defined in Equations (3)-(6) below, in this order; $\tau_X(V)$ (ms) and $X_\infty(V)$ are time constants of activation/inactivation and steady-state activation/inactivation functions, respectively, which are defined in Equations (7)-(16) below, in this order; and I_{app} is applied current.

$$I_{Na}(V,m,h) = g_{Na}m^3h(V - E_{Na}) \tag{3}$$

$$I_K(V,n) = g_K n^4 (V - E_K) \tag{4}$$

$$I_L(V) = g_L(V - E_L) \tag{5}$$

$$I_{Ca}(V,r,f) = g_{Ca}rf(V - E_{Ca}) \tag{6}$$

where g_{Na}, g_K, g_L, and g_{Ca} are the maximal conductances of a sodium current (the g_{Na} value was varied in the present study; the default value was 229 nS), a potassium current (14 nS), a leak current (1/11 nS), and a calcium current (the g_{Ca} value was varied in the present study; the default value was 65 nS), respectively; E_{Na}, E_K, E_L, and E_{Ca} are the reversal potentials of a sodium current (45 mV), a potassium current (−97 mV), a leak current (−29 mV), and a calcium current (61 mV), respectively.

$$\tau_m(V) = e^{-(V+286)/160} \tag{7}$$

$$\tau_h(V) = 0.51 + e^{-(V+26.6)/7.1} \tag{8}$$

$$\tau_n(V) = e^{-(V-67)/68} \tag{9}$$

$$\tau_r = 3.1 \tag{10}$$

$$\tau_f(V) = e^{-(V-444)/220} \tag{11}$$

$$m_\infty(V) = \frac{1}{1 + e^{-(V+35.2)/7.9}} \tag{12}$$

$$h_\infty(V) = \frac{1}{1 + e^{(V+62)/5.5}} \tag{13}$$

$$n_\infty(V) = \frac{1}{\left(1 + e^{-(V-14)/17}\right)^{0.25}} \tag{14}$$

$$r_\infty(V) = \frac{1}{1 + e^{-(V+25)/7.5}} \tag{15}$$

$$f_\infty(V) = \frac{1}{1 + e^{(V+260)/65}} \tag{16}$$

Detailed explanations of the model equations have been described previously [2].

The free and open source software Scilab (http://www.scilab.org/) was used to numerically solve ODEs (initial conditions: $V = -80$ mV, $m = 0.34$, $h = 0.045$, $n = 0.54$, $r = 0.01$, $f = 0.04$).

3. RESULTS

The time courses of the membrane potential of the SCN neuron model under different parameter values are shown in **Figure 1**. At default parameter values, the model showed spontaneous repetitive spiking (**Figure 1(a)**), as reported previously (Figure 2(a) in [2]). When only g_{Ca} was completely blocked, the model showed the stable steady state instead of the spontaneous spiking state (**Figure 1(b)**).

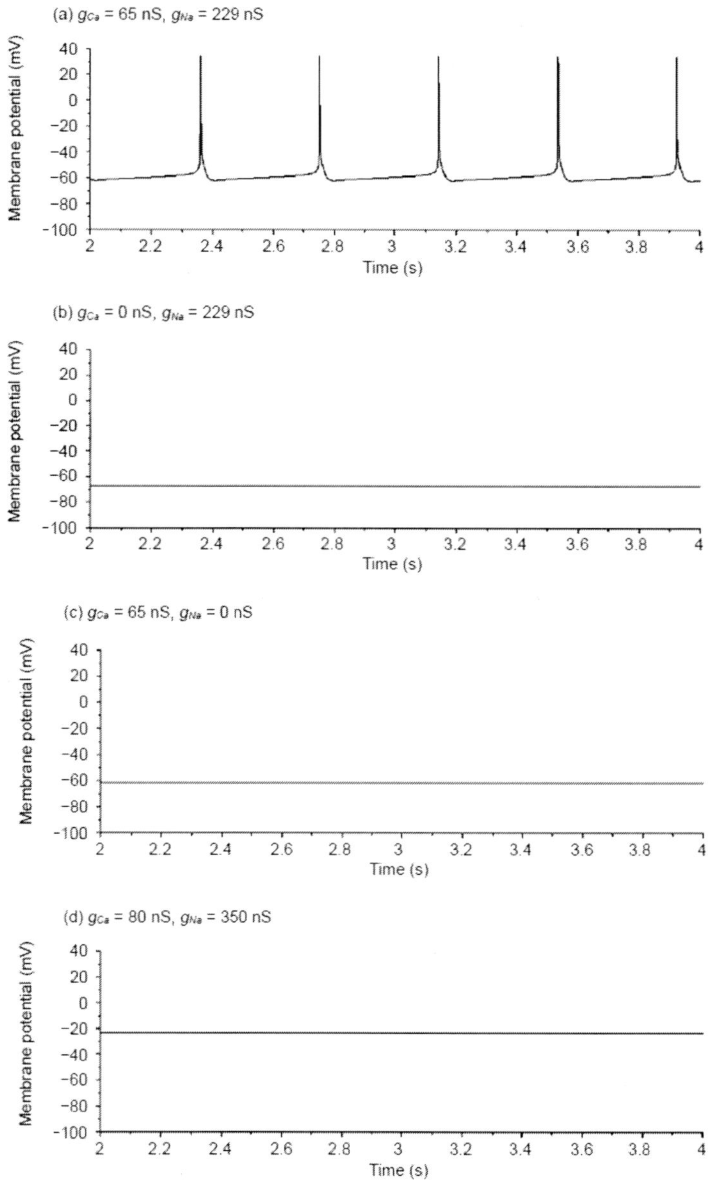

Figure 1. Time courses of the simulated membrane potential of circadian pacemaker neurons under several (g_{Ca}, g_{Na}) conditions. (a) The spontaneous repetitive spiking state at (g_{Ca}, g_{Na}) = (65 nS, 229 nS). (b) The hyperpolarized steady state at (g_{Ca}, g_{Na}) = (0 nS, 229 nS). (c) The hyperpolarized steady state at (g_{Ca}, g_{Na}) = (65 nS, 0 nS). (d) The depolarized steady state at (g_{Ca}, g_{Na}) = (80 nS, 350 nS).

When only g_{Na} was completely blocked, the model showed a similar pattern, as illustrated in **Figure 1(b)** (**Figure 1(c)**). When both g_{Ca} and g_{Na} were set to be values larger than the default values, the model showed the stable steady state (**Figure 1(d)**). However, the steady state membrane potential value shown in **Figure 1(d)** was more depolarized than that shown in **Figure 1(b)** and **Figure 1(c)**.

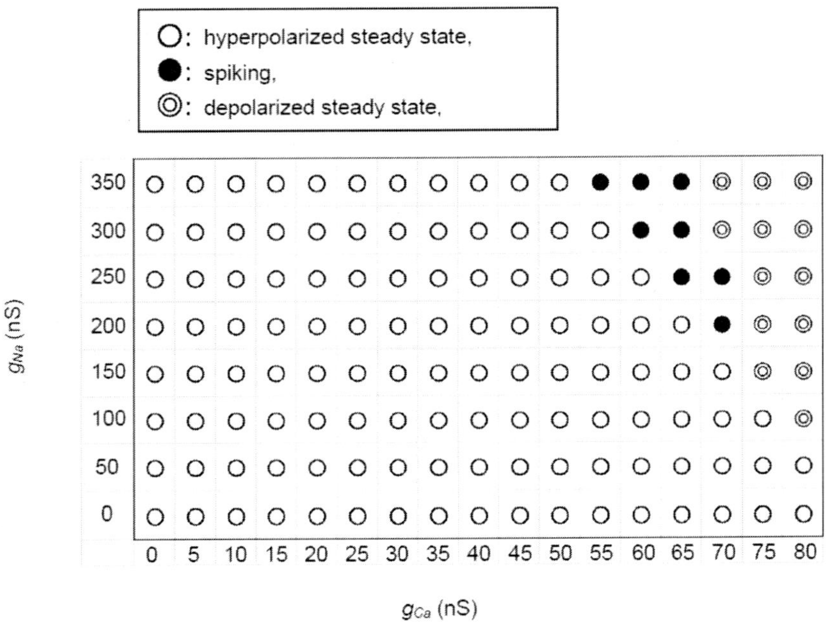

Figure 2. The sensitivity of the dynamical states of the circadian pacemaker neuron model to g_{Ca} and g_{Na} variations. Black circle: the spontaneous repetitive spiking state, white circle: the hyperpolarized steady state, white double circle: the depolarized steady state.

Figure 2 illustrates the phase diagram of the model in the (g_{Ca}, g_{Na})-parameter space. The dynamical states of the model were classified into three states: the hyperpolarized steady state (white circle), the spontaneous spiking state (black circle), and the depolarized steady state (double circle). A decrease in g_{Na} induced a decrease in the g_{Ca} range in which repetitive spiking occurred, and finally, the g_{Ca} range disappeared. Therefore, it is concluded that when g_{Na} is zero, repetitive spiking cannot occur, no matter how large the g_{Ca} value is. An increase in g_{Na} induced a decrease in the g_{Ca} threshold required to induce the transition from the hyperpolarized steady state to the spiking state. For example, even when

g_{Ca} greatly decreased to 30 nS, repetitive spiking occurred under conditions in which g_{Na} greatly increased to 1603 nS (a seven-fold increase from the default value) (**Figure 3(a)**). In addition, even when g_{Ca} was zero, repetitive spiking occurred under conditions in which g_{Na} was 1603 nS (**Figure 3(b)**).

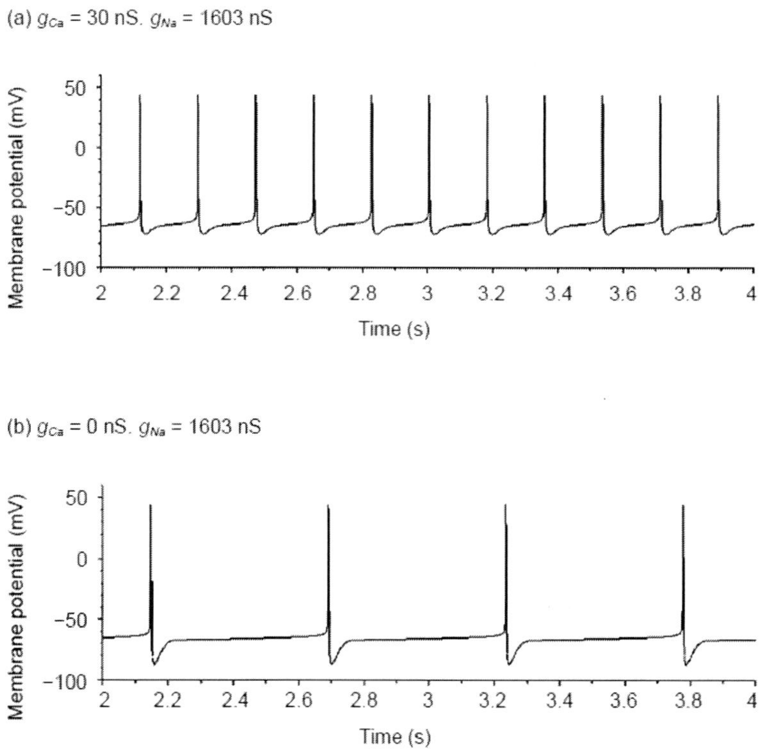

Figure 3. Time courses of the simulated membrane potential of circadian pacemaker neurons under (g_{Ca}, g_{Na}) conditions different from those in **Figure 1**. (a) The spontaneous repetitive spiking state at (g_{Ca}, g_{Na}) = (30 nS, 1603 nS). (b) The spontaneous repetitive spiking state at (g_{Ca}, g_{Na}) = (0 nS, 1603 nS).

4. DISCUSSION

The present study revealed the sensitivity of the model dynamics to variations in g_{Ca} and g_{Na} of circadian pacemaker neurons. Although Sim and Forger revealed the difference in time courses between the calcium and sodium currents (Figure 2(c) and Figure 2(d), respectively, in [2]), they did not clarify whether the calcium and sodium conductances are indispensable to repetitive spiking under the default condition. The present results (**Figures 1(a)-(c)**) resolved this issue; both g_{Ca} and g_{Na} were indispensable to repetitive spiking under the default condition. Interestingly, the present study also revealed the difference between the two conductances. When we considered repetitive spiking under the condition in which the g_{Ca} value was smaller than its default value but the g_{Na} value was much larger than its default value, the sodium conductance was indispensable but the calcium conductance was dispensable to this type of repetitive spiking (Figure 2 and Figure 3(a) and Figure 3(a)).

Previous studies have reported the relationship between ionic conductances and neuronal spiking in various types of mathematical models [4] -[7]. In particular, similar to the circadian pacemaker neuron model analyzed in the present study, a mathematical model of midbrain dopaminergic neurons included the sodium and calcium conductances [4]. In the midbrain dopaminergic neuron model (Figure 4 in [4]), (1) an increase in g_{Na} induced a decrease in the g_{Ca} threshold required to induce the transition from the nonpacemaking state to the pacemaking state and (2) the pacemaking occurred in the absence of the calcium conductance under conditions in which the sodium conductance was set to a large value. The present circadian pacemaker neuron model also exhibited these two characteristics. However, the present results revealed two important differences between these two models: (1) the (g_{Ca}, g_{Na})-parameter space was divided into only two states (the pacemaking and nonpacemaking states) in the midbrain dopaminergic neuron model (Figure 4 in [4]), whereas the (g_{Ca}, g_{Na})-parameter space as divided into three states (the depolarized steady state, the repetitive spiking state, and the hyperpolarized steady state) in the circadian pacemaker neuron model (**Figure 2**), and (2) pacemaking occurred in the absence of the sodium conductance under the condition in which the calcium conductance was set to a large value in the midbrain dopaminergic neuron model (Figure 4 in [4]), whereas pacemaking did not occur in the absence of the sodium conductance even when the calcium conductance was set to a large value in the circadian pacemaker neuron model (**Figure 2**).

5. CONCLUSION

The present investigation focused on a mathematical model of circadian pacemaker neurons, performed numerical simulation analysis, and compared this numerical result with that of the previous studies. The important and novel findings of the present study are as follows: 1) there was a difference in the contribution to repetitive spiking under certain conditions between the sodium and calcium conductances: for the generation of repetitive spiking under certain conditions, the sodium conductance played an indispensable role, whereas the calcium conductance was not necessarily essential, and 2) the (g_{Ca}, g_{Na})-parameter space of the circadian pacemaker neuron model showed a different pattern compared with that of the midbrain dopaminergic neuron model. These findings can contribute to our in-depth understanding of the influence of the sodium and calcium conductances on neuronal repetitive spiking.

ACKNOWLEDGEMENTS

The author would like to thank Enago (www.enago.jp) for the English language review.

REFERENCES

1. Ermentrout, G.B. and Terman, D. (2010) Mathematical Foundations of Neuroscience (Interdisciplinary Applied Mathematics). Springer, New York. http://dx.doi. org/ 10.1007/978-0-387-87708-2
2. Sim, C.K. and Forger, D.B. (2007) Modeling the Electrophysiology of Suprachiasmatic Nucleus Neurons. *Journal of Biological Rhythms*, **22**, 445-453. http://dx.doi.org/10. 1177/0748730407306041
3. Ashrafuzzaman, M. and Tuszynski, J. (2012) Membrane Biophysics (Biological and Medical Physics, Biomedical ngineering). Springer, New York.
4. Drion, G., Massotte, L., Sepulchre, R. and Seutin, V. (2011) How Modeling Can Reconcile Apparently Discrepant Experimental Results: The Case of Pacemaking in Dopaminergic Neurons. *PLoS Computational Biology*, **7**, e1002050. http://dx.doi.org/10. 1371/ journal.pcbi.1002050
5. Shirahata, T. (2011) The Effect of Variations in Sodium Conductances on Pacemaking in a Dopaminergic Retinal Neuron Model. *Acta Biologica Hungarica*, **62**, 211-214. http://dx. doi.org/10.1556/ABiol.62.2011.2.11
6. Shirahata, T. (2014) Effect of Sodium Conductance Variations on Electrical Behavior of a Neocortical Neuron Model. *Acta Biologica Hungarica*, **65**, 379-384. http://dx.doi.org/ 10.1556/ABiol.65.2014.4.2
7. Shirahata, T. (2015) Numerical Study of a Mathematical Model of Vibrissa Motoneurons: The Relationship between Repetitive Spiking and Two Types of Sodium Conductance. *International Journal of Theoretical and Mathematical Physics*, **5**, 48-52.

CHAPTER 4

Analysis of Cross-Correlations in Emerging Markets Using Random Matrix Theory

Thomas Chinwe Urama*, Patrick Oseloka Ezepue, Chimezie Peters Nnanwa

Statistics, Information Modelling and Financial Mathematics Research Group, Department of Engineering & Mathematics, Sheffield Hallam University, Sheffield, UK

ABSTRACT

This paper investigates the universal financial dynamics in two dominant stock markets in Sub-Saharan Africa, through an in-depth analysis of the cross-correlation matrix of price returns in Nigerian Stock Market (NSM) and Johannesburg Stock Exchange (JSE), for the period 2009 to 2013. The strength of correlations between stocks is known to be higher in JSE than that of the NSM. The stock price dynamics in the NSM is important for modeling Nigerian derivatives in the future, and in the light of this, it is pertinent to note that the interactions of other stocks with the oil sector are weak, whereas the banking stocks have strong positive interactions with the other sectors in the stock exchange. For the JSE, it is the oil sector and beverages that have greater sectorial correlations, instead of the banks which have a weaker one in relation to other assets in the stock exchange.

Keywords: Random Matrix Theory, Cross-Correlations, Emerging Markets, Option Pricing, Eigenvalues Eigenvectors, Inverse Participation Ratios and Implied Volatility

1. INTRODUCTION

This paper investigates and compares the spectral properties of correlation matrices of price fluctuations in Nigerian and South African Stock Markets, using the Random Matrix Theory (RMT). Alternative approaches, namely factor and principal components analysis for measuring the extent of correlations could be found as presented in [1] [2] [3] [4] [5]. In this research, we use RMT to compare the empirical correlation matrix with Wishart random matrix, which models normality and departures from which connote the existence of significant market information in the observed price fluctuations [6].

Pafka and Kondor [6] assert that correlation matrices of financial returns play a crucial role in various aspects of modern finance including investment theory, capital allocation, and risk management. Also, [7] declare that following the introduction of RMT into the financial markets by [8] and [9] , the concept has been used in the study of the statistical properties of cross-correlations in different financial markets [10] - [22]. Laurent Laloux *et al.* [8] opine that for financial assets, the study of the empirical correlation matrix is very relevant, since, from their finding, it is its estimation in the price movements of different assets that constitutes a significant and indispensable aspect of risk management. They declare that the probability of huge losses for a certain portfolio or option book is dominated by correlated moves of its different components and that a position which is simultaneously long in stocks and short in bonds will be risky as stocks and bonds usually move in opposite directions during crisis periods.

The interesting question that concerned investors need to answer is how volatility, which is a measure of market fluctuations, affects the dynamics of the market or vice versa. It is, therefore, expedient to explore the relationship between volatility and the coupling of stocks with one another, using correlation matrices [14]. Thus, correlations amongst the volatility of different assets are very useful, not only for portfolio selection, but also in pricing options and certain multivariate econometric models for price forecasting and volatility estimations. Engle and Figlewski [23] assert that with regards to Black-Scholes option pricing model the variance of the portfolio, ρ, of options exposed to Vega risk only is given by

$$Var(\rho) = \sum_{i,j,k,l} \frac{w_i w_l \Lambda_{ij} \Lambda_{lk} C_{jk}}{v_j v_k \sigma_j \sigma_k} \quad (1)$$

where w_i are the weights in the portfolio, C_{ij} is the correlation matrix for the implied volatility for the underlying assets and the Vega matrix Λ_{ij} is defined as

$$\Lambda_{ij} = \frac{\partial p_i}{\partial v_j} \quad (2)$$

with P_i as the price of option *i*, v_j is the implied volatility of asset underlying option *j* and ?*i* is the standard deviation of the implied volatility v_i.

Similarly, for investors using derivatives products as a hedge on the underlying assets and for risk management, it is advisable that such investors should buy call and put options respectively for assets whose returns move in opposing directions, as may be witnessed from the calculated empirical correlation matrix. Furthermore, an accurate quantification of correlations between the returns of various stocks is practically important in quantifying risks of stock portfolios, pricing options, and forecasting. [24] note that financial correlation matrices are the key input parameters for Markowitz [25] fundamental portfolio optimization problem aimed at providing a recipe for the selection of a portfolio of assets, such that the risk associated with the investment is minimized for a given expected return. Edelman Alan [26] asserts that RMT makes it possible for a comparison between the cross-correlation matrices obtained from a given number of empirical time series data for a period *T* with an entirely random matrix *W*, otherwise known as Wishart matrix of the same size with the empirical correlation matrix, to obtain some useful information about the market(s), which is necessary for portfolio optimization and risk management. RMT predictions represent an average over all possible interactions between the constituents of the assets in a given market under consideration. The deviations from universal predictions of RMT obtained from the Wishart matrix are used in identifying the system specific, non-random properties of the system under consideration and such variations provide information about the underlying interaction of the assets. In other words, we compare the statistics of the cross-correlation coefficients of price fluctuations of stock *i* and *j* against a random matrix having the same symmetric properties as that of the empirical matrix. The RMT is known to distinguish the random and non-random parts of the cross-correlation matrix *C*, the non-random parts of C which deviates from RMT results is known to provide information regarding the genuine collective behaviour of the stocks under consideration and indeed the entire market at large [27].

Theoretically, the comparative analyses of asset price fluctuations (hence correlation structures) between the JSE and NSM will enable us to calibrate suitable derivative models to be proposed for adoption in the NSM for portfolio optimization and risk management. This is because from the research visit embarked upon by the researchers to the Nigerian Stock Exchange (NSE) in 2014; policy makers in the NSE are taking a clue from the JSE in their proposed introduction of some pioneer derivative products and subsequently an appropriate pricing and valuation of such products in the NSM. The research into the correlations between price changes of different stocks is not only necessary for quantifying the risk in a given portfolio, but it is also of scientific interest to researchers in Economics and Financial Mathematics [28] [29]. Interestingly, interpreting the correlations between individual stocks-price changes in a given financial market can be likened to the difficulties experienced by physicists in the fifties, in interpreting the spectra of complex nuclei. Due to the enormous amounts of spectroscopic data on the energy levels that were available, which were too complicated to be analysed through model calculations, since the nature of the interactions were not known, Random Matrix Theory (RMT) was developed to take care of the Statistics of energy levels of the complex quantum systems [30] [31].

Similarly, for financial time series in a stock exchange, the nature of interactions among constituent stock are unknown, hence the need to adopt the RMT approach in exploring these interactions between individual pairs of stocks, for use in portfolio optimization and risk management. The estimation of risk and expected returns based on variance and expected returns in a given portfolio constitutes Markowitz's model [32]. In view of the fact that the statistical properties of correlations between different stocks seem to be less universal across different stock market, [17] , in this paper, we first demonstrate the validity of the general predictions of RMT for the eigenvalue statistics of the correlation matrix and subsequently calculate the deviations, if any, of the empirical data from the Wishart matrix predictions, to identify the nature of the correlations between the individual stocks and distinguish same from those of the deviations due to randomness, in the NSM and JSE. In doing this, the period T under consideration has to be relatively large enough when compared with the number of stocks or assets being considered to minimize the noise in the correlation matrix. The two sources of noise envisaged in the use of RMT in investigating the cross- correlations of stocks in a given financial market include (a) the noise from the period length T considered with respect to the number of stock and; (b) that resulting from the fact that financial time series of historical return itself is finite or bounded thereby introducing inadvertently estimation errors (noise) in the correlation matrix [6].

Szilard and Kondor [33] also observe that the effect of noise strongly depends on the ratio of stocks to the period considered, given by $r \frac{N}{T}$, where N is the number of stocks considered and T the length of the available time series. They note that for the ratio r = 0.6 and above, there will be a pronounced effect of noise on the empirical analysis as was discovered by [9] [21] [34] and that for a smaller value of r (r = 0.2 or less); the error due to noise drops to tolerable levels.

In our case for NSM $r = \frac{82}{1018} = 0.08$ and that of JSE we have $r = \frac{35}{1147} = 0.03$ thus both lying in the admissible region in the values of r. When this is done, if the eigenvalues of the empirical correlation matrix and that of the Wishart matrix lie in the same region without any significant deviations, then the stocks are said to be uncorrelated and therefore no information or deduction can be made about the nature of the market, since it is the deviations of the eigenvalues of the correlation matrix from that of the Wishart matrix that carries information about the entire market. However, if there exists at least one eigenvalue lying outside the theoretical predicted bound of the eigenvalues in the empirical correlation matrix obtained from the stock market returns, then the deviating eigenvalue(s) is(are) known to carry information about the market under consideration.

To the best of our knowledge, no such work on the comparison of stock market correlations has been carried out on African emerging markets, especially JSE and NSM which are major emerging markets in the Sub-Saharan Africa. Most of the work on such comparison has been carried out for developed markets or developed versus emerging markets, see, for instance: [10] [17] [19] [35] [36]. On the other hand, for some comparison for different stock exchanges within the same market environment, see [17].

In some sense, the JSE is gradually approaching a developed market whereas the NSM is an ideal African emerging market with no known trades on derivative products currently existing in the market, unlike the JSE where trade on derivatives has been in existence for over two decades. Option contracts were introduced in JSE in October 1992, agricultural commodity futures in 1995 and a fully automated trading system in May 1996, whereas in the NSM trade in derivative products are still at the formative stage, with a recently approved derivative trade on foreign exchange future under the auspices of Financial Market Derivative Quotations (FMDQ) in 2016. As the policy makers in the NSM are benchmarking themselves on the relevant trade on derivatives in JSE towards an effective take off of derivative trade in the NSM, it is pertinent to compare the asset return correlations between the two markets, to understand the similarities and differences in the statistical properties using random matrix theory.

2. DATA

The data set consists of the daily closing prices of 82 stocks listed in the Nigerian Stock Market, NSM from 3rd August 2009 to 26th August 2013, giving a total of 1019 daily closing returns after removing (a) assets that were delisted, (b) those that did not trade at all or (c) are partially in business for the period under review. The stocks considered for NSM are drawn from the Agriculture, Oil and Gas, Real Estates/Construction, Consumer Goods and Services, Health care, ICT, Financial Services, Conglomerates, Industrial Goods, and Natural Resources. For the JSE, we have a total in 35 stocks selected from Top 40 shares in the Industrial Metals and Mining, Banking, Insurance, Health care, Mobil Telecommunications, Oil and Gas, Financial services, Food and Drugs, Tobacco, Forestry and Paper, Real Estate, Media, Personal Goods and Beverages, covering the period 2nd January 2009 to 01st August 2013 covering a similar period as that of NSM (This period was chosen for the research because that was the period when we could get the complete market information for the two stock exchanges being considered).

For the values of the daily asset prices to be continuous and to minimize the effect of thin trading, we remove the public holidays in the period under consideration and to reduce noise in the analysis, market data for the present day is assumed to be the same with the previous day for cases where there are no information on trade for any particular asset on a given date. Also, we eliminate stocks that infrequently traded within the period under review. Let $P_i(t)$ be the closing price on a given day t, for stock i and define the natural logarithmic return of the index as

$$r_i(t) = \ln \frac{P_i(t+1)}{P_i(t)} \tag{3}$$

where $r_i(t)$ is the number of observations in the two stock exchanges, NSM and JSE.

3. THEORETICAL BACKGROUNDS

3.1. Computing Volatility

We calculate the price changes of assets in the two markets over a time scale Δt which is equivalent to one day and denote the price of i at a time t

as $s_i(t)$ with the corresponding price change or logarithmic returns $G_i(t)$ over time scale Δt as

$$G_i(t) = \ln[S_i(t+\Delta t)] - \ln[S_i(t)] \tag{4}$$

We quantify the volatility in the respective asset return as a local average of the absolute value of daily returns of indices in an appropriate time window of T days as

$$v = \frac{\sum_{t=1}^{T-1} |G_i(t)|}{T-1} \tag{5}$$

To standardize the values of obtained from Equation (4) above for all values of i, we normalize $G(t)_i$ as follows

$$g(t)_i = \frac{\prec G(t)_i - G(t)_i}{\sigma_i} \tag{6}$$

where $\sigma_i = \sqrt{G(t)_i^2 - G(t)_i^2}$ represents the average in the period studied.

From real time series data of the implied volatility surface, we can calculate the element of $N \times N$ correlation matrix C as follows

$$C_{ij} = \langle g_i(t)\, g_j(t) \rangle = \frac{\langle [G_i(t) - \langle G_i \rangle][G_j(t) - \langle G_j \rangle] \rangle}{\sqrt{[\langle G_i^2 \rangle - \langle G_i^2 \rangle][\langle G_j^2 \rangle - \langle G_j^2 \rangle]}} \tag{7}$$

C_{ij} lies in the range of the closed interval $-1 \le C_{ij} \le 1$, with C_{ij}=0 means there is no correlation, C_{ij}=-1implies anti-correlation and C_{ij}=1 means perfect correlation for the empirical correlation matrix.

3.2. Eigenvalue Spectrum of the Correlation Matrix

As stated earlier, our aim is to extract information about the cross-correlation from the empirical correlation matrix C. To this end, we are going to compare the properties of C with those of a random matrix; see, [9] [11] [21] [22] [37]. It can be shown from [38] that the empirical correlation matrix C can be expressed as

$$C = \frac{1}{L} GG^T \tag{8}$$

where G is the normalized $N \times L$ matrix and G^T is the transpose of G. This empirical matrix will be compared with a random Wishart matrix R given by:

$$R = \frac{1}{L} AA^T \tag{9}$$

to classify the information and noise in the system [22] [37] , where A is an $N \times L$ matrix whose entries are independent identically distributed random variables that are normally distributed and have zero mean and unit variance.

In our bid to use the random matrix theory in portfolio optimization and (derivative) assets risk management, we should be conversant with the universal properties of random matrices. Wilcox et al. [16] assert that there are four underlying properties of random matrices which include (a) Wishart distribution eigenvalues from the correlation matrix, (b) Wigner surmise for eigenvalue spacing (c) the distribution of eigenvector components of the corresponding eigenvalues and finally (d) Inverse participation ratio for Eigenvector components of the resulting correlation matrix. Authors like [26] [30] [40] , assert that the statistical properties of Rare known and that in particular for the limit as $N \to \infty$ we have that $Q = \frac{L}{N} (\geq 1)$ is fixed. The probability function $P_{rm}(\lambda)$ of eigenvalues λ of the random correlation matrix R is given by

$$P(\lambda) = \frac{Q}{2\pi\sigma^2} \frac{\sqrt{(\lambda_{\max} - \lambda)(\lambda - \lambda_{\min})}}{\lambda} \tag{10}$$

for λ such that $\lambda_{\min} \leq \lambda \leq \lambda_{\max}$, where σ^2 is the variance of the elements of A. Here $\sigma^2=1$and $\lambda_{\min}$ and $\lambda_{\max}$ satisfy

$$\lambda_{\max/\min} = \sigma^2 \left(1 + \frac{1}{Q} \mp 2\sqrt{1/Q} \right) \tag{11}$$

The values of lambda from Equation (10) that satisfy (11) and (12) are called the Wishart distribution of eigenvalues from the correlation matrix. These values of lambda obtained from Equation (11) as stated before determine the bounds of theoretical eigenvalue distribution. When the eigenvalues of empirical correlation matrix *C* are beyond these bounds, they are said to deviate from the random matrix bounds and are therefore supposed to carry some useful information about the market, [12].

The distribution of eigenvalue spacing was introduced as the required test for the case when there are not significant deviations of the empirical eigenvalue distribution to that of the random matrix prediction Wilcox et al. [16]. When the eigenvalues so obtained from the correlation matrix do not deviate significantly from the predictions of the RMT we apply the so-called Wigner surmise for eigenvalue spacing otherwise called Gaussian orthogonal ensemble [11] and is given by

$$P(s) = \frac{s}{2\pi} \exp\left(-\frac{s\pi^2}{4} \right) \tag{12}$$

where $(\lambda_{i+1} - \lambda_i)/d$ and *d* denotes the average of the differences $\lambda_{i+1} - \lambda_i$ as *i* varies.

3.3. Distribution of Eigenvector Component

The concept that low lying eigenvalues are really random can also be verified by studying the statistical structure of the corresponding eigenvectors. The *j*th component of the eigenvector corresponding to each eigenvalue λ_α will be denoted by, v_{aj} and then normalized such that $\sum_{j=1}^{N} v_{\alpha,j}^2 = N$. Plerou *et al.* [9] assert that if there is no information contained in the eigenvector, v_{aj} , one expects that for a fixed *α*, the distribution of $u = v_{aj}$ *(as j is varired)* is a maximum entropy. This, therefore, leads to what is called Porter-Thomas distribution in the theory of random matrices written as

$$p(u) = \frac{1}{\sqrt{2\pi}} \exp\left(-\frac{u^2}{2} \right) \tag{13}$$

In line with the assumption of pure randomness and independence, the distribution of the components, $u_a(l)$for l=1,2,3---N of an eigenvector u_a of a random correlation matrix, R should obey the standard normal distribution with zero mean and unit variance, [41]. The distribution so obtained from (13) above are expected to fit well the histogram of the eigenvector except for those corresponding to the highest eigenvalues which lie beyond the theoretical value of, $\lambda_{\max}$, [9].

3.4. Inverse Participation Ratio

Guhr, T. *et al.* [41] assert that to quantify the number of components that participates significantly in each eigenvector, we use inverse participation ratio (IPR). This (IPR) shows the degree of deviation of the distribution of eigenvectors from RMT results and distinguishes one eigenvector with approximately equal components with another that has a small number of huge components. For each eigenvector, *va*, [11] defined the inverse participation ratio as

$$I_a = \sum_{j=1}^{N} \left(v_{aj}\right)^4 \tag{14}$$

where N is the number of the time series (the number of implied volatility considered) and hence the number of eigenvalue components and v_a^j is the j th component of the eigenvector,v_a. There are two limiting cases of $I_a(i)$; If an eigenvector v_a has an identical component, $v_a(j)=\frac{1}{\sqrt{N}}$ then $I_a=\frac{1}{N}$ and (ii) For the case when the eigenvector v_a has one element with v_a=(j)=1 and the remaining components zero, then I_a=1 Therefore, the IPR can be illustrated as the inverse of the number of elements of an eigenvector that are different from zero that contribute significantly to the value of the eigenvector. [42] in their study of the RMT assert that the expectation of the IPR is given by

$$\langle I_a \rangle = N\int_{-\infty}^{\infty} \left[v_a(j)\right]^4 \frac{1}{\sqrt{2\pi N}} \exp\left(-\frac{\left[v_a(j)\right]^2}{2N}\right) dv_a(j) = \frac{3}{N} \tag{15}$$

since the kurtosis (extreme deviations) for a distribution of eigenvector components s 3.

4. EMPIRICAL RESULT AND DATA ANALYSIS

4.1. Eigenvalue Analysis

We took a sample study of eighty-two (N = 82) stocks from the Nigerian stock exchange which gave rise to L = 1019 daily closing prices. For the Johannesburg stock exchange, JSE we had a sample study of thirty-five (N' = 35) stocks with a total of L' = 1148. The theoretical eigenvalue bounds in the NSM are respectively

$\lambda_- = 0.51$ and $\lambda_+ = 1.65$ as minimum and maximum values with $Q = \frac{L}{N} = 12.41$.

Further from the calculation, the market value shows that the largest eigenvalue $\lambda_1 = 4.87$ which is approximately three times larger than the predicted RMT of value (1.64). Similarly for the JSE, the theoretical eigenvalue bounds of the correlation matrix are $\lambda_- = 0.21$ and $\lambda_+ = 2.37$ as minimum and maximum eigenvalues respectively, with $Q' = \frac{L'}{N'} = 32.77$ .A high percentage of the eigenvalues obtained from the empirical correlation matrix of stock market price returns lie below $\lambda_{min}(\lambda_-)$, just as obtained by [16] and this is attributable to the fact that many of the liquid stocks behave independently when compared with the rest of the market. The empirical market value calculations show that the largest eigenvalue $\lambda_1 = 11.86$ which is five times larger than the predicted RMT value of 2.37 above. If there were no correlations between the stocks in NSM and JSE, the eigenvalues derived from the market data would have been bounded between $\lambda_- = 0.51$ and $\lambda_+ = 1.65$ for NSM and $\lambda_- = 0.21$ and $\lambda_+ = 2.37$ for JSE respectively. In NSM 7.3% of the eigenvalue lie outside the theoretical value and therefore contain information about the market whereas in JSE 8.57% of the total eigenvalue carry information about the entire market (see Figure 1 and Figure 2). With these significant deviations in the empirical eigenvalue distribution from the RMT predictions, the test for Wigner surmises for eigenvalue spacing are not relevant in this case.

The average $\langle C_{ij} \rangle$of the elements of the market correlation matrix for the NSM is 0.041, and that of the JSE is 0.168, showing that even though the two markets are both emerging the JSE is about four times more correlated than that of the NSM. Thus, this shows that the Johannesburg market is much more emerging than the Nigerian market, [10]. It, therefore, means that since many assets in JSE are more correlated than that of the NSM, perhaps different macroeconomic forces are driving the two markets, [19]. It is also worthy of mention that the empirical correlation matrices obtained from the two markets are positive definite since all the eigenvalues obtained are all positive.

The comparable informative indices (7.3% and 8.6%) for NSM and JSE, respectively, suggest a similarity between the market microstructures in the system.

Figure 3 above represent the distribution of eigenvectors for the various eigenvalues in the empirical correlation matrix of the NSM. The eigenvector labelled U1 and U82 represents an eigenvector for deviating eigenvalue in the theoretical (hypothetical) region whereas the other 4 diagrams are the eigenvector components of the eigenvalue within the regions predicted from the Random Matrix Theory.

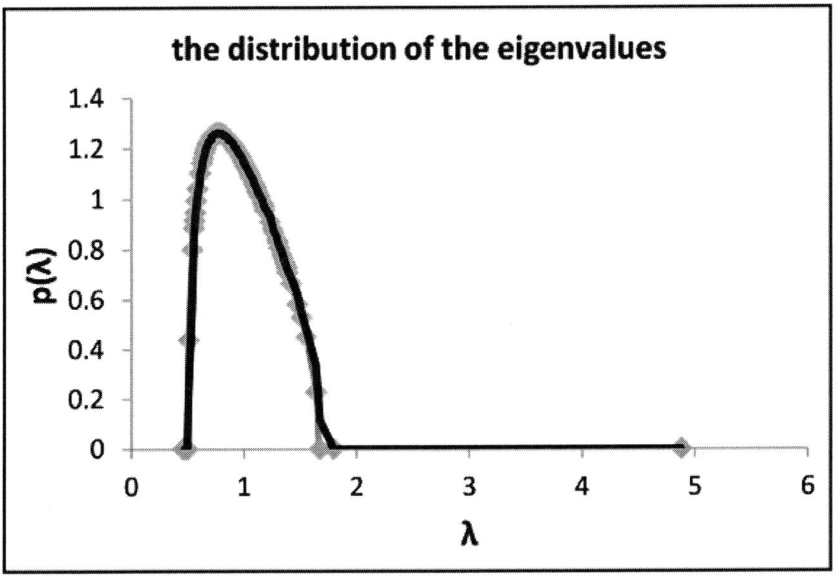

Figure 1. Theoretical (Marcenko-Pastur) empirical eigenvalues for NSM (source: Nigerian Stock Market price return 2009-2013).

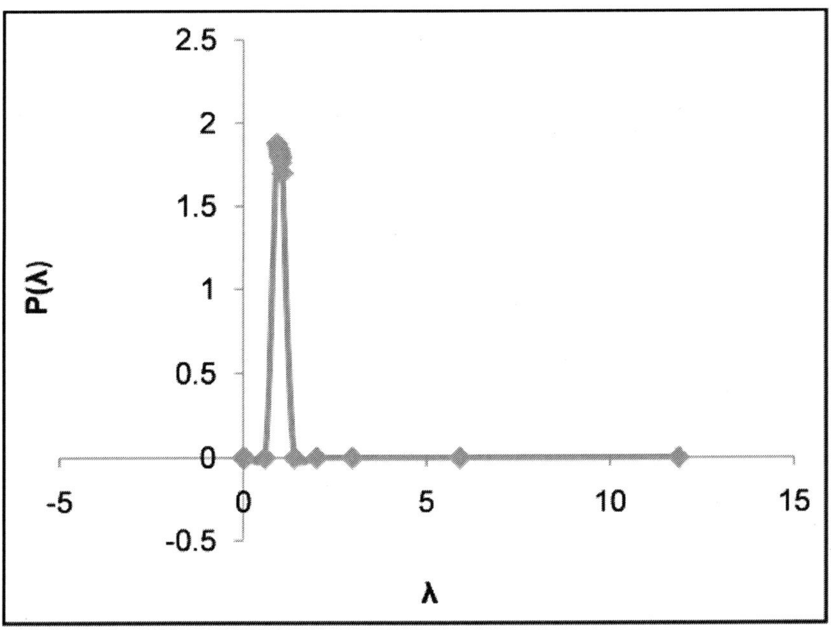

Figure 2. Theoretical (Marcenko-Pastur) empirical eigenvalues for JSE (source: Johannesburg Stock Exchange price return 2009-2013).

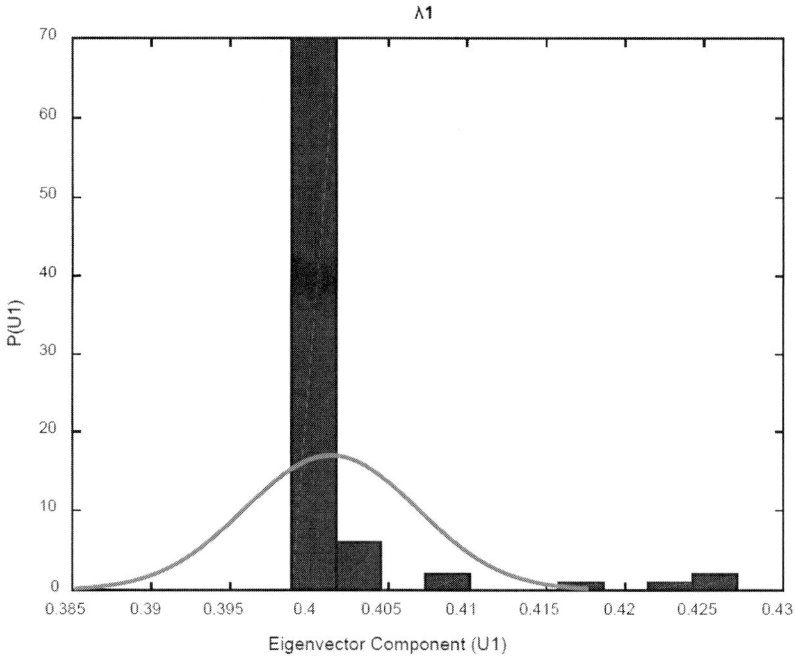
λ1
P(U1)
70
60
50
40
30
20
10
0
0.385
0.39
0.395
0.4
0.405
0.41
0.415
0.42
0.425
0.43
Eigenvector Component (U1)

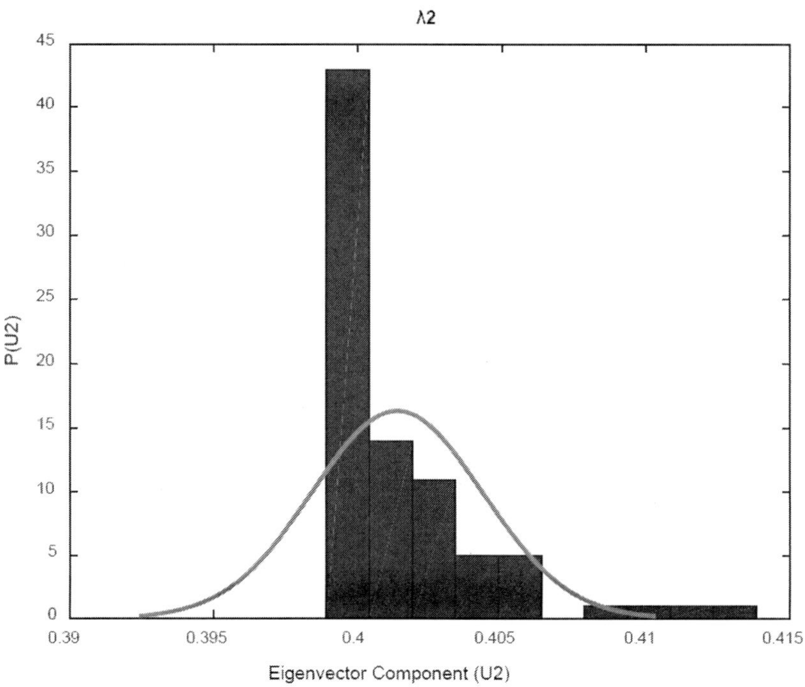
λ2
P(U2)
45
40
35
30
25
20
15
10
5
0
0.39
0.395
0.4
0.405
0.41
0.415
Eigenvector Component (U2)

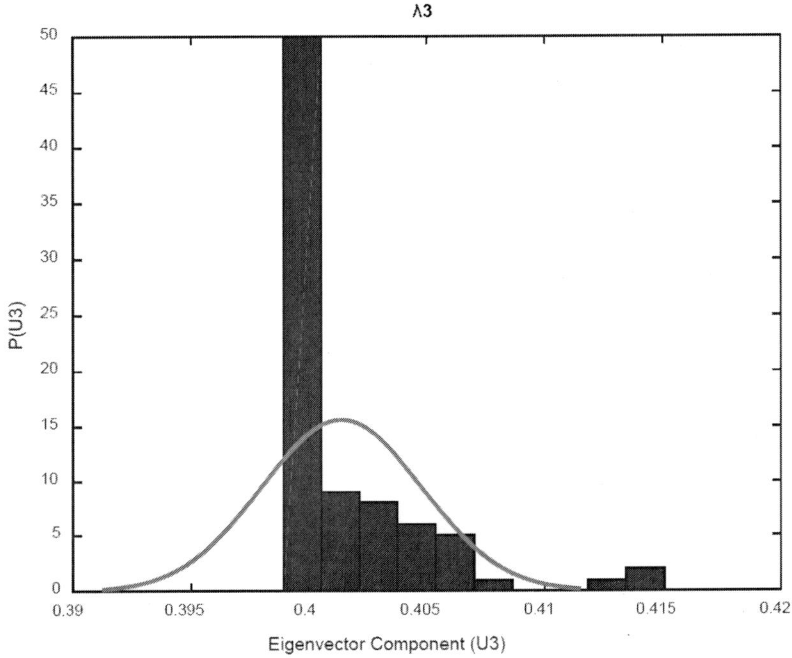
λ3
P(U3)
50
45
40
35
30
25
20
15
10
5
0
0.39
0.395
0.4
0.405
0.41
0.415
0.42
Eigenvector Component (U3)

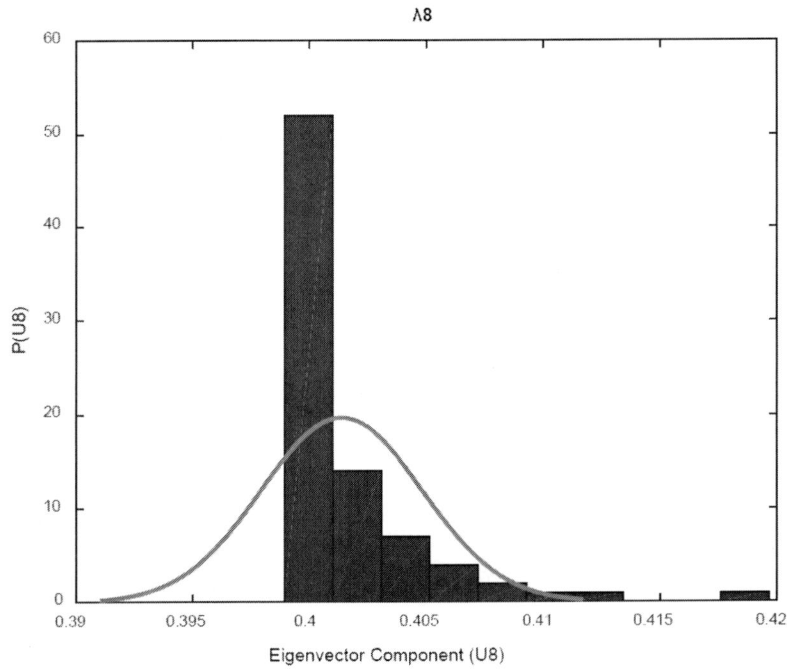
λ8
P(U8)
60
50
40
30
20
10
0
0.39
0.395
0.4
0.405
0.41
0.415
0.42
Eigenvector Component (U8)

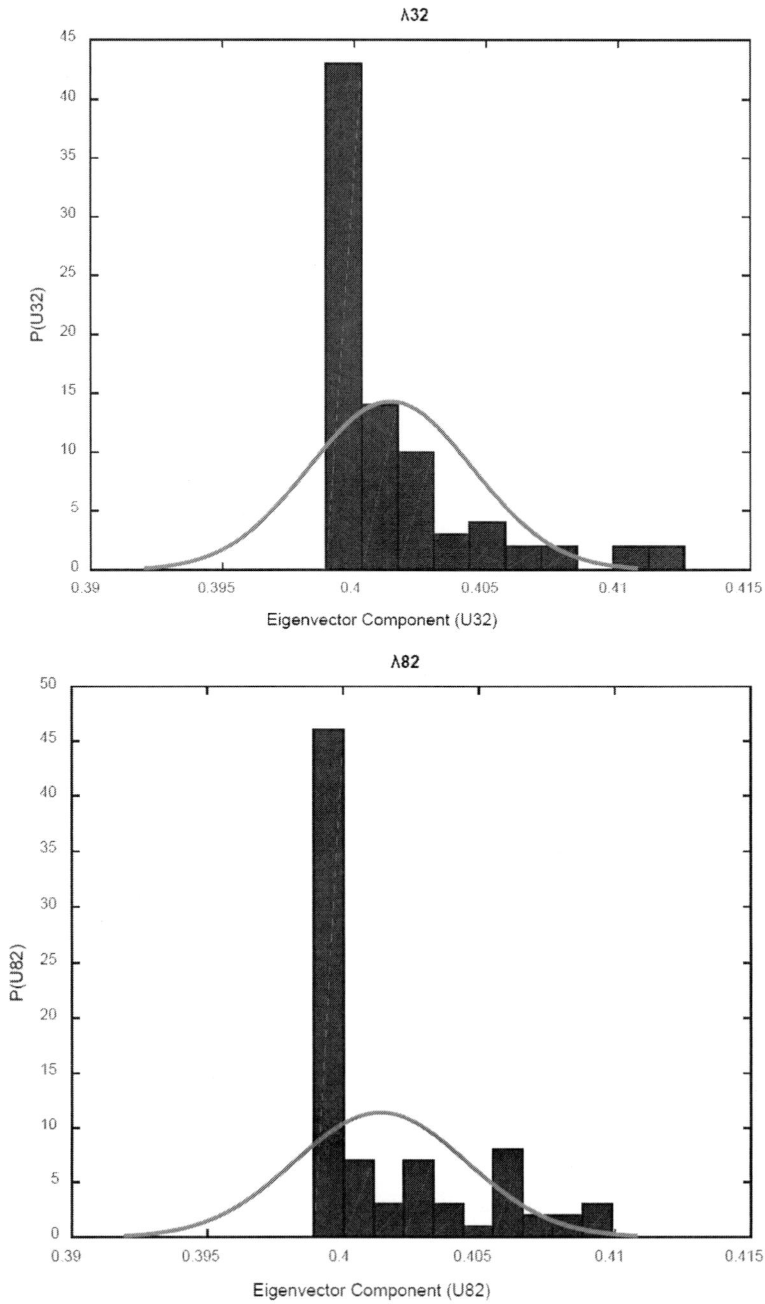

Figure 3. Distribution of eigenvector components of stocks in NSM. Source [43]

The overwhelming non-informativeness of the remaining 92.7% and 91.4% of the overall markets, further suggests typical random behaviour of the two markets. Typically, the distribution of the first three eigenvectors indicates the key features (mean, standard deviation and kurtosis) of a market. A look at these first three distributions for the NSM shows compared to the normal distribution, they are skewed and leptokurtic in mean and standard deviations, but fairly symmetric in kurtosis. The JSE versions portray similar non-symmetric behaviours, but fairly symmetric in kurtosis. The NSM distributions would seem to follow a beta-gamma family of distribution while the JSE ones are mostly negatively skewed, as opposed to the first two NSM distributions which are positively skewed. In general, higher-order distributions are examined for a more detailed understanding of market-dynamics, for example, market microstructure.

These distributions present the same profiles as the first three distributions in the two markets, which suggest persistence of market features and the driving economic forces. Given the fact the distributions reveal the presence of market information outside the noisy RMT range; the results suggest potential market inefficiency and ability to make money from the markets. We cannot, however, say more that this regarding the stylised fats and market features, without a detailed examination of the key financial economics features typically explored in empirical finance, namely market efficiency, volatility, bubbles, anomalies, valuations and predictability.

Figure 4 shows the eigenvector distribution for some eigenvalues within and outside the theoretical region of the Random Matrix Theory. The last diagrams

V34 and V35 represent the eigenvectors corresponding to an eigenvalue outside the region predicted by RMT which contain the information about the market. The other eigenvectors correspond to the eigenvalues due to noise as they lie in the region predicted by RMT.

The key interest in this paper is to assess how similar the NSM and JSE are, to facilitate future modelling of as yet non-existent derivative prices in the NSM using available information on existing derivative prices in the JSE. For this, a comparative look at the two sets of eigenvector distributions suggest a flipping over or reverse dynamics in the JSE in comparison with the NSM. For example, the U2 and U_3(NSM) versus V2 and V_3(JSE) eigenvalue distributions are mirror reflections of each other. The practical implication of this reveals that different market forces seem to drive the NSM and JSE. This result is intuitively meaningful because the NSM is an oil-dependent and erratic in its price dynamics and market microstructure unlike the JSE which is mining dependent, and is therefore relatively stable in nature. Consequently, attempts to model, say, non-existent derivative prices in Nigeria using existing prices in the JSE have to be taken cautiously.

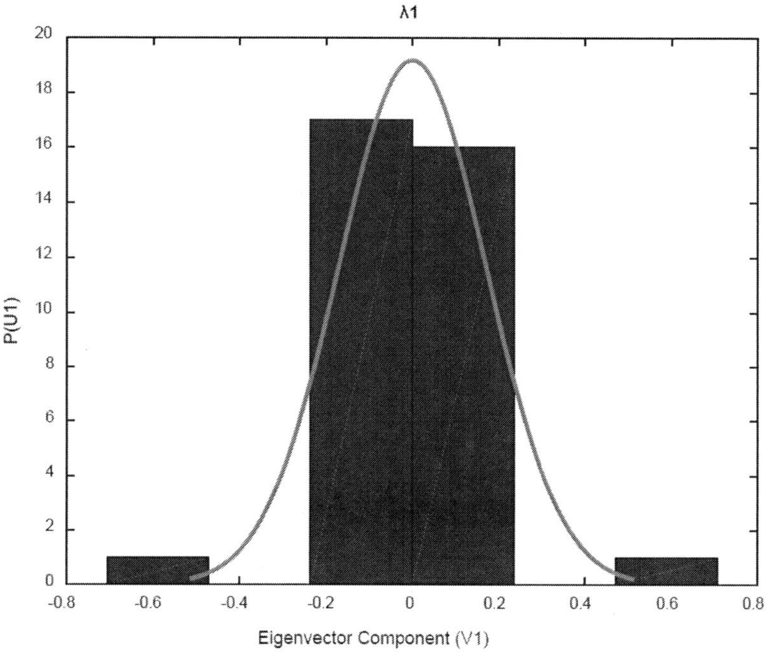
λ1
P(U1)
Eigenvector Component (V1)

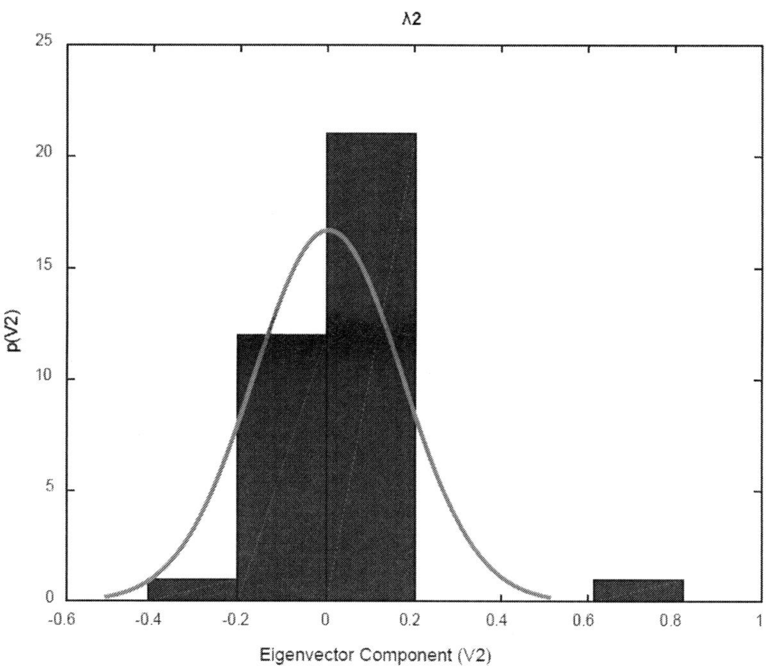
λ2
p(V2)
Eigenvector Component (V2)

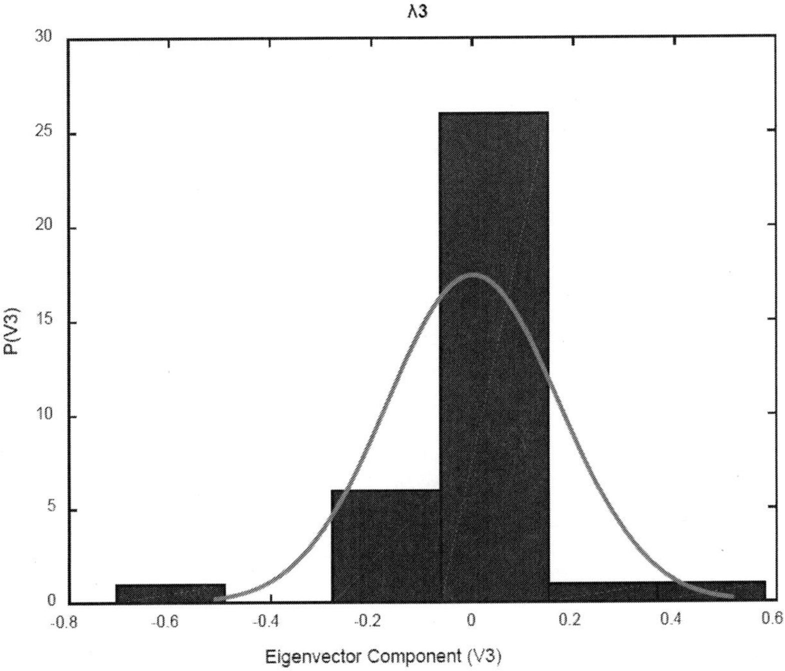
λ3
30
25
20
15
10
5
0
P(V3)
-0.8
-0.6
-0.4
-0.2
0
0.2
0.4
0.6
Eigenvector Component (V3)

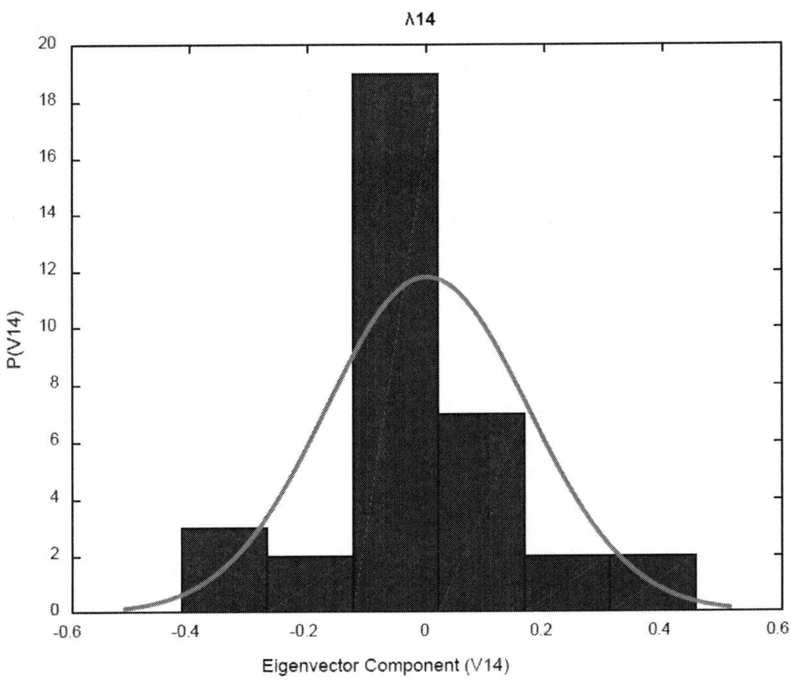
λ14
20
18
16
14
12
10
8
6
4
2
0
P(V14)
-0.6
-0.4
-0.2
0
0.2
0.4
0.6
Eigenvector Component (V14)

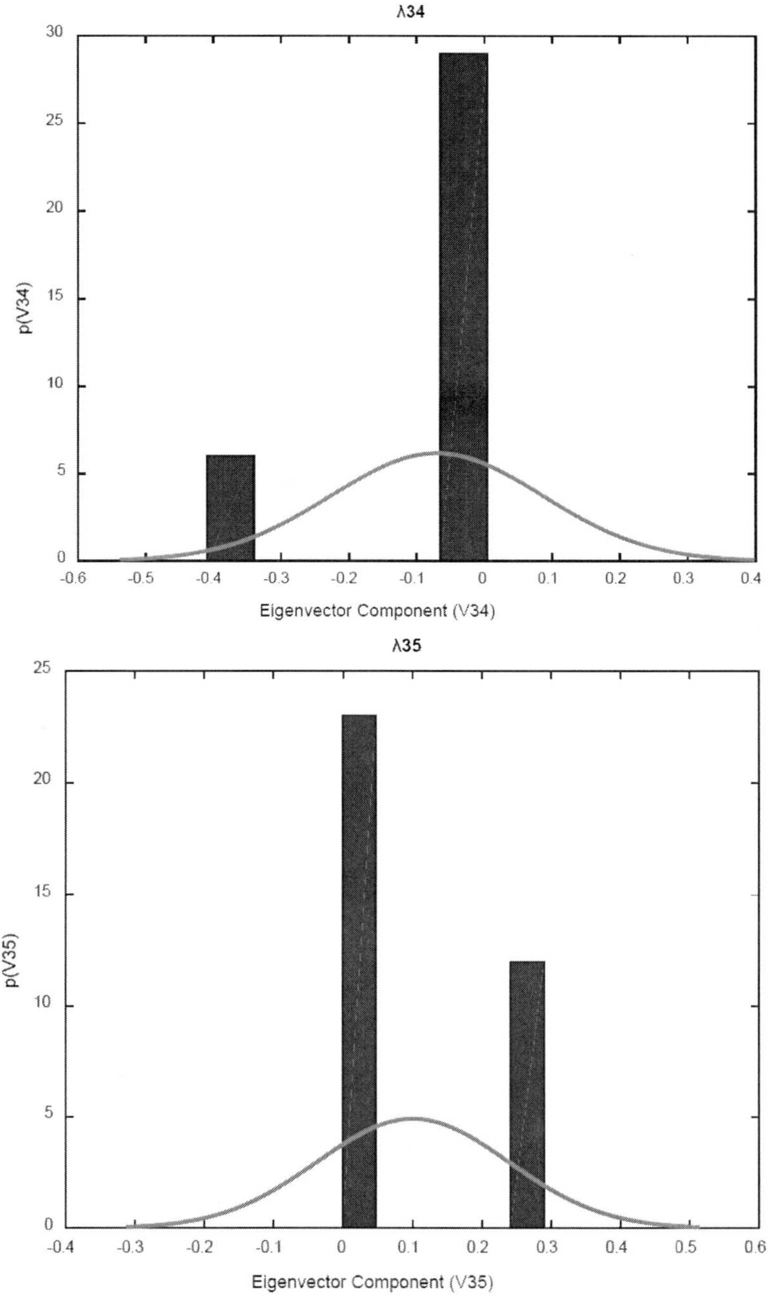

Figure 4. Distribution of eigenvector components of stocks in JSE. Source: [44].

That said, the flipping-over features suggest that including NSM and JSE stocks in an African Emerging Markets portfolio would achieve reasonable portfolio diversification and corresponding Markowitz-style mean-variance portfolio optimization. These insights reveal the power of statistical physics tools such as RMT in peering through complex market dynamics which may not manifest with traditional mathematical finance techniques.

4.2. Inverse Participation Ratios (IPRs)

The inverse participation ratio (IPR) is the multiplicative inverse of the number of eigenvector components that contribute significantly to the eigenmode, [11]. For the largest eigenvalue nine deviating from the RMT bounds, almost all the stocks contribute to the corresponding eigenvector thereby justifying treating this eigenvector as the market factor. The eigenvector corresponding to other deviating eigenvalues also exhibits that their corresponding stocks contribute slightly to the overall market features in the two exchanges, NSM and JSE.

The average IPR value is around 3/82 for NSM & 1/35 for JSE respectively larger than would be expected $1/N = 1/82 = 0.01$ for NSM & $1/35 = 0.03$ for JSE, if all components contributed to each eigenvector, [37]. The remaining eigenvectors appear to be random with some deviations from the predicted value of $3/N = 0.04$ and 0.09 respectively for NSM and JSE possibly as a result of the existence of fat tails and high kurtosis of the return distributions.

The lower end of JSE and the higher end of the eigenvalues for both exchanges (NSM and JSE) show deviations suggesting the existence of localized modes. It is noticeable from Figure 5 and Figure 6 that these deviations are fewer in number for JSE than that of the NSM, which implies that distinct groups whose members are mutually correlated in their price movements are witnessed in both markets although they are more noticeable in JSE.

4.3. Contributions to Knowledge

This paper stems from a doctoral research which aims to model yet non-existing derivative prices in the NSM, using existing prices in the JSE. The underpinning heuristics (not developed in detail in this paper) is to backtrack from measures of similarity or dissimilarity between the stylized facts and other empirical correlates of the two market dynamics, one of which is the random matrix correlation structures. The paper, therefore, is novel in foregrounding the modeling of non-existing financial derivatives for the first time known to the authors.

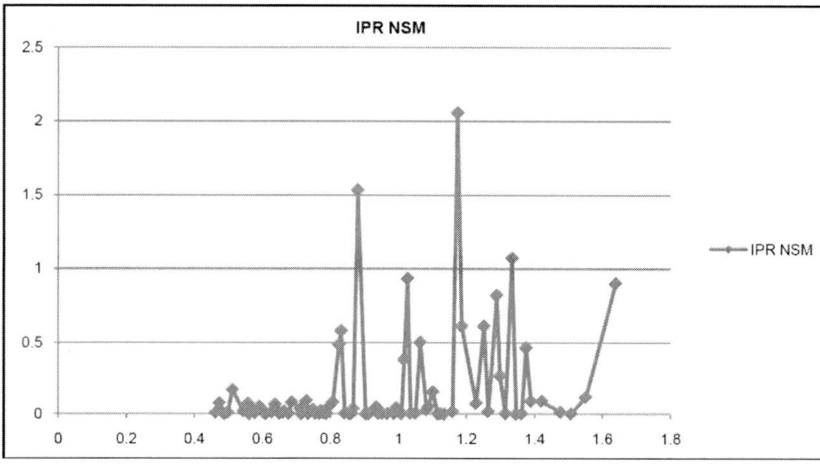

Figure 5. Inverse participation ratio and their ranks for NSM. Source: [43].

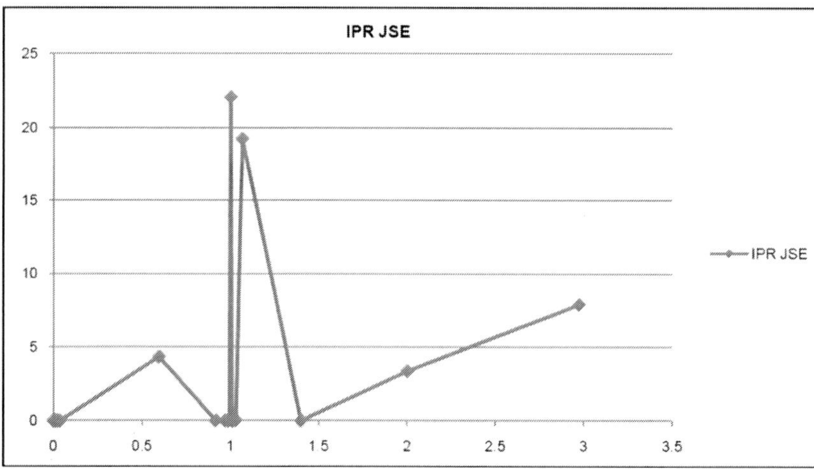

Figure 6. Inverse participation ratio and their ranks for JSE. Source: [44].

4.4. Limitations of the Study

It would have been preferable to use up to date data (2009-2016) for the two markets to accommodate the recent impact of oil price fluctuation on the market dynamics. This was not possible since for the NSM available data from the Nigerian Stock Exchange when this research was being carried out range from 2009-2013. The authors therefore, used this range that was available for the analysis. Strictly speaking from the point of using the results in derivative pricing, this limitation is not severe as one can forecast parts of the data that are not available or simulate alternative impact scenarios for the revealed price paths of crude oil between 2013 and 2016, for example.

5. CONCLUSION AND HINTS ON FUTURE WORK

The analysis of the correlation and structure of stock market returns for the two most dominant markets in the Sub-Saharan Africa, NSM and JSE, was carried out in this paper using RMT. Marcenko-Pastur eigenvalue distribution predicted that the theoretical eigenvalues should be in the range of 0.52 and 1.65 for NSM and 0.21 and 2.37 for JSE respectively. While for NSM it was observed that 6 out of 82 stocks considered that have their corresponding eigenvalues lie outside this theoretical bound of eigenvalues, in JSE 3 out of the 35 stocks has their eigenvalues outside the predicted eigenvalue bounds.

Therefore, 89% of the information from the return distributions is purely random thereby leaving us with the alternative hypothesis of the RMT which states that the information on the market lies on the deviating eigenvalues which imply then that for NSM the true market characteristic lies with only 11% of the assets examined. Similarly, for JSE, only 9% of the stocks considered have information about the market which can be used in constructing portfolios with better stable returns and optimal risk management. As stated earlier, these correlation matrices contain some relevant information for option pricing and hedging [45].

We noted earlier in the literature review that random matrix theory could be very useful in options trading, hedging and in the management of risks associated with a portfolio of investment. In this regard, we intend to use the RMT results in this paper to construct suitable investment portfolios from overall market and sector-based results, for given weights and implied volatilities of the stocks in the respective portfolios under consideration.

As Nigeria is yet to commence trade on derivative products, we will carry out heuristic analyses of the option price data for NSM, and execute same for JSE using obtained data from the Johannesburg Stock Exchange, to adjudicate the relative performances of different derivative pricing models in the two markets.

REFERENCES

1. Gentle, J. (1998) Numerical Linear Algebra with Applications in Statistics. Statistics and Computing, Springer, Berlin. https://doi.org/10.1007/978-1-4612-0623-1
2. Gentle, J. (1998) Singular Value Factorization. Springer, Berlin
3. Morrison, D.F. (1990) Multivariate Statistical Methods. 3rd Edition, McGraw-Hill, Inc., New York, 1-85
4. Jackson, E. (2003) A User's Guide to Principal Components. John-Wiley and Sons Inc., New York, 1-65
5. Cont, R. and Fonseca, J. (2002) Dynamics of Implied Volatility Surfaces. Quantitative Finance, 2, 45-60. https://doi.org/10.1088/1469-7688/2/1/304
6. Pafka, S. and Kondor, I. (2004) Estimated Correlation Matrices and Portfolio Optimization. Physica A, 343, 623-634
7. Wang, G., Xie, C., Chen, S., Yang, J. and Yang, M. (2013) Random Matrix Theory Analysis of Cross-Correlations in the US Stock Market: Evidence from Pearson's Correlation Coefficient and Detrended Cross-Correlation Coefficient. Physica A, 392, 3715-3730
8. Laloux, L., Cizeau, P., Bouchaud, J. and Potters, M. (1999) Noise Dressing of Financial Correlation Matrices. Physical Review Letters, 83, 1467-1470. https://doi.org/10.1103/PhysRevLett.83.1467
9. Plerou, V., Gopikrishnan, P., Rosenow, B., Amaral, L.A.N. and Stanley, H.E. (1999) Universal and Non-Universal Properties of Cross-Correlations in Financial Time Series. Physical Review Letters, 83, 1471-1474. https://doi.org/10.1103/PhysRevLett.83.1471
10. Shen, J. and Zheng, B. (2009) Cross-Correlation in Financial Dynamics. A Letters Journal Exploring the Frontiers of Physics, 86, 48005. https://doi.org/10.1209/0295-5075/86/48005
11. Plerou, V., Gopikrishnan, P., Rosenow, B., Amaral, L.A.N., Guhr, T. and Stanley, H.E. (2002) Random Matrix Approach to Cross-Correlations in Financial Data. Physical Review E, 65, Article ID: 066126. https://doi.org/10.1103/physreve.65.066126
12. Cukur, S., Eryigit, M. and Eryigit R. (2007) Cross-Correlation in an Emerging Market Financial Data. Physica A, 376, 555-564
13. Junior, L. and Franca, I. (2012) Correlation of Financial Markets in Times of Crisis. Physica A, 391, 187-208
14. Kulkarni, V. and Deo, N. (2007) Correlation and Volatility in an Indian Stock Market: A Random Matrix Approach. EuroPhysics Journal B, 60, 101-109
15. Sharma, C. and Banerjee, K. (2015) A Study of Correlation in the Stock Market. Physica A: Statistical Mechanics and Its Applications, 321-330
16. Wilcox, D. and Gebbie, T. (2007) An Analysis of Cross-Correlations in an Emerging Market. Physica A, 375, 584-598
17. Kumar, R. and Sinha, S. (2007) Collective Behaviour of Stock Price Movements in an Emerging Market. Physical Review E, 76, Article ID: 046116
18. Kim, M., Lee, S., Hwang, D., Kim, S. and Ko, I. (2010) Dynamics of Implied Volatility Surfaces from Random Matrix Theory. Physica A, 389, 2762-2769
19. Fenn, D.J., Porter, M., Williams, S., McDonald, M., Johnson, N. and Jones, N. (2011) Temporal Evolution of Financial Market Correlations. Physical Review E, 84, Article ID: 026109. https://doi.org/10.1103/physreve.84.026109
20. Nobi, A., Maeng, S., Gyun, G. and Lee, J. (2013) Random Matrix Theory and Cross-Correlations in Global Financial Indices and Local Stock Market Indices. Journal of the Korean Physical Society, 62, 569-574. https://doi.org/10.3938/jkps.62.569
21. Laloux, L., Cizeau, P. and Potters, M.J. (2000) Bouchaud, Random Matrix Theory and Financial Correlations. International Journal of Theoretical and Applied Finance, 3, 1-7. https://doi.org/10.1142/S0219024900000255
22. Gopikrishnan, P., Rosenow, B., Plerou, V. and Stanley, E. (2001) Quantifying and Interpreting Collective Behaviour in Financial Markets. Physical Review E, 64, Article ID: 035106

23. Engle, R. and Figlewski, S. (2014) Modeling the Dynamics of Correlation among Implied Volatilities. REV Finance Group, Inc., Arcadia, 1-28.
24. Plerou, V., Gopikrishnan, P., Rosenow, B., Amaral, L.A.N. and Stanley, H.E. (2000) A Random Matrix Theory Approach to Financial Cross-Correlations. Physica A, 287, 374-382
25. Markowitz, H. (1952) Portfolio Selection. Journal of Finance, 2, 77-91. https://doi.org/10.1111/j.1540-6261.1952.tb01525.x
26. Edelman, A. (1988) Eigen Values and Condition Numbers of Random Matrices. SIAM Journal on Matrix Analysis and Applications, 9, 543-560. https://doi.org/ 10.1137/0609045
27. Plerou, V., Gopikrishnan, P., Rosenow, B., Amaral, L.A.N. and Stanley, H.E. (2001) Collective Behaviour of Stock Price Movements: A Random Matrix Theory Approach. Physica A, 299, 175-180
28. Gee-Rae, K. and Markowitz, H.M. (1989) Investment Rules, Margin and Market Volatility. The Journal of Portfolio Management, 6, 45-52
29. Palmer, R.G., Arthur, W.B., Holland, J.H. and LeBaron, B. (1994) Artificial Economic Life: A Simple Model of a Stock Market. Physica D, 75, 264-274
30. Kondor, I. and Kertesz, J. (1999) EconoPhysics: An Emerging Science. Kluwer, Dordrecht
31. Chatterjee, A. and Chakrabarti, B.K. (2006) Econophysics of Stock and Other Markets. Springer, Berlin. https://doi.org/10.1007/978-88-470-0502-0
32. Markowitz, H. (1959) Portfolio Selection: Efficient Diversification of Investment. John Wiley and Sons, New York
33. Pafka, S. and Kondor, I. (2003) Noisy Covariance Matrices and Portfolio Optimization 2. Physica A, 319, 487-494
34. Galluccio, S., Bouchaud, J. and Potters, M. (1998) Rational Decisions, Random Matrices and Spin Glasses. Physica A, 259, 449-456
35. Wang, D., Podobrink, B., Horvatic, D. and Stanley, H.E. (2011) Quantifying and Modeling Long-Range Cross-Correlations in Multiple Time Series with Applications to World Stock Indices. Physical Review E, 83, Article ID: 046121. https://doi.org/ 10. 1103/PhysRevE.83.046121
36. Sensoy, A., Yuksel, S. and Erturk, M. (2013) Analysis of Cross-Correlations between Financial Markets after the 2008 Crisis. Physica A, 392, 5027-5045
37. Colon, T., Ruskin, H.J. and Crane, M. (2007) Random Matrix Theory and Fund of Funds Portfolio Optimization. Physica A, 382, 565-576
38. Sharifi, S., Crane, M., Shamaie, A. and Ruskin, H. (2004) Random Matrix Theory for Portfolio Optimization: A Stability Approach. Physica A, 335, 629-643.
39. Sengupta, A.M. and Mitra, P.P. (1999) Distributions of Singular Values for Some Random Matrices. Physical Review E, 60, 1-4. https://doi.org/10.1103/PhysRevE.60.3389
40. Bai, Z.D. (1999) Methodologies in Spectral Analysis of Large Dimensional Random Matrices: A Review. Statistica Sinica, 9, 611-677
41. Guhr, T., Muller, A. and Weidenmuller, H. (1998) Random Matrix Theories in Quantum Physics: Common Concepts. Physics Reports, 299, 189-425
42. Utsugi, A., Ino, K. and Oshikawa, M. (2004) Random Matrix Theory Analysis of Cross-Correlations in Financial Markets. Physical Review E, 70, Article ID: 026110. https://doi.org/10.1103/physreve.70.026110
43. Closing Prices of Some Stocks from the Nigerian Stock Market Official Data Obtained through the Nigerian Stock Exchang
44. Closing Prices of Some Stocks Selected from the 40 Most Actively Traded Stocks on the Johannesburg Stock Exchange, South Africa, Obtained from the Johannesburg Stock Exchange
45. Hull, J. (1997) Options, Futures and Other Derivative Securities. Prince Hall, Englewood Cliffs

CHAPTER 5

Analysis on a Mathematical Model for Tumor Induced Angiogenesis

Richard C. Ewool, Zachariah Sinkala

Department of Computational Science & Mathematics, Middle Tennessee State University, Murfreesboro, USA

ABSTRACT

Tumor-induced angiogenesis is the process by which unmetastasized tumors recruit red blood vessels by way of chemical stimuli to grow towards the tumor for vascularization and metastasis. We model the process of tumor-induced angiogenesis at the tissue level using ordinary and partial differential equations (ODEs and PDEs) that have a source term. The source term is associated with a signal for growth factors from the tumor. We assume that the source term depends on time, and a parameter (time parameter). We use an explicit stabilized Runge-Kutta method to solve the partial differential equation. By introducing a source term into the PDE model, we extend the PDE model used by H. A. Harrington et al. Our results suggest that the time parameter could play some role in understanding angiogenesis.

Keywords: Explicit Stabilized Runge-Kutta Methods, Angiogenesis, Inhibitors, Rat Cornea, Source Term

1. INTRODUCTION

Tumor Angiogenesis is the process by which cancerous cells signal surrounding blood capillaries to initiate the formation of new red blood cells. These new capillaries gradually form a dendritic structure and spread through the extra cellular matrix. The red blood vessels grow progressively over time towards the local tumor mass under the influence of chemical stimuli such as vascular endothelial fibroblast growth factors (VEGF) or basic fibroblast growth factors (BFGF) [1] -[4]. The chemical agents that contribute to the process of angiogenesis come from a source, located usually in the necrotic, proliferating or both regions of the local tumor [5] [6]. It is a well established fact that unvascularized tumors cannot grow beyond 2 mm and are not likely to metastasize. This has been proved experimentally by Judah Folkman et al. [7]. Since the publication by Judah Folkaman several hypothesis and research have been conducted to investigate the possibility of controlling angiogenesis in the case of cancerous cells. Our research can play a role in understanding the challenge that angiogenesis presents by studying the influence of the source of growth factors in angiogenesis.

By introducing a source term into the PDE model, we extend the model proposed by H. A. Harrington et al. [8]. The model describes how tumor angiogenic factors from a source influences the growth of red blood cells in the direction of the tumor. It also assesses the interplay between tumor angiogenic inhibitor and the tumor angiogenic factors.

We assume that the source term depends on time and a location in the tumor (most likely the necrotic region, proliferating cells or both) which induces tumor angiogenic growth factors (TAF), such as vascular endothelial growth factors (VEGF) or basic fibroblast growth factors (BFGF). The proposed source term $Q(t,T_0)$ is a function of the form;

$$Q(t,T_0) = \begin{cases} 0, \text{ if } (0 \le t < T_0) \\ Q_p, \text{ if } (t \ge T_0) \end{cases}$$

T_0 represents the time it takes the tumor to initiate new signaling growth factors. For T_0 large enough, our model reduces to H. A. Harrington's model. The threshold needed to signal growth factors from the source term, $Q_p$is assumed to be a constant value for simplicity and lack of experimental data. Figure 1 is a cartoon diagram of the model as described in the previous paragraphs.

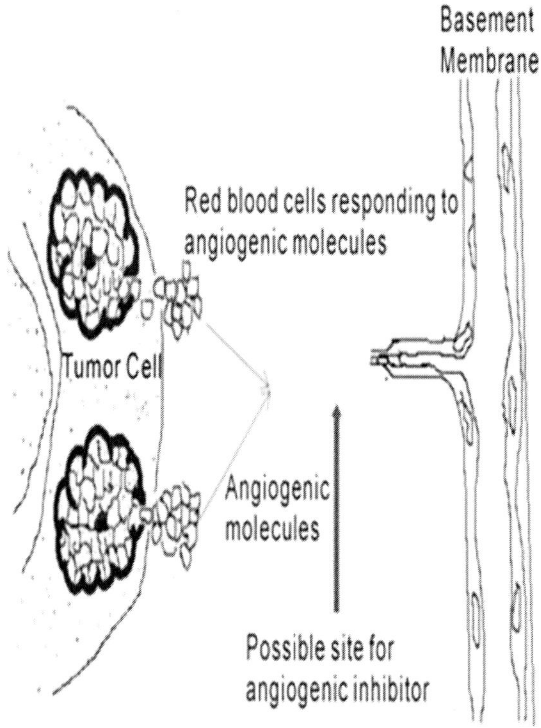

Figure 1. Diffusion of tumor angiogeneic factors and response of endothelial cells in the direction of the tumor angiogenic factors. The vertical arrow shows a possible location where an inhibitor can be introduced. The Tumor cells on the right hand side represent the source from which the signaling factors are induce from. The signaling factors are induced at some given time intervals.

In order to mimic the experimental work by Folkman et al., we adapted Figure 2 as observed in [8][9]. Our model as mentioned earlier is a modified form of Harrington's model. We proposed a similar model by adding a source term which depends on a threshold parameter, time. We described the diffusion and inhibition of tumor angiogenic factors using partial differential equations on a two dimensional plane. In Figure 3, the diamond region represents the source of the tumor angiogenic factors whose initiation depends on the threshold parameter, time.

We shall take a look at a hybrid model for tumor induced angiogenesis at the tissue level. As described previously, our model describes the diffusion of tumor angiogenic factors within the extra cellular matrix and how red blood vessels respond to these factors by growing towards the tumor along the gradient of the chemical stimuli.

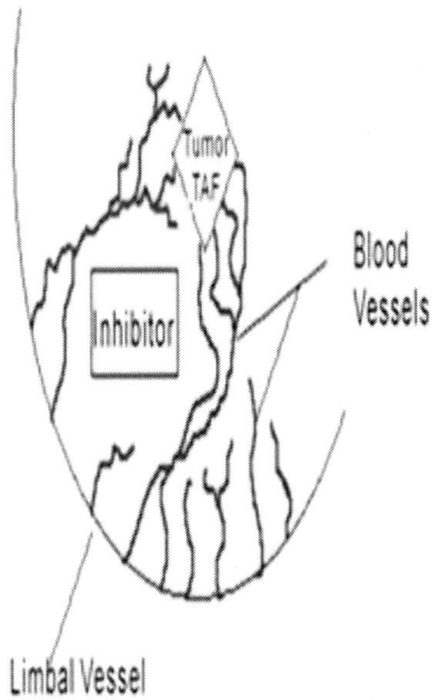

Figure 2. The following diagram mimics the interaction between tumor induced angiogenic factors and an inhibitor in the cornea. Proliferating blood vessels from the limbal vessel are avoid the path where the inhibitor has been introduced. Blood vessels growing towards tumor in response to tumor angiogeneic factors.

The contents of the paper are divides as follows; section 2 describes the general form of the model and its parameters. In section 3 we discuss the numerical methods used in solving the system. In section 4 we compute numerically the PDEs arising from the model and conclude in section 5 with a summary of our results and some future directions.

2. MODEL DESCRIPTION

We assume that the process of angiogenesis is spatially homogeneous, thus we can model the spread and interaction of the coupled tumor angiogenic and inhibitor factors within the extracellular matrix with a source term using ordinary differential equations.

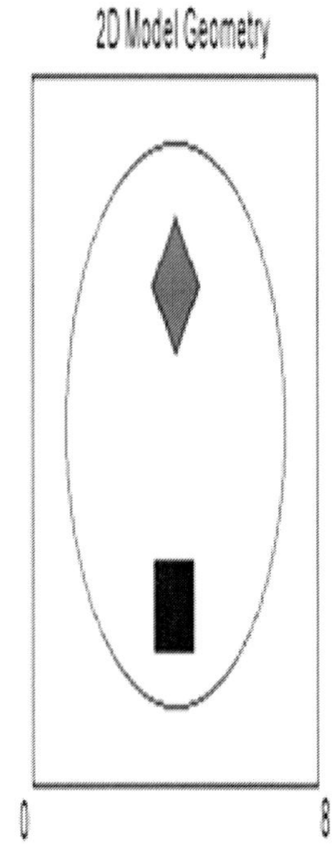

Figure 3. The above diagram is an 8 × 8 square which represents the region where the interaction between tumor angiogenic factors and inhibitors occur. The shaded rectangle in the square, represents the initial concentration of tumor angiogenic factors and location of source term. The diamond region represents the initial concentration of inhibitor.

Although the ordinary differential equation model does not capture all the details of angiogenesis, it is a good starting point to understand its dynamics. The ordinary differential equation model is as follows;

$$\frac{\mathrm{d}C}{\mathrm{d}t} = -kC - \mu IC - \mathrm{kon}IC + Q(t, T_0) \quad (1)$$

$$\frac{\mathrm{d}I}{\mathrm{d}t} = -\mathrm{kon}IC \quad (2)$$

with initial conditions;

$$C(0) = C_0 \quad (3)$$

Table 1 . Parameter values for the PDE and ODE model. These values were Taken from previous literature [4] and [5].

Description	Constant	Value
Diffusion Coefficient of bFGF (TAF)	D_1	$0.5 \times 10^{-6}\ cm^2 \cdot s^{-1}$
Rate Constant of bFGF degradation	k	$2.89 \times 10^{-2}\ h^{-1}$
Rate constant of uptake	μ	$2000\ \mu m \cdot h^{-1}$
Interplay constant	kon	$3.45 \times 60\ M^{-1} \cdot h^{-1}$
Diffusion Coefficient of Inhibitor	D_2	$10^{-6}\ cm^2 \cdot s^{-1}$

$$I(0) = I_0 \tag{4}$$

where C and I are state variables representing the concentrations of tumor angiogenic factor (TAF) and inhibitor respectively, and

$$Q(t,T_0) = \begin{cases} 0, \text{ if}\left(0 \leq t < T_0\right) \\ Q_p, \text{ if}\left(t \geq T_0\right) \end{cases}$$

represents a source term which depends on time. C_0 and I_0 represent initial concentration of tumor angiogenic factor and inhibitor respectively. The parameters in the ordinary differential equations (1) and (2) are constants obtained from previous literature [8] [9] are described in detail in Table 1 The function Q is obtained artificially due to the lack of experimental data. The following graphs displays our results from solving equations (1)-(4).

Ordinary differential equation models as mentioned earlier are not enough to capture the dynamics of spatially heterogeneous biological behavior. We therefore include a spatial term and study the dynamic of the system within this frame work.

The partial differential equations are modeled from the results of Harrington et al. by adding a source term to their equations. We impose Dirichlet boundary and initial conditions on the system as follows;

$$\frac{\partial C}{\partial t} = D_1\left(\frac{\partial^2 C}{\partial x^2} + \frac{\partial^2 C}{\partial y^2}\right) - kC - \mu IC - \text{kon}IC + Q(t,T_0) \tag{5}$$

$$\frac{\partial I}{\partial t} = D_2\left(\frac{\partial^2 I}{\partial x^2} + \frac{\partial^2 I}{\partial y^2}\right) - \text{kon}IC \tag{6}$$

where the parameters D_1 and D_2 represents the diffusion rate constant for TAFs and inhibitors respectively. Similarly the parameter k is the natural inactivation rate of the TAF's while the parameters μ and I is the capillary indicator function which depends on time and the location where the tumor is source is located.

Lastly the parameter kon is the rate constant controlling the relationship between the TAF's and the inhibitors.

3. DISCUSSION OF NUMEICAL METHOD—EXPLICIT STABILIZED RUNGE-KUTTA METHODS

Some recent and older publications in [10] -[13] showed that explicit stablized Runge Kutta methods possess extended stability domains and have less restriction on the choice of their step size as compared to the more classical Runge-Kutta methods. This property thus makes explicit stabilized runge kutta methods good options to consider when solving partial differential equations numerically, much as implicit methods are the more desirable choice. It has been shown by Assyr Abdulle [10] that the explicit stabilized Runge-Kutta schemes work very well for mildly stiff system. The model under consideration is mildly stiff and thus it is reasonable to use explicit stabilized methods.

An extensive discussion on the construction and how to implement the methods can be found in [10] -[14]. We however provide here a more concise overview to motivate the implementation of this method to our model. This discussion is not a substitute to reading on the more elaborate technical report by Assyr Abdulle on the subject and other text book material. The general idea for the construction of the methods can be divided into two steps:

- Construct Stability polynomials bounded in a long strip around the negative real axis.
- Develop a numerical methods which has such favorable stability functions.

The first step to achieving the steps mentioned above will be to generate an optimal stability polynomial of order P and degrees S of the form

$$R_s(z) = 1 + z + \cdots + \frac{z^p}{p!} + \alpha_{p+1} z^{p+1} + \cdots + \alpha_s z^s$$

such that $|R_s(z)| \le 1$ and $z \in [-l_s, 0]$.

It has been shown by [14] that polynomials of this type for arbitrary P and S values exists and are unique. Riha's approach to constructing such polynomials is very analytical in nature and is not used very much in practice. A more friendly approach to constructing these polynomials is the use of numerical approximations. Some numerical approximations which have been used to construct some of these polynomials include DUMKA method, Runge-Kutta Cheybyshev method and the optimal Runge-Kutta Chebyshev method (ROCK). These methods are what we seek to employ in solving our system and consequently test how well they perform as they have not been widely used as compared to the more traditional implicit methods. We will discuss in the next subsections the DUMKA method and the optimal Runge-Kutta Chebyshev method. We use these methods in solving our model numerically.

3.1. Dumka Method

DUMKA method is computed through an iterative procedure which is based on the zeros of the optimal stability polynomials. Optimal stability polynomials can be thought of as the generalized Chebyshev polynomials. There are a lot of literature discussion on this subject a very good source can be found in [15]. Every step leading to the solution under this method consists of a two stage scheme.

This approach helps reduce error propagation that can arise when solving differential equations numerically. One general approach for solving time-dependent partial differential equations is to discretize the space variables to obtain a system of ordinary differential equations of the form $y' = f(t, y), y(t_0) = y_0$, where, $y, y_0 \in R^n, t \geq 0$ and $f(t,y)$ has a value in R^n. The general form of the DUMKA method for solving the resulting system of ordinary differential equation is given as follows;

$$g_i = g_{i-1} + \Delta t \alpha_i f(g_{i-1})$$

$$g_{i+1}^* = g_i + \Delta t \alpha_i f(g_i)$$

$$g_{i+1} = g_{i+1}^* - \Delta t \left(\alpha_i - \frac{\beta_i}{\alpha_i} \right) \left(f(g_i) - f(g_{i-1}) \right)$$

where $g_0 = y_0$ is an initial condition, Δt is the time step, a_i are the zero's of the optimal stability polynomials and β_i are the squares of the zero's of the optimal stability polynomials.g_i is called the internal stages and are constructed by composition. This can be composition of t explicit Euler methods.

The interested reader should refer to [14] -[16] for more in depth discussion on the method.

3.2. Rock Method

The ROCK method is built on a recurrence relation and has been derived for orders P = 2 and P = 4. In [16] , the author proved that the optimal stability polynomials of even orders have exactly P complex zeros. In the construction of the ROCK method, the first thing we do is to look for some P from the family of polynomials, such that the approximation to the optimal stability polynomial is given as;

$$R_s(Z) = W_p(Z)P_{s-p}(Z) \tag{7}$$

$$R_s(Z) = 1 + Z + \cdots + \frac{Z^z}{Z!} + O\left(Z^{z+1}\right) \tag{8}$$

where P_{s-p} is a member of a family of polynomials $\{P_j(x)\}_{j\geq 0}$, which are orthogonal to the weight function $\frac{W_p(z)^2}{\sqrt{1-z^2}}$ and the zeros of $W_p(z)$ are closely or approximately equal to the P complex zeros of equation (2).

The theoretical frame work of equation (1) is based on Bernstein's Theorem.

Due to the recurrence relation of orthogonal polynomials $\{P_j(x)\}_{j\geq 0}$, a method based on recurrence formula can be constructed.

Next we shall look at the construction of the approximate polynomial to Equation (1) for P = 2 such that

$$R_s(z) = W_2(z)P_{s-2}(z)$$

The three-term recurrence formula associated with the polynomials $\{P_j(x)\}_{j\geq 0}$ where

$$P_j(z) = \left(\alpha_j z - \beta_j\right)P_{j-1}(z) - \alpha_j P_{j-2}$$

can be used to determine the internal stages of the ROCK method defined as

$$g_1 = y_0 + \alpha_1 \Delta f\left(g_0\right) \tag{9}$$

$$g_i = y_0 + \alpha_i \Delta f\left(g_{i-1}\right) - \beta_i g_{i-1} - \gamma_i g_{i-2} \qquad (10)$$

and

$$W_2\left(z\right) = 1 + 2\delta z + \tau z^2$$

The final structure of the method is as follows

$$g_{s-1} = g_{s-2} + h\delta f\left(g_{s-2}\right)$$

$$g_s^* = g_{s-1} + h\delta f\left(g_{s-1}\right)$$

$$g_s = g_s^* - h\delta\left(1 - \frac{\tau}{\delta^2}\right)\left(f\left(g_{s-1}\right) - f\left(g_{s-2}\right)\right)$$

In a similar fashion of derivation, the fourth order ROCK method is given as follows

$$g_0 = y_0$$

$$g_1 = y_0 + h\mu_1 f\left(g_0\right)$$

$$g_j = h\mu_i f\left(g_{j-1}\right) - v_j g_{j-1} - k_j g_{j-2}$$

$$y_1 = g_{s-4}$$

In this paper we apply the schemes discussed above that is the DUMKA method and the second and fourth order optimal Runge-Kutta Chebyshev method (ROCK-2, ROCK-4) on the hybrid model for tumor induced angiogenesis.

As mentioned earlier the above discussion is not a substitution for reading the references [14] -[16] and additional books. The reader who is interested in explicit stabilized numerical methods should read further on the subject as noted.

4. RESULTS AND DISCUSSION

We now take a look at the results obtained from the application of the explicit stabilized Runge-Kutta methods on the 2-dimensional partial differential equation model for tumor angiogenesis. The model together with its initial and Dirichlet boundary conditions as mentioned earlier are given as follows;

$$\frac{\partial C}{\partial t} = D_1\left(\frac{\partial^2 C}{\partial x^2} + \frac{\partial^2 C}{\partial y^2}\right) - kC - \mu IC - \text{kon}IC + Q(t, T_0) \tag{11}$$

$$\frac{\partial I}{\partial t} = D_2\left(\frac{\partial^2 I}{\partial x^2} + \frac{\partial^2 I}{\partial y^2}\right) - \text{kon}IC \tag{12}$$

with initial conditions

$$C(x,y,0) = \begin{cases} 1, & \text{if } \max\left(|x-x_0|, |y-y_0|\right) \le 0.4 \\ 0, & \text{otherwise} \end{cases}$$

$$I(x,y,0) = \begin{cases} 10, & \text{if } \left(|x-z_0|, |y-z_0|\right) \le 0.4 \\ 0, & \text{otherwise} \end{cases}$$

We note that the parameter values 4.0, 2.5, 0.4, 1.0, 10.0 and 5.5 together with the domain are arbitrarily chosen. We would have liked, to have chosen these values from experimental data but none was immediately available to us.

The following graphs are solutions of the system of partial differential equations (11) and (12) together with the initial conditions. These solutions represent the diffusion of the tumor angiogenic factors and the tumor angiogenic factor inhibitors within the extra cellular matrix. We realized that the behavior of the system closely matches what the literature suggests. That is in the presence of the inhibitors, the concentration of the tumor angiogenic factors is suppressed whiles the inhibitors remain within the system under consideration.

We also looked at the case where the inhibitor is placed at a different position by changing the initial condition slightly and check its effect on the tumor angiogenic factors. The location of the inhibitor relative to the tumor angiogenic factors is important. This is because it appears the inhibitor has to be located close enough to the tumor source in order to have an effect on the tumor angiogenic factors.

Table 2. Computational times for the different numerical methods for varying time steps tau is reported below.

Method	*tau*	Number of Functional Evaluation	CPU Run Time (Seconds)
DUMKA	0.003	118	1.34
	0.01	172	1.44
	0.1	439	1.93
ROCK2	0.003	23	0.08
	0.01	50	0.19
	0.1	165	0.58
ROCK4	0.003	41	0.16
	0.01	82	0.27
	0.1	261	0.87

Lastly, we considered the effect of a source term $Q(t,T_0)$; which is a function of time and a threshold value. As mentioned earlier the threshold value determines the effect that the source term has on the model under consideration. The plots below represent the L1 and L2 norms respectively of the solutions of the system (11)-(12) with respect to varying threshold values.

Finally the computational times of the three explicit stabilized Runge-Kutta methods is reported in table 2.

From the results reported in Table 2 from the numerical computation, the ROCK2 out performs the ROCK 4 method in terms of computational speed and number of functional evaluations. This is consistent with the theory since ROCK2 has less stages than the ROCK4 method as mentioned previously in the discussion section.

5. CONCLUSIONS AND FUTURE WORK

We extended the model of Harringtionby adding a time dependent source term of the form

$$Q(t,T_0) = \begin{cases} 0, \text{ if}\left(0 \le t < T_0\right) \\ Q_p, \text{ if}\left(t \ge T_0\right) \end{cases}$$

The source term models the signaling of growth factors from the tumor, to recruit red blood vessels: which depends on a threshold parameter, T_0 (represents the time it takes the tumor to initiate signaling growth factors) to facilitate the process of angiogenesis.

We believe our results will provide some answers to some questions arising from angiogenesis. For example, experimentalist can use the threshold parameter, T_0 to characterized aggressive tumors and non-aggressive tumors. This is because our results indicate that if T_0 is large enough, there is a long delay of initiating the growth factors. Thus we categorize these tumors to be non-aggressive. On the other hand, if T_0 is small, angiogenesis is initiated quickly and we categorize this as aggressive tumors (For example in Figures 4-8, small threshold implies high tumor angiogenic factors).

We used three explicit stabilized Runge-Kutta methods for solving the extended mathematical model for tumor induced angiogenesis in the cornea.

In our model we used a simple geometry as shown in Figure 3. We will like to explore into more complex geometries. In the future, we will like to use Splines instead of methods of lines (which is used for the discrete **Table2** Computational times for the different numerical methods for varying time steps tau is reported below.

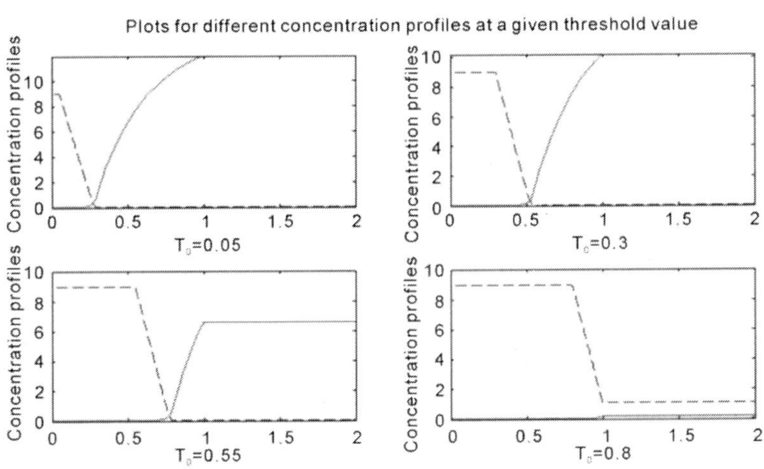

Figure 4. Represents the solution profile for the ode model at 4 different threshold values. The threshold values T_0 are 0.05, 0.3, 0.55 and 0.8 respectively. The dashed lines represents the solution profile of the inhibitor and the solid line represents the profile of the tumor angiogenic factors.

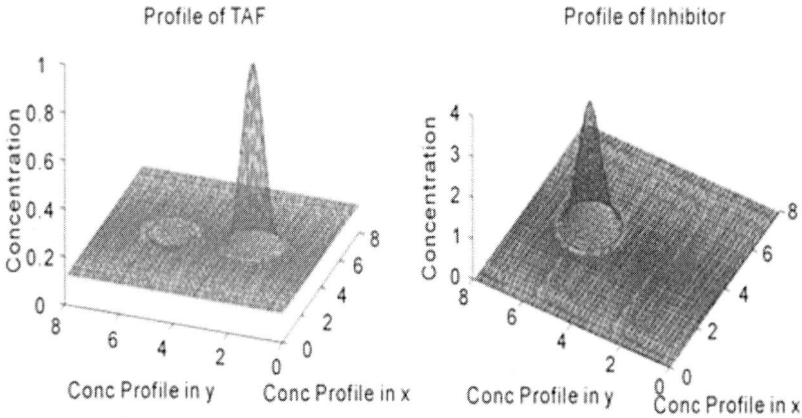

Figure 5. Concentration Profiles of tumor angiogenic factors (TAF) and Inhibitor. The following profile is a snap shot of the diffusion of TAF and inhibitor concentration at the end of the time interval.

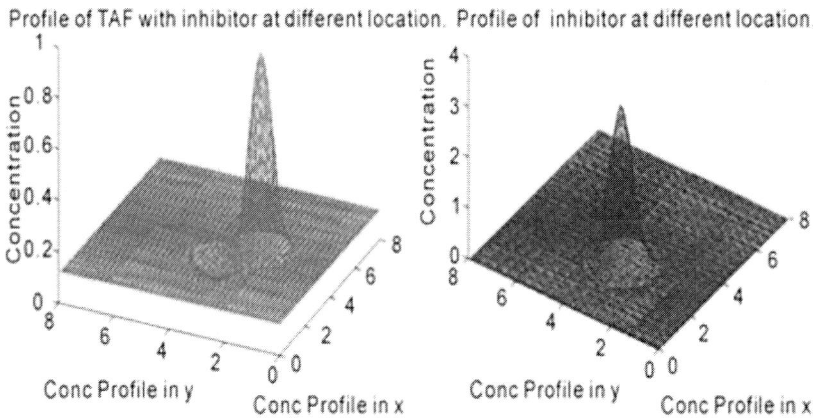

Figure 6. The following profile is a snap shot of the diffusion of TAF and inhibitor concentration. In this model, the initial position (initial condition) of the tumor angiogenic factors and the inihibitor profile was changed.

Richard C. Ewool, Zachariah Sinkala (PDE) to include both cellular (ODE) and sub-cellular dynamics (individual based model). In the extension we will like to consider as well, functional coefficients for the diffusion parameter.

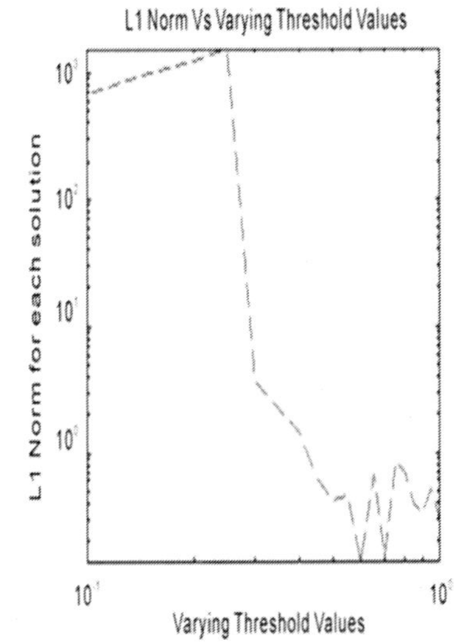

Figure 7. L1 norm Vs varying threshold values. This plot show that for a small threshold value, there is a high concentration of the tumor angioegenic factors.

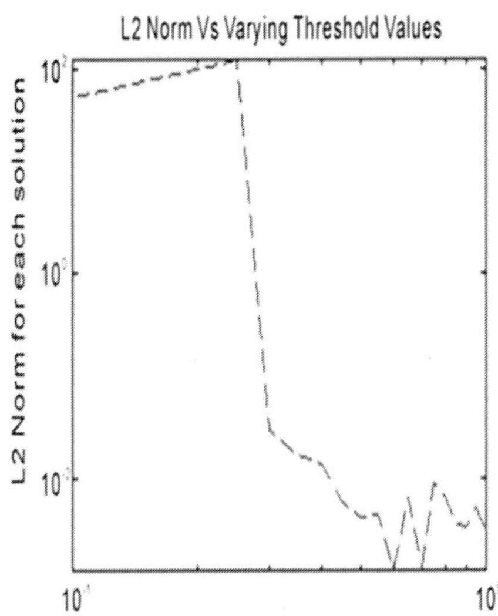

Figure 8. L2 norm Vs varying threshold values. The plots of the L2-norm suggests a similar phenomenon. That is the threshold parameter has a strong influence on the behavior of the solution.

REFERENCES

1. Benjamin Ribba et al. (2006) A Multiscale Mathematical Model of Avascular Tumor Growth to Investigate the Therapeutic Benefit of Anti-Invasive Agents. Elsevier Science, Preprint.
2. Pamuk, S. (2007) A Mathematical Model for Tumor Angiogenesis. University of Idaho, Moscow, Idaho.
3. Anderson, A. and Chaplain, M. (1998) Continuous and Discrete Mathematical Models of Tumor-Induced Angiogenesis. Bulletin of Mathematical Biology, 60, 857-899. http://dx.doi.org/10.1006/bulm.1998.0042
4. Levine, H., Tucker, A. and Nilsen-Hamilton, M. (2002) A Mathematical Model for the Role of Cell Signal Transduction in the Initiation and Inhibition of Angiogenesis. Growth Factors, 20, 155-175. http://dx.doi.org/10.1080/0897719031000084355
5. Good, D., Polverini, P., Rastinejad, F., Le Beau, M., Lemons, R., Frazier, W. and Bouck, N. (1990) A Tumor Suppressor-Dependent Inhibitor of Angiogenesis Is Immunologically and Functionally Indistinguishable from a Fragment of Thrombospondin. Proceedings of the National Academy of Sciences, 87, 6624-6628. http://dx.doi.org/10.1073/ pnas. 87. 17.6624
6. Bickel, S.T., Juliano, J.D. and Nagy, J.D. (2014) Evolution of Proliferation and the Angiogenic Switch in Tumors with High Clonal Diversity. PLoS ONE, 9, Article ID: e91992.
7. Folkman, J. (1971) Fundamental Concept of the Angiogenic Process. The New England Journal of Medicine, 285, 1182-1186.
8. Harrington, H.A., Maier, M., Naidoo, L., Whitaker, N. and Kevrekidis, P.G. (2006) A Hybrid Model for Tumor-Induced Angiogenesis in the Cornea in the Presence of Inhibitors. Mathematical and Computer Modelling, 46, 513-524.http://dx.doi.org/10.1016/ j.mcm. 2006.11.034
9. Tong, S. and Yuan, F. (2001) Numerical Simualtion of Angiogeneis in the Cornea. Microvascular Research, 61, 14-27.
10. Abdulle, A. (2002) Fourth Order Chebyshev Methods with Recurrence Relation. SIAM Journal on Scientific Computing, 23, 2041-2054. http://dx.doi.org/10.1137/ S106482 7500379549
11. Van der Houwen, S.B, (1980) On the Internal Stage Runge-Kutta Methods for Large m-Values. Zeitschrift für Angewandte Mathematik und Mechanik, 60, 479-485.
12. Verwer, J.G. (1996) Explicit Runge-Kutta Methods for Parabolic Differential Equations. Applied Numerical Mathematics, 22, 359-379. http://dx.doi.org/10.1016/S0168-9274(96)00022-0
13. Sommeijer, B., Shampine, L. and Verwer, J. (1998) RKC: An Explicit Solver for Parabolic PDEs. Journal of Computational and Applied Mathematics, 88, 326.
14. Riha, W. (1972) Optimal Stability Polynominals. Computing, 9, 37-43. http://dx.doi.org/ 10.1007/BF02236374
15. Abdulle, A. (2011) Explicit Stabilized Runge-Kutta Methods. MATHICSE Technical Report, Nr. 27.2011.
16. LeVeque, R.J. (2007) Finite Difference Methods for Ordinary and Partial Differential Equations. Society for Industrial and Applied Mathematics (SIAM), Philadelphia, 175-179. http://dx.doi.org/10.1007/BF02236374

CHAPTER 6

Solution of System of Linear Fractional Differential Equations with Modified Derivative of Jumarie Type

Uttam Ghosh[1], Susmita Sarkar[2], Shantanu Das[3, 4, 5]

[1]*Department of Mathematics, Nabadwip Vidyasagar College, Nabadwip, Nadia, West Bengal, India*

[2]*Department of Applied Mathematics, University of Calcutta, Kolkata, India*

[3]*Reactor Control Systems Design Section E & I Group B.A.R.C Mumbai India*

[4]*Department of Physics, Jadavpur University Kolkata*

[5]*Department of Appl. Mathematics, University of Calcutta*

ABSTRACT

Solution of fractional differential equations is an emerging area of present day research because such equations arise in various applied fields. In this paper we have developed analytical method to solve the system of fractional differential equations in-terms of Mittag-Leffler function and generalized Sine and Cosine functions, where the fractional derivative operator is of Jumarie type. The use of Jumarie type fractional derivative, which is modified Rieman-Liouvellie fractional derivative, eases the solution to such fractional order systems. The use of this type of Jumarie fractional derivative gives a conjugation with classical methods of solution of system of linear integer order differential equations, by usage of Mittag-Leffler and generalized trigonometric functions. The ease of this method and its conjugation to classical method to solve system of linear fractional differential equation is appealing to researchers in fractional dynamic systems. Here after developing the method, the algorithm is applied in physical system of fractional differential equation. The analytical results obtained are then graphically plotted for several examples for system of linear fractional differential equation.

Keywords: fractional calculus, Jumarie fractional derivative, Mittag-Leffler function, generalized sine and cosine function, fractional differential equations

1. INTRODUCTION

The fractional calculus is a current research topic in applied sciences such as applied mathematics, physics, mathematical biology and engineering. The rule of fractional derivative is not unique till date. The definition of fractional derivative is given by many authors. The commonly used definition is the Riemann-Liouvellie (R-L) definition [1, 2, 3, 4, 5]. Other useful definition includes Caputo definition of fractional derivative (1967) [1, 2, 3, 4, 5]. Jumarie's left handed modification of R-L fractional derivative is useful to avoid nonzero fractional derivative of a constant functions [7]. Recently in the paper [8]Ghosh *et al* proposed a theory of characterization of non-differentiable points with Jumarie type fractional derivative with right handed modification of R-L fractional derivative. The differential equations in different form of fractional derivatives give different type of solutions [1-5][1]. Therefore, there is no standard algorithm to solve fractional differential equations. Thus the solution and its interpretation of the fractional differential equations is a rising field of Applied Mathematics. To solve the linear and non-linear differential equations recently used methods are Predictor-Corrector method [9], Adomain decomposition method [2, 10, 11], Homotopy Perturbation Method [12] Variational Iteration Method [13], Differential transform method [14]. Recently in [15] Ghosh *et al* developed analytical method for solution of linear fractional differential equations with Jumarie type derivative [7] in terms of Mittag-Leffler functions and generalized sine and cosine functions. This new finding of [15] has been extended in this paper to get analytical solution of system of linear fractional differential equations. In section 1.0 we have defined some important definitions of fractional derivative that is basic Riemann-Liouvellie (RL) fractional derivative, the Caputo fractional derivative, the Jumarie fractional derivative, the Mittag-Leffler function and generalized Sine and Cosine functions. In section 2.0

solution of system of fractional differential equations has been described and in section 3.0 an application of this method to physical system has been discussed.

1.1. The Basic Definitions of Fractional Derivatives and Some Higher Transcendental Functions:

a) Basic definitions of fractional derivative:

i) Riemann- Liouvellie (R-L) definition

The R-L definition of the left fractional derivative is,

$$ {}_aD_x^{\alpha} f(x) = \frac{1}{\Gamma(m+1-\alpha)} \left(\frac{d}{dx}\right)^{m+1} \int_a^x (x-\tau)^{m-\alpha} f(\tau)d\tau \tag{1.1}$$

Where:1, $m \leq \alpha < m + 1$, m:is positiveinteger.

In particular when$0 \leq \alpha < 1$then

$$ {}_aD_x^{\alpha} f(x) = \frac{1}{\Gamma(1-\alpha)} \frac{d}{dx} \int_a^x (x-\tau)^{-\alpha} f(\tau)d\tau. \tag{1.2}$$

The definition (1.1) is known as the left R-L definition of the fractional derivative. The corresponding right R-L definition is

$$ {}_xD_b^{\alpha} f(x) = \frac{1}{\Gamma(m+1-\alpha)} \left(-\frac{d}{dx}\right)^{m+1} \int_x^b (\tau-x)^{m-\alpha} f(\tau)d\tau \tag{1.3}$$

Where $m \leq \alpha < m + 1$

The derivative of a constant is obtained as non-zero using the above definitions (1.1)-(1.3) which contradicts the classical derivative of the constant, which is zero. In 1967 Prof. M. Caputo proposed a modification of the R-L definition of fractional derivative which can overcome this shortcoming of the R-L definition.

ii) Caputo definition

M. Caputo defines the fractional derivative in the following form [6]

$$ {}_{a}^{C}D_{x}^{\alpha} f(x) = \frac{1}{\Gamma(n-\alpha)} \int_{a}^{x} (x-\tau)^{n-\alpha-1} f^{(n)}(\tau) d\tau \tag{1.4} $$

Where n - 1 ≤ a < n

In this definition first differentiate $f(x)$, n-times then integrate n-⍰ times. The disadvantage of this method is that $f(x)$, must be differentiable n-times then the α -th order derivative will exist, where n-1 ≤ ⍰ < n . If the function is non-differentiable then this definition is not applicable. Two main advantages of this method are (i) fractional derivative of a constant is zero (ii) the fractional differential equation of Caputo type has initial conditions of classical derivative type but the R-L type differential equations has initial conditions fractional type i.e. $\lim_{x \to a} {}_{a}D_{x}^{\alpha-1}[f(x)] = b_1$.

This means that a fractional differential equation composed with RL fractional derivatives require concept of fractional initial states, sometimes they are hard to interpret physically [2].

iii) Modified definitions of fractional derivative:

To overcome the fractional derivative of a constant, non-zero, another modification of the definition of left R-L type fractional derivative of the function $f(x)$, in the interval $[a, b]$ was proposed by Jumarie [7] in the form, that is following.

$$ f_{L}^{(\alpha)}(x) = {}_{a}^{J}D_{x}^{\alpha} f(x) = \begin{cases} \dfrac{1}{\Gamma(-\alpha)} \displaystyle\int_{a}^{x} (x-\tau)^{-\alpha-1} f(\tau) d\tau, & \alpha < 0. \\ \dfrac{1}{\Gamma(1-\alpha)} \dfrac{d}{dx} \displaystyle\int_{a}^{x} (x-\tau)^{-\alpha} \left(f(\tau) - f(a) \right) d\tau, & 0 < \alpha < 1 \\ \left(f^{(\alpha-m)}(x) \right)^{(m)}, & m \le \alpha < m+1. \end{cases} \tag{1.5} $$

We consider that $f(x) - f(a) = 0$ for $x < a$. In (1.5), the first expression is just fractional integration; the second line is RL derivative of order $0 < \alpha < 1$ of offset function that is $f(x) - f(a)$. For $\alpha < 1$, we use the third line; that is first we differentiate the offset function with order $0 < (\alpha - m) < 1$, by the formula of second line, and then apply whole m order differentiation to it.

Here we chose integer m, just less than the real number α; that is $m \leq \alpha < m + 1$.

The logic of Jumarie fractional derivative is that, we do RL fractional derivative operation on a new function by forming that new function from a given function by offsetting the value of the function at the start point. Here the differentiability requirement as demanded by Caputo definition is not there. Also the fractional derivative of constant function is zero, which is non-zero by RL fractional derivative definition.

We have recently modified the right R-L definition of fractional derivative of the function $f(x)$, in the interval $[a, b]$ in the following form [8],

$$f_R^{(\alpha)}(x) = {}_x^J D_b^{\alpha} f(x)$$

$$= \begin{cases} -\dfrac{1}{\Gamma(-\alpha)} \displaystyle\int_x^b (\tau - x)^{-\alpha-1} f(\tau)d\tau, & \alpha < 0. \\ -\dfrac{1}{\Gamma(1-\alpha)} \dfrac{d}{dx} \displaystyle\int_x^b (\tau - x)^{-\alpha} \left(f(b) - f(\tau) \right) d\tau, & 0 < \alpha < 1 \\ \left(f^{(\alpha-m)}(x) \right)^{(m)}, & m \leq \alpha < m+1. \end{cases} \tag{1.6}$$

In the same paper [8], we have shown that both the modifications (1.5) and (1.6) give fractional derivatives of non-differentiable points their values are different, at that point, but we get finite values there, of fractional derivatives. Whereas in classical integer order calculus, where we have different values of right and left derivatives at non differentiable points in approach limit from left side or right side, but infinity (or minus infinity) at that point, where function is non-differentiable. But in case of Jumarie fractional derivative and right modified RL fractional derivative [8], there is no approach limit at the non-differentiable points, but a finite value is obtained at that non differentiable point of the function. The difference is that integer order calculus returns infinity or minus infinity at non-differentiable points, where as the Jumarie fractional derivative returns a finite number indicating the character of otherwise non-differentiable points in a function, in left sense or right sense. This has a significant application in characterizing otherwise non-differentiable but continuous points in the function. However, the finite value of the non differentiable point after fractional differentiation depends on the interval length. In the rest of the paper $^{J}D^{\alpha}$ will represent Jumarie fractional derivative.

b) Mittag-Leffler function and the generalized Sine and Cosine functions

The Mittag-Leffler function was introduced by the Swedish mathematician Gösta Mittag-Leffler [17, 18,19, 20] in 1903. It is the direct generalization of exponential functions. The one parameter Mittag-Leffler function is defined (in series form) as:

$$E_\alpha(z) \overset{\text{def}}{=} \sum_{k=0}^{\infty} \frac{z^k}{\Gamma(1+\alpha k)}, \quad z \in \mathbb{C}, \quad \operatorname{Re}(\alpha) > 0.$$

In the solutions of FDE we use this series definition in MATLAB plots. One parameter Mittag-Leffler function in relation to few transcendental functions is as follows

The integral representation of the Mittag-Leffler function [17, 18, 19, 20] is,

$$E_\alpha(z) = \frac{1}{2\pi} \int_C \frac{t^{\alpha-1} e^t}{t^\alpha - z} dt, \qquad z \in \mathbb{C}, \quad \operatorname{Re}(\alpha) > 0$$

Here the path of the integral C is a loop which starts and ends at $-\infty$ and encloses the circles of disk |t|≤|z|^1/⍰ in positive sense:|arg(t)|≤ ⍰⍰on C [17, 18, 19, 20].

The two parameter Mittag-Leffler function (in series form) and its relation with few transcendental functions are as following

$$E_{\alpha,\beta}(z) \overset{\text{def}}{=} \sum_{k=0}^{\infty} \frac{z^k}{\Gamma(\beta+\alpha k)}, \qquad z,\beta \in \mathbb{C}, \qquad \operatorname{Re}(\alpha) > 0$$

$$E_{\alpha,1}(z) = E_\alpha(z) = e^z \qquad z \in \mathbb{C}, \qquad \operatorname{Re}(\alpha) > 0$$

$$E_{1,2}(z) = \frac{e^z - 1}{z}$$

$$E_{2,2}(z) = \frac{\sinh(\sqrt{z})}{\sqrt{z}}.$$

The corresponding integral representation [17, 18, 19, 20] of the two parameter Mittag-Leffler function is,

$$E_{\alpha,\beta}(z) = \frac{1}{2\pi} \int_C \frac{t^{\alpha-\beta} e^t}{t^\alpha - z} dt, \qquad z \in \mathbb{C}, \qquad \operatorname{Re}(\alpha) > 0$$

where the contour C is already defined, in the above paragraph.

Using the modified definition, of fractional derivative of Jumarrie type, [7, 8] we get

$$ {}^{J}D^{\alpha}\left[1\right]=0, \qquad 0<\alpha<1. $$

The Jumarie fractional derivative of any constant function is zero, unlike a non-zero value of fractional RL derivative of a constant.

We now find Jumarie fractional derivative of Mittag-Leffler function $E\alpha$ (at $^{\alpha}$)

$$ E_{\alpha}(at^{\alpha})\overset{\text{def}}{=}1+\frac{at^{\alpha}}{\Gamma(1+\alpha)}+\frac{a^{2}t^{2\alpha}}{\Gamma(1+2\alpha)}+\frac{a^{3}t^{3\alpha}}{\Gamma(1+3\alpha)}+\dots\dots\dots $$

Using Jumarie derivative of order α, with $0 \le 1 < \alpha$ with start point as $a = 0$ for $f(t) = t^{na}$, [15] that is

$$ {}_{0}^{J}D_{t}^{\alpha}\left[t^{n\alpha}\right]=\frac{\Gamma(n\alpha+1)}{\Gamma\left(\alpha(n-1)+1\right)}(t)^{\alpha(n-1)} $$

for $n = 1,2,3,\ldots$; and also using Jumarie derivative of constant as zero ${}^{J}D^{?}[1] = 0$, we get the following very useful identity.

$$ \begin{aligned} {}_{0}^{J}D_{t}^{\alpha}\left[E_{\alpha}(at^{\alpha})\right] &= D^{\alpha}\begin{pmatrix} 1+\dfrac{at^{\alpha}}{\Gamma(1+\alpha)}+\dfrac{a^{2}t^{2\alpha}}{\Gamma(1+2\alpha)} \\ +\dfrac{a^{3}t^{3\alpha}}{\Gamma(1+3\alpha)}+\dots\dots\dots\dots\dots\dots \end{pmatrix} \\ &= 0+\frac{\Gamma(1+\alpha)a}{\Gamma(1)\Gamma(1+\alpha)}+\frac{\Gamma(1+2\alpha)a^{2}t^{\alpha}}{\Gamma(1+2\alpha)\Gamma(1+\alpha)} \\ &\quad +\frac{\Gamma(1+3\alpha)a^{3}t^{2\alpha}}{\Gamma(1+3\alpha)\Gamma(1+2\alpha)}+\dots\dots\dots \\ &= a\left(1+\frac{at^{\alpha}}{\Gamma(1+\alpha)}+\frac{a^{2}t^{2\alpha}}{\Gamma(1+2\alpha)}+\frac{a^{3}t^{3\alpha}}{\Gamma(1+3\alpha)}+\dots\dots\dots\dots\right) \\ &= aE_{\alpha}(at^{\alpha}) \end{aligned} $$

Thus

$$ {}_0^J D_t^\alpha \left[E_\alpha(at^\alpha) \right] = aE_\alpha(at^\alpha) $$

This shows that $AE_\alpha(at^\alpha)$ is a solution is a solution of the fractional differential equation [15]

$$ {}_0^J D^\alpha y = ay $$

Where A is arbitrary constant.

Therefore

$$ {}_0^J D^\alpha y = ay $$

with $y(0) = 1$ has solution

$$ y = E_\alpha(at^\alpha). $$

The fractional Sine and Cosine functions are expressed as following [16],

$$ E_\alpha(ix^\alpha) \overset{\text{def}}{=} \cos_\alpha(x^\alpha) + i\sin_\alpha(x^\alpha) $$
$$ \cos_\alpha(x^\alpha) \overset{\text{def}}{=} \sum_{k=0}^{\infty} (-1)^k \frac{x^{2k\alpha}}{\Gamma(1+2\alpha k)} $$
$$ \sin_\alpha(x^\alpha) \overset{\text{def}}{=} \sum_{k=0}^{\infty} (-1)^k \frac{x^{(2k+1)\alpha}}{\Gamma(1+(1+2k)\alpha)} $$

The above series form of fractional Sine and Cosine are used to plots, in solutions. It can be easily shown that [15, 16]

$$ {}^J D^\alpha \left[\sin_\alpha(x^\alpha) \right] = \cos_\alpha(x^\alpha) $$
$$ {}^J D^\alpha \left[\cos_\alpha(x^\alpha) \right] = -\sin_\alpha(x^\alpha). $$

This has been proved by the following term by term differentiation. The series presentation of $\cos_\alpha(x^\alpha)$ is [15],

$$\cos_{\alpha}(x^{\alpha}) = 1 - \frac{x^{2\alpha}}{\Gamma(1+2\alpha)} + \frac{x^{4\alpha}}{\Gamma(1+4\alpha)} - \frac{x^{6\alpha}}{\Gamma(1+6\alpha)} + ...$$

Taking its term by term Jumarie fractional derivative of order α we get,

$$\begin{aligned}
&{}_{0}^{J}D_{x}^{\alpha}\left[\cos_{\alpha}(x^{\alpha})\right] \\
&= 0 - \frac{\Gamma(1+2\alpha)x^{2\alpha-\alpha}}{\Gamma(1+2\alpha)\Gamma(1+\alpha)} + \frac{\Gamma(1+4\alpha)x^{4\alpha-\alpha}}{\Gamma(1+4\alpha)\Gamma(1+3\alpha)} \\
&\quad - \frac{\Gamma(1+6\alpha)x^{6\alpha-\alpha}}{\Gamma(1+6\alpha)\Gamma(1+5\alpha)} + ... \\
&= -\left[\frac{x^{\alpha}}{\Gamma(1+\alpha)} - \frac{x^{3\alpha}}{\Gamma(1+3\alpha)} + ...\right] \\
&= -\sin_{\alpha}(x^{\alpha})
\end{aligned}$$

Similarly we can get the expression for fractional derivative of Jumarie type of order α for $\sin_{\alpha}(x^{\alpha})$.

2. SYSTEM OF LINEAR FRACTIONAL DIFFERENTIAL EQUATIONS

Before considering the system of fractional differential equations we state the results [15] which arises in solving a single linear fractional differential equations composed by Jumarie derivative; we will be using the following theorems.

Theorem 1: The fractional differential equation

$$\left({}^{J}D^{\alpha} - a\right)\left({}^{J}D^{\alpha} - b\right)y(t) = 0$$

has solution of the form

$$y = AE_{\alpha}(at^{\alpha}) + BE_{\alpha}(bt^{\alpha})$$

where A and B are constants.

Theorem 2: The fractional differential equation

$$^{J}D^{2\alpha}[y]-2a\left(^{J}D^{\alpha}[y]\right)+a^{2}y=0$$

has solution of the form

$$y=(At^{\alpha}+B)E_{\alpha}(at^{\alpha})$$

where A and B are constants.

Theorem 3: Solution of the fractional differential equation

$$^{J}D^{2\alpha}[y]-2a\left(^{J}D^{\alpha}[y]\right)+(a^{2}+b^{\alpha})y=0$$

is of the form

$$y=\left(E_{\alpha}(at^{\alpha})\right)\left(A\cos_{\alpha}(bt^{\alpha})+B\sin_{\alpha}(bt^{\alpha})\right)$$

where A and B are constants.

Consider the system of linear fractional differential equations

$$\left.\begin{aligned} ^{J}D^{\alpha}[x]=ax+by \\ ^{J}D^{\alpha}[y]=cx+dy \end{aligned}\right\}. \tag{2.1}$$

Here a, b, c and d are constants, the operator $^{J}D^{\alpha}$ is the Jumarie fractional derivative operator, call it for convenience $^{J}D^{\alpha}\equiv\frac{d^{\alpha}}{dt^{\alpha}}$, and x and $y$are functions of t. In matrix form we write the (2.1) in following way

$$\begin{bmatrix} ^{J}D^{\alpha}x \\ ^{J}D^{\alpha}y \end{bmatrix}=\begin{bmatrix} a & b \\ c & d \end{bmatrix}\begin{bmatrix} x \\ y \end{bmatrix} \text{also} \frac{d^{\alpha}}{dt^{\alpha}}\begin{bmatrix} x \\ y \end{bmatrix}=\begin{bmatrix} a & b \\ c & d \end{bmatrix}\begin{bmatrix} x \\ y \end{bmatrix}.$$

The above system (2.1) can be written in the following form

$$\left.\begin{array}{l}\left({}^{J}D^{\alpha}-a\right)x-by=0\\-cx+\left({}^{J}D^{\alpha}-d\right)y=0\end{array}\right\}. \tag{2.2}$$

Operating $({}^{J}D^{\alpha}-\alpha)$ on both sides of the second equation of (2.2) we get the following steps.

$$\begin{array}{l}-c({}^{J}D^{\alpha}-a)x+({}^{J}D^{\alpha}-a)({}^{J}D^{\alpha}-d)y=0\\-cby+({}^{J}D^{\alpha}-a)({}^{J}D^{\alpha}-d)y=0\\{}^{J}D^{2\alpha}[y]-(a+d)\left({}^{J}D^{\alpha}[y]\right)+(ad-bc)y=0\end{array} \tag{2.3}$$

Equation (2.3) is a linear fractional order differential equation (with order 2⍰), with Jumarie derivative operator.

Let

$$a+d=\lambda_1+\lambda_2 \qquad \text{and} \qquad ad-bc=\lambda_1\lambda_2$$

then the equation (2.3) can be re-written as,

$${}^{J}D^{2\alpha}[y]-(\lambda_1+\lambda_2)\left({}^{J}D^{\alpha}[y]\right)+\lambda_1\lambda_2 y=0 \tag{2.4}$$

For λ_1, λ_2 real and distinct, solution of the equation (2.4) can be written in the form (from Theorem 1)

$$y=A_1E_{\alpha}(\lambda_1t^{\alpha})+B_1E_{\alpha}(\lambda_2t^{\alpha}) \tag{2.5}$$

Again from second equation of (2.1) we get after putting the value of y the following,

$$
\begin{aligned}
&{}^{J}D^{\alpha}\left[y\right]=cx+dy\\
&{}^{J}D^{\alpha}\left[A_1E_{\alpha}(\lambda_1t^{\alpha})+B_1E_{\alpha}(\lambda_2t^{\alpha})\right]\\
&=cx+d\left[A_1E_{\alpha}(\lambda_1t^{\alpha})+B_1E_{\alpha}(\lambda_2t^{\alpha})\right]\\
&cx=\left[A_1\lambda_1E_{\alpha}(\lambda_1t^{\alpha})+B_1\lambda_2E_{\alpha}(\lambda_2t^{\alpha}\right]\\
&-d\left[A_1E_{\alpha}(\lambda_1t^{\alpha})+B_1E_{\alpha}(\lambda_2t^{\alpha})\right]\\
&cx=A_1(\lambda_1-d)E_{\alpha}(\lambda_1t^{\alpha})+B_1(\lambda_2-d)E_{\alpha}(\lambda_2t^{\alpha})
\end{aligned}
$$

From above we obtain the following

$$
\begin{aligned}
x&=\tfrac{1}{c}\left[A_1(\lambda_1-d)E_{\alpha}(\lambda_1t^{\alpha})+B_1(\lambda_2-d)E_{\alpha}(\lambda_2t^{\alpha})\right]\\
&=A_2E_{\alpha}(\lambda_1t^{\alpha})+B_2E_{\alpha}(\lambda_2t^{\alpha})\\
A_2&=\frac{A_1(\lambda_1-d)}{c};\qquad B_2=\frac{B_1(\lambda_2-d)}{c}
\end{aligned}
\tag{2.6}
$$

A_1, B_1 are arbitrary constants in above derivation, and $\lambda_1+\lambda_2=a+d$, $\lambda_1\lambda_2=ad-bc$.

Thus the solution can be written in the form

$$
\begin{bmatrix}x\\y\end{bmatrix}=\begin{bmatrix}A_2\\A_1\end{bmatrix}E_{\alpha}(\lambda_1t^{\alpha})+\begin{bmatrix}B_2\\B_1\end{bmatrix}E_{\alpha}(\lambda_2t^{\alpha})
\tag{2.7}
$$

Again to solve the system of fractional differential equation (2.1) we use the method similar to as used in classical differential equations.

Since ${}^{J}D^{\alpha}[\xi(\tau)]=\lambda x(t)$ has solution in the form $x(t)=AE_{?}(\lambda t^{?})$, A is arbitary constant [15], putting

$$
x=AE_{\alpha}(\lambda t^{\alpha})\quad\text{and}\quad y=BE_{\alpha}(\lambda t^{\alpha})
$$

in (2.1) we get the following

$$
\begin{aligned}
&{}^{J}D^{\alpha}x - ax - by = 0 \\
&{}^{J}D^{\alpha}\left[AE_{\alpha}(\lambda t^{\alpha})\right] - aAE_{\alpha}(\lambda t^{\alpha}) - bBE_{\alpha}(\lambda t^{\alpha}) = 0 \\
&A\lambda E_{\alpha}(\lambda t^{\alpha}) - AaE_{\alpha}(\lambda t^{\alpha}) - bBE_{\alpha}(\lambda t^{\alpha}) = 0 \\
&A(\lambda - a) - Bb = 0
\end{aligned}
$$

$$
\begin{aligned}
&{}^{J}D^{\alpha}y - cx - dy = 0 \\
&{}^{J}D^{\alpha}\left[BE_{\alpha}(\lambda t^{\alpha})\right] - cAE_{\alpha}(\lambda t^{\alpha}) - dBE_{\alpha}(\lambda t^{\alpha}) = 0 \\
&B\lambda E_{\alpha}(\lambda t^{\alpha}) - AcE_{\alpha}(\lambda t^{\alpha}) - dBE_{\alpha}(\lambda t^{\alpha}) = 0 \\
&-Ac + B(\lambda - d) = 0
\end{aligned}
$$

$$
\left.\begin{aligned}
A(\lambda - a) - bB = 0 \\
-cA + (\lambda - d)B = 0
\end{aligned}\right\} \tag{2.8}
$$

Eliminating A and B from (2.8) we get,

$$
\begin{vmatrix} a-\lambda & b \\ c & d-\lambda \end{vmatrix} = 0 \text{ giving } \lambda^2 - (a+d)\lambda + (ad - bc) = 0 \tag{2.9}
$$

which is known as the characteristic equation with roots λ_1 and λ_2, also termed as eigen-values. Three cases may arises

i) The roots λ_1 and λ_2 are real and distinct.

ii) The roots are real and equal i.e. $\lambda_1 = \lambda_2 = \lambda$ (say).

iii) The roots are complex i.e. of the form $\lambda_1, \lambda_2 = p \pm iq$ (say).

Case –I

For real and distinct roots λ_1 and λ_2, we write

$$
\begin{aligned}
&x_1 = A_1E_{\alpha}(\lambda_1 t^{\alpha}),\ y_1 = B_1E_{\alpha}(\lambda_1 t^{\alpha}) \\
&\text{and } x_2 = A_2E_{\alpha}(\lambda_2 t^{\alpha}), y_2 = B_2E_{\alpha}(\lambda_2 t^{\alpha})
\end{aligned}
$$

The solution of (2.1) is

$$
\begin{aligned}
x = x_1 + x_2 = A_1E_{\alpha}(\lambda_1 t^{\alpha}) + A_2E_{\alpha}(\lambda_2 t^{\alpha}) \\
y = y_1 + y_2 = B_1E_{\alpha}(\lambda_1 t^{\alpha}) + B_2E_{\alpha}(\lambda_2 t^{\alpha})
\end{aligned}
$$

Thus the solution of the system of fractional differential equation can be written as,

$$\begin{bmatrix} x \\ y \end{bmatrix} = c_1 \underset{\sim}{A} E_\alpha(\lambda_1 t^\alpha) + c_2 \underset{\sim}{B} E_\alpha(\lambda_2 t^\alpha)$$

Where $\underset{\sim}{A} = \begin{bmatrix} A_1 \\ B_1 \end{bmatrix}$ and $\underset{\sim}{B} = \begin{bmatrix} A_2 \\ B_2 \end{bmatrix}$ are the arbitry constants.

Example: 1

$$\left.\begin{aligned} {}^J D^\alpha x = 2x + y \\ {}^J D^\alpha y = x + 2y \end{aligned}\right\},$$
$$0 < \alpha \le 1 \text{ with } x(0) = 2,\ y(0) = 1.$$

Let

$$x = AE_\alpha(\lambda t^\alpha) \text{ and } y = BE_\alpha(\lambda t^\alpha).$$

be the solution of the above differential equation. Substituting this in the above equations we get,

$$\left.\begin{aligned} A(2-\lambda) + B = 0 \\ A + (2-\lambda)B = 0 \end{aligned}\right\} \tag{2.10}$$

The corresponding characteristic equation is,

$$\begin{vmatrix} 2-\lambda & 1 \\ 1 & 2-\lambda \end{vmatrix} = 0 \textit{ giving } \lambda = 1, 3.$$

Putting λ = 1 in (2.8) we get A + B = 0, taking A = 1 we get B = -1 and putting λ = 3 in (2.8) we get A = B, taking A = 1 we get B = 1. Hence the solutions are,

$$x_1 = E_\alpha(t^\alpha), y_1 = -E_\alpha(t^\alpha)$$

And $x_2 = E\alpha\,(3t^\alpha), y_2 = E\alpha\,(3t^\alpha)$.

Thus the general solution is

$$x = c_1 E_\alpha(t^\alpha) + c_2 E_\alpha(3t^\alpha)$$
$$y = -c_1 E_\alpha(t^\alpha) + c_2 E_\alpha(3t^\alpha).$$

Where c_1, c_2 are arbitrary constants.

Using the initial condition

$$x(0) = 2, y(0) = 0.$$

we get $c_1 = c_2 = 1$.

Thus the required solution is,

$$x = E_\alpha(t^\alpha) + E_\alpha(3t^\alpha)$$
$$y = -E_\alpha(t^\alpha) + E_\alpha(3t^\alpha)$$

Figure 1 represents the graphical presentation of $x(t)$ and $y(t)$ when the eigen-values of the system of differential equations are positive. Numerical simulation shows that $x(t)$ and $y(t)$ both grow rapidly with decrease of order of derivative i.e. as α decreases from 1 towards 0.

Example: 2

$$\left.\begin{aligned} {}^J D^\alpha x &= -2x + y \\ {}^J D^\alpha y &= x - 2y \end{aligned}\right\},\ 0 < \alpha \le 1 \text{ with } x(0) = 2,\ y(0) = 1$$

Let $x = AE_\alpha(\lambda t^\alpha)$ and $y = BE_\alpha(\lambda t^\alpha)$.

be the solution of the above differential equation. Substituting this in the above equations we get,

$$\left.\begin{aligned} A(-2-\lambda) + B &= 0 \\ A + (-2-\lambda)B &= 0 \end{aligned}\right\}.$$

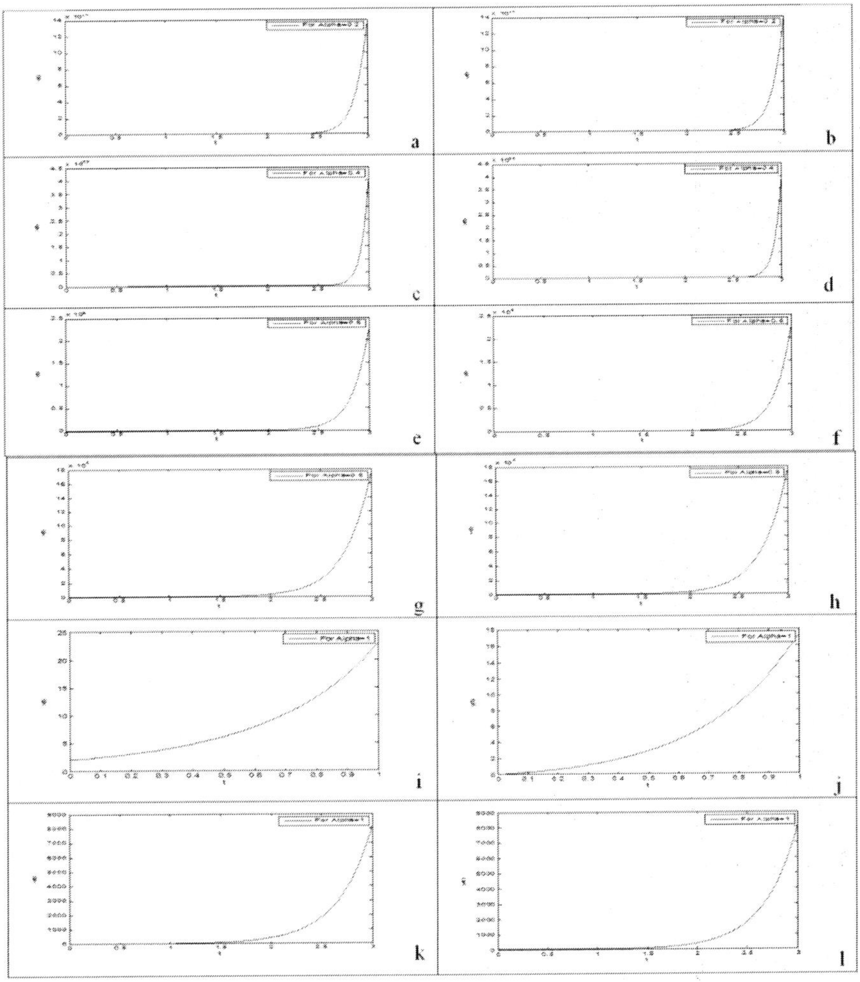

Figure 1. Numerical simulation of the solutions of fractional differential equation in Example-1 for x(t) and y(t) for different values of α = 0.2, 0.4, 0.6, 0.8 and 1.0. Figure (i) & (k) and figure (j) & (l) represents the same figure only length of x-axis is changed here to represent the prominent initial values

The corresponding characteristic equation is,

$$\begin{vmatrix} -2-\lambda & 1 \\ 1 & -2-\lambda \end{vmatrix} = 0 \quad \text{giving} \quad \lambda = -1, -3.$$

Putting λ = -1 in (2.8) we get A – B = 0, taking A = 1 we get B = 1 and putting λ = -3 in (2.8) we get A = -B, taking A = 1 we get B = -1. Hence the solutions are,

$$x_1 = E_\alpha(-t^\alpha), y_1 = E_\alpha(-t^\alpha)$$

and $x_2 = E\alpha(-3t^\alpha)$, $y_2 = -E\alpha(-3t^\alpha)$.

Thus the general solution is

$$x = c_1 E_\alpha(-t^\alpha) + c_2 E_\alpha(-3t^\alpha)$$
$$y = c_1 E_\alpha(-t^\alpha) - c_2 E_\alpha(-3t^\alpha).$$

Where c_1, c_2 are arbitrary constants.

Using the initial condition $x(0) = 2, y(0) = 0$ we get

$$c_1 = c_2 = 1.$$

Thus the required solution is,

$$x = E_\alpha(-t^\alpha) + E_\alpha(-3t^\alpha), y = E_\alpha(-t^\alpha) - E_\alpha(-3t^\alpha).$$

Figure 2 represents the graphical presentation of $x(t)$ and $y(t)$ when the eigen-values of the system of differential equations are negative. Numerical simulation shows that $x(t)$ and $y(t)$ are both decaying rapidly with decrease of order of derivative i.e. as decreases from 1 to 0.6. For negative eigen-values the solutions are decaying asymptotically to zero.

Case –II

The roots of the equation (2.3) are complex and are of the form 1, 2 = $p \pm iq$ then the solution $x(t)$ and $y(t)$ can be written in the form (from Theorem 3),

$$x = (A_1 + iA_2)\left[E_\alpha(p+iq)t^\alpha\right]$$

and $y = (B_1 + iB_2)\left[E_\alpha(p+iq)t^\alpha\right].$

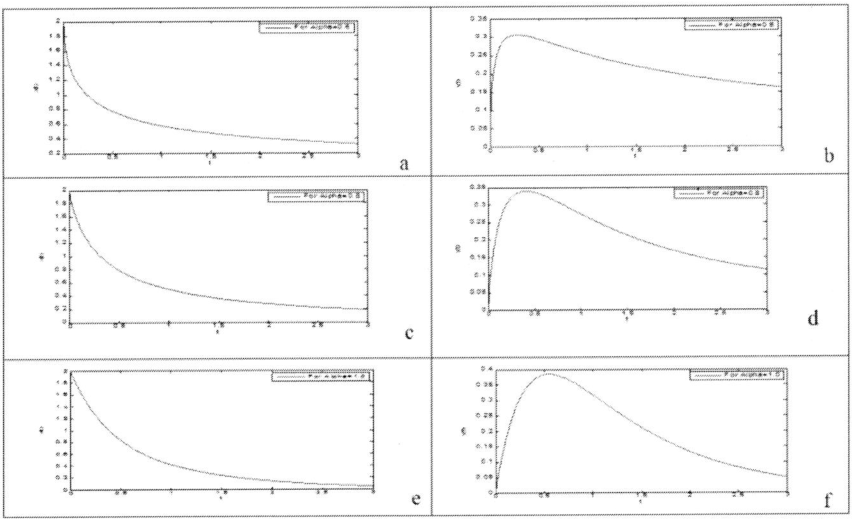

Figure 2. Numerical simulation of the solutions of fractional differential equation in Example-2 for $x(t)$ and $y(t)$ for different Values of α = 0.6, 0.8 and 1.0

Using the definition of Mittag-Leffler function and fractional cosine and sine functions, that is

$$E_\alpha(ix^\alpha) \overset{\text{def}}{=} \cos_\alpha(x^\alpha) + i\sin(x^\alpha)$$

we get,

$$\begin{aligned} x &= (A_1 + iA_2)\left[E_\alpha(p+iq)t^\alpha\right] \\ &= (A_1 + iA_2)\left[E_\alpha(pt^\alpha)\right]\left[E_\alpha(iq)t^\alpha\right] \\ &= (A_1 + iA_2)\left[E_\alpha(pt^\alpha)\right]\left[\cos_\alpha(qt^\alpha) + i\sin_\alpha(qt^\alpha)\right] \\ &= \left[E_\alpha(pt^\alpha)\right]\begin{bmatrix}\left(A_1\cos_\alpha(qt^\alpha) - A_2\sin_\alpha(qt^\alpha)\right) \\ +i\left(A_2\cos_\alpha(qt^\alpha) + A_1\sin_\alpha(qt^\alpha)\right)\end{bmatrix}. \end{aligned}$$

Similarly we get by repeating the above steps for y as follows

$$y=\left[E_{\alpha}(pt^{\alpha})\right]\begin{bmatrix}\left(B_1\cos_{\alpha}(qt^{\alpha})-B_2\sin_{\alpha}(qt^{\alpha})\right)\\+i\left(B_2\cos_{\alpha}(qt^{\alpha})+B_1\sin_{\alpha}(qt^{\alpha})\right)\end{bmatrix}.$$

In above obtained expressions for x and y, we have complex quantity as $u = iv$. This may be also considered as linear combination of $u$and v considering i a constant. Therefore, we can say that x is linear combination of x_1 (the real part of obtained complex x), and x_2 (the imaginary part of obtained complex x). Similarly we have y as linear combination of y_1 and y_2 [21]. With this argument we write the following

$$x_1=\left[E_{\alpha}(pt^{\alpha})\right]\left[A_1\cos_{\alpha}(qt^{\alpha})-A_2\sin_{\alpha}(qt^{\alpha})\right]$$
$$x_2=\left[E_{\alpha}(pt^{\alpha})\right]\left[A_2\cos_{\alpha}(qt^{\alpha})+A_1\sin_{\alpha}(qt^{\alpha})\right]$$
$$y_1=\left[E_{\alpha}(pt^{\alpha})\right]\left[B_1\cos_{\alpha}(qt^{\alpha})-B_2\sin_{\alpha}(qt^{\alpha})\right]$$
$$y_2=\left[E_{\alpha}(pt^{\alpha})\right]\left[B_2\cos_{\alpha}(qt^{\alpha})+B_1\sin_{\alpha}(qt^{\alpha})\right]$$

It can be shown that (x_1, y_1), $(x_2, y_2)$are solutions of the given equations (2.1).

Thus the general solution in this case can be written in the form as in classical integer order differential equation [[21], pp.305].

The linear combination of x_1 and x_2, gives x and linear combination of y_1 and y_2 gives y, which is represented as following

$$x=\left[E_{\alpha}(pt^{\alpha})\right]\begin{bmatrix}M\left(A_1\cos_{\alpha}(qt^{\alpha})-A_2\sin_{\alpha}(qt^{\alpha})\right)\\+N\left(A_2\cos_{\alpha}(qt^{\alpha})+A_1\sin_{\alpha}(qt^{\alpha})\right)\end{bmatrix}$$

$$y=\left[E_{\alpha}(pt^{\alpha})\right]\begin{bmatrix}M\left(B_1\cos_{\alpha}(qt^{\alpha})-B_2\sin_{\alpha}(qt^{\alpha})\right)\\+N\left(B_2\cos_{\alpha}(qt^{\alpha})+B_1\sin_{\alpha}(qt^{\alpha})\right)\end{bmatrix}.$$

With M, N as arbitrary constants, determined from initial states. We demonstrate by following examples.

Example: 3

$$\left.\begin{aligned} {}^{J}D^{\alpha}x &= 3x + 2y \\ {}^{J}D^{\alpha}y &= -5x + y \end{aligned}\right\},$$
$$0 < \alpha \le 1 \text{ with } x(0) = 2,\ y(0) = 1.$$

Let

$$x = AE_{\alpha}(\lambda t^{\alpha}) \text{ and } y = BE_{\alpha}(\lambda t^{\alpha})$$

be the solution of the above differential equation, then putting in the above equation we get,

$$\left.\begin{aligned} A(3-\lambda) + 2B &= 0 \\ -5A + (1-\lambda)B &= 0 \end{aligned}\right\} \qquad (2.11)$$

The corresponding characteristic equation is,

$$\begin{vmatrix} 3-\lambda & 2 \\ -5 & 1-\lambda \end{vmatrix} = 0 \quad \text{giving} \quad \lambda = 2 \pm 3i\,.$$

Putting $\lambda = 2 + 3i$ in (2.8) by putting $a = 3$, $b = 2$, $c = -5$, $d = 1$ we get the following

The (2.8) returns the same answer that is $2B = (3i-1)A$.We choose here $A = 2$, so $B = 3i-1$, as we obtained one equation with two unknowns. Thus $B = 3i-1$ and $A = 2$ can be taken as one of the trial solution of the above. Hence the solution is,

$$\begin{aligned} x &= 2E_{\alpha}\left((2+3i)t^{\alpha}\right) \\ &= 2\left[E_{\alpha}(2t^{\alpha})\right]\left[\cos_{\alpha}(3t^{\alpha}) + i\sin_{\alpha}(3t^{\alpha})\right] \end{aligned}$$

Thus

$$x_1 = 2\left[E_{\alpha}(2t^{\alpha})\right]\left[\cos_{\alpha}(3t^{\alpha})\right]$$
$$\text{and } x_2 = 2\left[E_{\alpha}(2t^{\alpha})\right]\left[\sin_{\alpha}(3t^{\alpha})\right]$$

Similarly the solution for y can be written in the form

$$\begin{aligned} y &= (-1+3i)\left[E_\alpha (2+3i)t^\alpha\right] \\ &= (-1+3i)\left[E_\alpha (2t^\alpha)\right]\left[\cos_\alpha (3t^\alpha) + i\sin_\alpha (3t^\alpha)\right] \\ &= \left[E_\alpha (2t^\alpha)\right]\begin{bmatrix} \left(-\cos_\alpha (3t^\alpha) - 3\sin_\alpha (3t^\alpha)\right) \\ +i\left(3\cos_\alpha (3t^\alpha) - \sin_\alpha (3t^\alpha)\right) \end{bmatrix}. \end{aligned}$$

Hence

$$\begin{aligned} y_1 &= \left[E_\alpha (2t^\alpha)\right]\left[-\cos_\alpha (3t^\alpha) - 3\sin_\alpha (3t^\alpha)\right] \\ \text{and } y_2 &= \left[E_\alpha (2t^\alpha)\right]\left[3\cos_\alpha (3t^\alpha) - \sin_\alpha (3t^\alpha)\right]. \end{aligned}$$

Therefore the general solution is linear combination of x_1, x_2 for x and linear combination of y_1, y_2 for y, and we write the following

$$\begin{aligned} x &= \left[E_\alpha (2t^\alpha)\right]\left[2M\cos_\alpha (3t^\alpha) + 2N\sin_\alpha (3t^\alpha)\right] \\ y &= \left[E_\alpha (2t^\alpha)\right]\begin{bmatrix} M\left(-\cos_\alpha (3t^\alpha) - 3\sin_\alpha (3t^\alpha)\right) \\ +N\left(3\cos_\alpha (3t^\alpha) - \sin_\alpha (3t^\alpha)\right) \end{bmatrix}. \end{aligned}$$

where M and N are arbitrary constants. Using initial conditions $x(0) = 2$, $y(0) = 1$ we get $2M = 2$ and $3N - M = 1$ giving $M = 1$, $N = 2$.

Hence the required solution is,

$$\begin{aligned} x &= \left[E_\alpha (2t^\alpha)\right]\left[2\cos_\alpha (3t^\alpha) + 4\sin_\alpha (3t^\alpha)\right] \\ y &= \left[E_\alpha (2t^\alpha)\right]\left[\cos_\alpha (3t^\alpha) - 5\sin_\alpha (3t^\alpha)\right]. \end{aligned}$$

Numerical simulation in Figure 3 shows that for $\alpha = 0.6$, 0.8 and 1.0 after the initiation($t = 0$) of the system $x(t)$ and $y(t)$ both oscillate, Period of oscillation changes with decrease of α.

Case-III

In this case roots of the equations (2.9) being equal, that is $\lambda_1 = \lambda_2 = \lambda$. Then one solution will be of the form

$$x_1 = AE_\alpha(\lambda t^\alpha) \quad \text{and} \quad y_1 = BE_\alpha(\lambda t^\alpha)$$

and the other solution will be

$$x_2 = (A_1 t^\alpha + A_2)E_\alpha(\lambda t^\alpha) \text{ and } y_2 = (B_1 t^\alpha + B_2)E_\alpha(\lambda t^\alpha).$$

Hence the general solution is,

$$x = c_1 AE_\alpha(\lambda t^\alpha) + c_2(A_1 t^\alpha + A_2)E_\alpha(\lambda t^\alpha)$$
$$y = c_1 BE_\alpha(\lambda t^\alpha) + c_2(B_1 t^\alpha + B_2)E_\alpha(\lambda t^\alpha)$$

where $A, A_1, A_2, B, B_1, B_2, C_1, C_2$ are arbitry constants

Example 4:

$$\left.\begin{array}{l} {}^J D^\alpha x = 4x - y \\ {}^J D^\alpha y = x + 2y \end{array}\right\}, 0 < \alpha \le 1 \text{ with } x(0) = 2,\ y(0) = 1 \quad (2.12)$$

Let

$$x = AE_\alpha(\lambda t^\alpha) \quad \text{and} \quad y = BE_\alpha(\lambda t^\alpha)$$

be the solutions of the above differential equation, then putting in the above equation we get

$$\left.\begin{array}{l} A(4-\lambda) - B = 0 \\ A + (2-\lambda)B = 0 \end{array}\right\}.$$

Eliminating A and B as in previous examples, we get,

$$\begin{vmatrix} 4-\lambda & 1 \\ 1 & 2-\lambda \end{vmatrix} = 0 \quad \text{giving} \quad \lambda = 3, 3.$$

For $\lambda = 3$ from equation (2.4) we get $A = B = 1$.

Thus

$$x_1 = E_\alpha(3t^\alpha) \quad \text{and} \quad y_1 = E_\alpha(3t^\alpha)$$

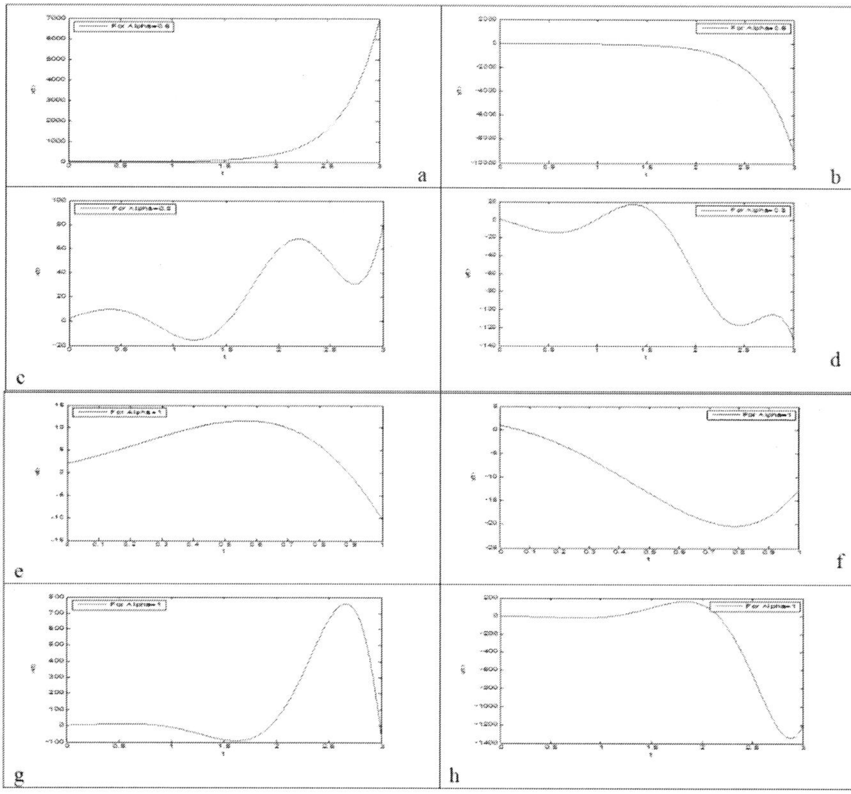

Figure 3. Numerical simulation of the solutions of fractional differential equation in Example-3 for $x(t)$ and $y(t)$ for different Values of α = 0.6, 0.8 and 1.0. Figure (e) & (f) and Figure (g) & (h) represents the same figure only length of x-axis is changed here to represent the prominent initial values

is one solution of the equation. The second solution is as in classical integer order differential equation [[21], pp.307]

$$x_2 = (A_1 t^{\alpha} + A_2)\left[E_{\alpha}(3t^{\alpha})\right]$$

$$\text{and } y_2 = (B_1 t^{\alpha} + B_2)\left[E_{\alpha}(3t^{\alpha})\right]$$

α-th order differentiating for above $x$and y we obtain the following

$$^{J}D^{\alpha}x = 3(A_1 t^{\alpha} + A_2)E_{\alpha}(3t^{\alpha}) + \Gamma(1+\alpha)A_1 E_{\alpha}(3t^{\alpha})$$

and

$$ {}^{J}D^{\alpha} y = 3(B_1 t^{\alpha} + B_2) E_{\alpha}(3t^{\alpha}) + \Gamma(1+\alpha) B_1 E_{\alpha}(3t^{\alpha}). $$

Putting the above obtained result in the given equation (2.12) we get,

Comparing the coefficients and simplifying we get

$$ A_1 = B_1,\, A_2 - B_2 = A_1\Gamma(1+\alpha) = B_1\Gamma(1+\alpha) $$

for simple non-zero values we take

$$ A_1 = B_1 = 1,\; A_2 - B_2 = \Gamma(1+\alpha), $$

for simplicity take $B_2 = 0$ then $A_2 = \Gamma(1+\alpha)$

Thus the other solution is

$$ x = \left[t^{\alpha} + \Gamma(1+\alpha)\right]\left[E_{\alpha}(3t^{\alpha})\right] \text{ and } y = t^{\alpha}\left[E_{\alpha}(3t^{\alpha})\right]. $$

Hence the general solution can be written in the form as in classical integer order differential equation [[21], pp.307]

$$ x = c_1 E_{\alpha}(3t^{\alpha}) + c_2\left[t^{\alpha} + \Gamma(1+\alpha)\right] E_{\alpha}(3t^{\alpha}) $$

$$ \textit{and}\;\; y = c_1 E_{\alpha}(3t^{\alpha}) + c_2 t^{\alpha} E_{\alpha}(3t^{\alpha}). $$

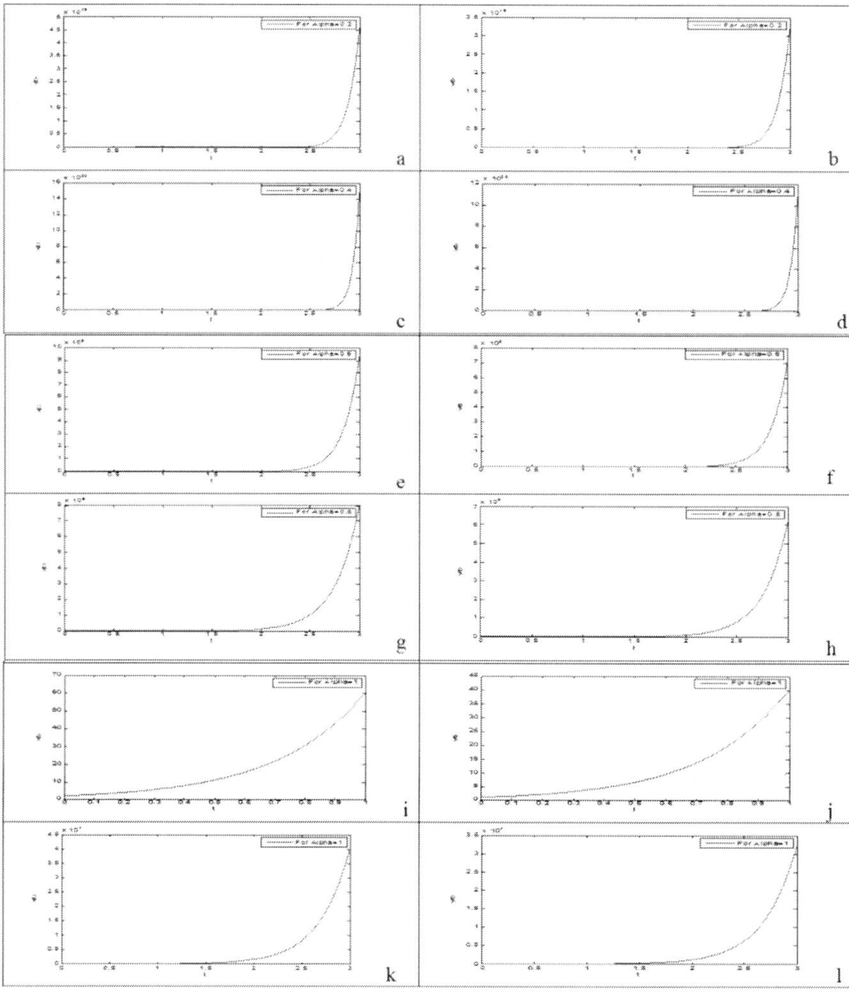

Figure 4. Numerical simulation of the solutions of fractional differential equation in Example-3 for $x(t)$ and $y(t)$ for different Values of $\alpha = 0.2, 0.4, 0.6, 0.8$ and 1.0. Figure (i) & (k) and Figure (j) & (l) represents the same figure only length of x-axis is changed here to represent the prominent initial values

Putting the initial condition.

$$x(0) = 2 \text{ and } y(0) = 1$$

and solving we get

$$c_1 = 1 \text{ and } c_2 = \frac{1}{\Gamma(1+\alpha)}.$$

Hence the solution is,

$$x = E_\alpha(3t^\alpha) + \frac{1}{\Gamma(1+\alpha)}\left[t^\alpha + \Gamma(1+\alpha)\right]\left[E_\alpha(3t^\alpha)\right]$$
$$\text{and } y = E_\alpha(3t^\alpha) + \frac{1}{\Gamma(1+\alpha)} t^\alpha E_\alpha(3t^\alpha).$$

Numerical simulation in Figure 4 shows that $x(t)$ and $y(t)$ start from 2 and 1 respectively; and both of them start to grow. This time interval to grow decreases as α increases. Moreover growth of $x(t)$ and $y(t)$ is higher for is for lower α values. As α increases growth rate of the solution decreases. This implies for lower values of α, $x(t)$ and $y(t)$ grow initially slowly. Once they start to grow, their growth rate is very high whereas for higher α values $x(t)$ and $y(t)$ start to grow sooner but their growth rate is low.

From the above discussion of the three cases one can state a theorem in the following form

Theorem 4:

eThe solutions of the system of differential equations

$${}^{J}D^\alpha X = AX, \text{ Where } A = \begin{bmatrix} a & b \\ c & d \end{bmatrix}, X = \begin{bmatrix} x \\ y \end{bmatrix}$$

are

Case (i)

$$X = c_1 \underset{\sim}{A} E_\alpha(\lambda_1 t^\alpha) + c_2 \underset{\sim}{B} E_\alpha(\lambda_2 t^\alpha),$$

Where $\underset{\sim}{A}$, $\underset{\sim}{B}$, c_1 and c_2 are the arbitry constants,

$$2\lambda = a + d,$$
$$\lambda^2 = ad - bc.$$

Case (ii)

$$X = c_1 \underset{\sim}{A} E_\alpha(\lambda t^\alpha) + c_2(\underset{\sim}{B_1} t^\alpha + \underset{\sim}{B_2}) E_\alpha(\lambda t^\alpha),$$

Where A, B_1, B_2, c_1 and c_2 are the arbitry constants,

$$2\lambda = a + d,$$
$$\lambda^2 = ad - bc.$$

Case (iii)

$$X = \left[E_\alpha(pt^\alpha) \right] \begin{bmatrix} A_1 \cos_\alpha(qt^\alpha) - A_2 \sin_\alpha(qt^\alpha) \\ +(A_2 \cos_\alpha(qt^\alpha) + A_1 \sin_\alpha(qt^\alpha)) \\ B_1 \cos_\alpha(qt^\alpha) - B_2 \sin_\alpha(qt^\alpha) \\ +(B_2 \cos_\alpha(qt^\alpha) + B_1 \sin_\alpha(qt^\alpha)) \end{bmatrix}$$

Where A_1, A_2, B_1, B_2 are the arbitry constants,

$$2p = a + d,$$
$$p^2 + q^2 = ad - bc.$$

3. APPLICATION OF THE ABOVE FORMULATION IN REAL LIFE PROBLEM:

Consider the following fractional damped oscillator, formulated by Jumarie fractional derivative

$$^J D^{2\alpha}[x] + 2a\left({}^J D^\alpha[x]\right) + bx = 0. \tag{3.1}$$

Let ${}^J D^\alpha[x] \equiv \frac{d^\alpha x}{dt^\alpha} = y$ then the given equation reduce to the following system of equation

$$\left. \begin{aligned} {}^J D^\alpha[y] &= \frac{d^\alpha y}{dt^\alpha} = -2ay - bx \\ {}^J D^\alpha[x] &= \frac{d^\alpha x}{dt^\alpha} = y \end{aligned} \right\}. \tag{3.2}$$

The above system of equation can be written in the form

$$\begin{bmatrix} {}^{J}D^{\alpha}x \\ {}^{J}D^{\alpha}y \end{bmatrix} = \begin{bmatrix} \frac{d^{\alpha}x}{dt^{\alpha}} \\ \frac{d^{\alpha}y}{dt^{\alpha}} \end{bmatrix} = \begin{bmatrix} 0 & 1 \\ -b & -2a \end{bmatrix} \begin{bmatrix} x \\ y \end{bmatrix}.$$

Let

$$x = AE_{\alpha}(\lambda t^{\alpha}) \text{ and } y = BE_{\alpha}(\lambda t^{\alpha})$$

be solutions of the differential equations.

Then

$$\left.\begin{aligned} A(0-\lambda) + B &= 0 \\ -bA + (-2a-\lambda)B &= 0 \end{aligned}\right\}. \tag{3.3}$$

For the above system of equation the auxiliary equation is,

$$\begin{vmatrix} 0-\lambda & 1 \\ -b & -2a-\lambda \end{vmatrix} = 0 \quad \text{giving} \quad \lambda^2 + 2a\lambda + b = 0.$$

Here the discriminant is $4(a^2 - b)$ We consider the case when $a^2 - b < 0$ then the eigen-values are

$$\lambda_1, \lambda_2 = p \pm iq \text{ where } p = -a,\ q = \sqrt{b-a^2}.$$

Then from (3.3) putting $\lambda = p + iq$ we get $A(p + iq) = B$, we can take the solution in the form $A = 1$, $B = (p + iq)$. The general solution will be of the form,

$$\begin{aligned} x &= E_{\alpha}\left[(p+iq)t^{\alpha})\right] \\ &= \left[E_{\alpha}(pt^{\alpha})\right]\left[\cos_{\alpha}(qt^{\alpha}) + \sin_{\alpha}(qt^{\alpha})\right] \\ x_1 &= \left[E_{\alpha}(pt^{\alpha})\right]\left[\cos_{\alpha}(qt^{\alpha})\right] \\ \text{and } x_2 &= \left[E_{\alpha}(pt^{\alpha})\right]\left[\sin_{\alpha}(qt^{\alpha})\right]. \end{aligned}$$

Thus the general solution can be written in the form as in classical integer order differential equation [[21], pp.305]

$$x = \left[E_\alpha (pt^\alpha) \right] \left[C_1 \cos_\alpha (qt^\alpha) + C_2 \sin_\alpha (qt^\alpha) \right]$$

Where C_1 and C_2 are arbitrary constants.

Here $x(0) = 2$ and ${}^J D^\alpha x(0) = 1$and solving we get $C_1 = 1$ and $C_2 = (1\text{-}2p)/q$ Hence the solution is

$$x = \left[E_\alpha (pt^\alpha) \right] \left[\cos_\alpha (qt^\alpha) + \frac{1-2p}{q} \sin_\alpha (qt^\alpha) \right].$$

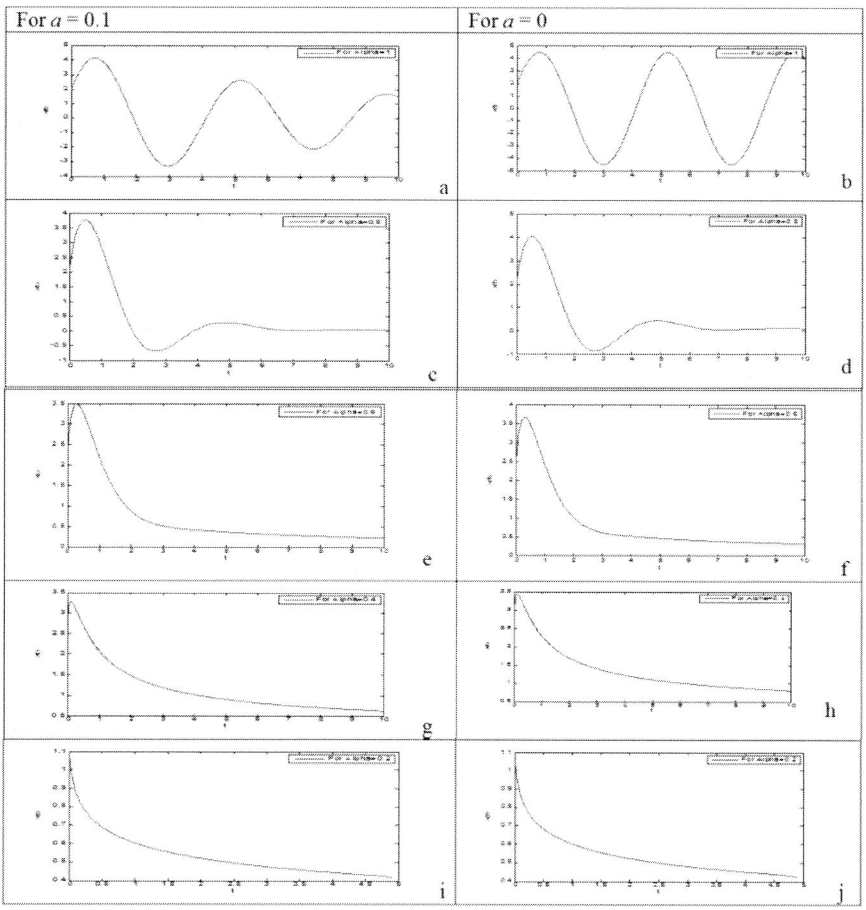

Figure 5. Numerical simulation of the solutions of fractional differential equation in equation (3.1) for different Values α = 0.2, 0.4, 0.6, 0.8, and 1.0. for b=2 and for a=0.1 & a=0

The numerical simulation of the solutions of the differential equation (3.1) has shown in Figure 5, for b = 2, left hand figures for a = 0.1 and right hand figures for a = 0. Figures 4 (a) and (b) are drawn for a = 1, it is clear from the figure in presence of damping the amplitude of the oscillation decreases with time. Figures 4 (c) and (d) are drawn for a = 0.8, it is clear from the figure in both the cases the amplitude of the oscillation decreases with time and ultimately amplitude tends to zero. Figures 4 (e) & (f) and (g) & (h) and (i) & (j) are drawn for a = 0.6, 0.4 and 0.2 respectively, it is clear from the figures with decrease of order of derivative the oscillator losses the oscillating behavior.

4. CONCLUSIONS

The system of fractional differential equation arises in different applications. Here we develop an algorithm to solve the system of fractional differential equations with modified fractional derivative (with Jumarie's fractional derivative formulation) in terms of Mittag-Leffler function and the generalized Sine and Cosine functions. From the numerical simulations it is observed that the growing or decaying of the solutions is fast in fractional order derivative case compare to the integer order derivative. The use of this type of Jumarie fractional derivative gives a conjugation with classical methods of solution of system of linear integer order differential equations, by usage of Mittag-Leffler and generalized trigonometric functions that we have demonstrated here in this paper. The ease of this method and its conjugation to classical method to solve system of linear fractional differential equation is appealing to researchers in fractional dynamic systems.

ACKNOWLEDGEMENT

Acknowledgments are to **Board of Research in Nuclear Science** (BRNS), Department of Atomic Energy Government of India for financial assistance received through BRNS research project no. 37(3)/14/46/ 2014-BRNS with BSC BRNS, title "Characterization of unreachable (Holderian) functions via Local Fractional Derivative and Deviation Function".

REFERENCES

1. K. S. Miller, B. Ross. An Introduction to the Fractional Calculus and Fractional Differential Equations.John Wiley & Sons, New York, NY, USA; 1993.
2. S. Das. Functional Fractional Calculus 2nd Edition, Springer-Verlag 2011.
3. Podlubny. Fractional Differential Equations, Mathematics in Science and Engineering, Academic Press, San Diego, Calif, USA. 1999; 198.
4. K. Diethelm. The analysis of Fractional Differential equations. Springer-Verlag, 2010.

a. Kilbas, H. M. Srivastava, J.J. Trujillo. Theory and Applications of Fractional Differential Equations. North-Holland Mathematics Studies, Elsevier Science, Amsterdam, the Netherlands, 204. 2006.

5. M. Caputo, "Linear models of dissipation whose q is almost frequency independent," *Geophysical Journal of the Royal Astronomical Society*, 1967. vol. 13, no. 5, pp. 529-539.
6. G. Jumarie. Modified Riemann-Liouville derivative and fractional Taylor series of non-differentiable functions Further results, Computers and Mathematics with Applications, 2006. (51), 1367-1376.
7. U. Ghosh, S. Sengupta, S. Sarkar and S. Das. Characterization of non-differentiable points of a function by Fractional derivative of Jumarie type. European Journal of Academic Essays 2(2): 70-86, 2015.
8. K. Diethelm, N. J. Ford and A. D. Freed, A predictor-corrector approach for the numerical solution of fractional differential equations, Nonlinear Dynamics, 29, 3-22, 2002.
9. Y. Hua, Y. Luoa, Z. Lu, Analytical solution of the linear fractional differential equation by Adomian decomposition method. Journal of Computational and Applied Mathematics. 215 (2008) 220-229.
10. S. S. Ray, R.K. Bera. An approximate solution of a nonlinear fractional differential equation by Adomian decomposition method. Applied Mathematics and Computation. 167 (2005) 561-571.
11. O. Abdulaziz, I. Hashim, S. Momani, Application of homotopy-perturbation method to fractional IVPs, J. Comput. Appl. Math. 216 (2008) 574-584.
12. R. Yulita Molliq, M.S.M. Noorani, I. Hashim, R.R. Ahmad. Approximate solutions of fractional Zakharov–Kuznetsov equations by VIM. Journal of Computational and Applied Mathematics. 233(2). 2009. 103-108.
13. V.S. Ertürk, S. Momani. Solving systems of fractional differential equations using differential transform method. Journal of Computational and Applied Mathematics 215 (2008) 142-151.
14. U. Ghosh., S. Sengupta, S. Sarkar and S. Das. Analytic solution of linear fractional differential equation with Jumarie derivative in term of Mittag-Leffler function. American Journal of Mathematical Analysis 3(2). 2015. 32-38, 2015.
15. G. Jumarie. Fourier's Transformation of fractional order via Mittag-Leffler function and modified Riemann-Liouville derivatives. J. Appl. Math. & informatics. 2008. 26. 1101-1121.
16. Mittag-Leffler, G. M.. Sur la nouvelle fonction $E\alpha$ (x), *C. R. Acad. Sci. Paris*, (Ser. II) 137, 554-558 (1903).
17. Erdelyi, A. Asymptotic expansions, Dover (1954).
18. Erdelyi, A. (Ed). Tables of Integral Transforms. vol. 1, McGraw-Hill, 1954.
19. Erdelyi, A. On some functional transformation Univ Potitec Torino 1950.
20. S. L. Ross. Differential equations (3rd edition). John Wiley & Sons. 1984.

CHAPTER 7

On the full bosonic string from Minkowski space to Riemannian manifolds

Volker Branding

University of Vienna, Faculty of Mathematics, Oskar-Morgenstern-Platz 1, 1090 Vienna, Austria

ABSTRACT

We study the action of the full bosonic string for the domain being two-dimensional Minkowski space and the target a Riemannian manifold. Its critical points couple the wave map equation to a scalar and a two-form potential. Besides investigating their basic features we establish existence results for the latter.

Keywords: Wave maps with scalar and two-form potential; Full bosonic string; Cauchy problem

1. INTRODUCTION AND RESULTS

The nonlinear σ-model is one of the most prominent models studied in quantum field theory. In mathematical terms it is described by the Dirichlet energy of a map between two manifolds. In the case that the domain has a Riemannian metric its critical points lead to the *harmonic map equation*, which is a semilinear elliptic equation, see [7] for a recent survey. However, if the domain has a Lorentzian metric its critical points are given by the *wave map equation*, which is a semilinear hyperbolic equation, see [13]; [16] ; [14] and references therein for an introduction to wave maps.

In this article we will focus on the nonlinear σ-model coupled to two external potentials, which arises as the full action of the bosonic string, see [11], p. 108 for the physical background. More precisely, we will study the following action

$$S(\phi) = \int_M \Big(\frac{1}{2}|d\phi|^2 + \phi^* B + V(\phi)\Big) d\mu \tag{1.1}$$

Here, $\phi\colon M \to N$ is a map between two manifolds M and N with $\dim M = 2$ and $\dim N \geq 3$. In addition, B is a two-form on N, which we pullback via the map ϕ. Finally, $V\colon N \to \mathbb{R}$ is a scalar function. For the sake of completeness, we want to mention that in the physics literature one often finds that the scalar potential $V(\phi)$ in (1.1) gets multiplied by the scalar curvature of the domain. The action functional (1.1) models a bosonic string under the influence of an external magnetic and scalar field.

This article is a sequel to [2], where the action functional (1.1) has been investigated in the case that the domain is a closed Riemannian surface and the target a Riemannian manifold. In this case an existence result could be obtained by the heat-flow method, which requires the target manifold to have negative curvature.

Here, we consider the situation that the domain M is two-dimensional Minkowski space and the target a Riemannian manifold. It turns out that contrary to the Riemannian case we do not have to impose a curvature condition on the target to obtain an existence result.

For wave maps from two-dimensional Minkowski space to Riemannian manifolds there exist two classical existence results. The first one is due to Gu [6], who showed that the Cauchy problem for wave maps from two-dimensional Minkowski space has a global smooth solution for smooth initial data. Using the method of a priori estimates the existence of a global weak solution for special target manifolds was shown in [5], which was later extended to arbitrary targets in [3].

A problem similar to the one studied in this article has been investigated in [10] ; [4]: The authors study timelike minimal surfaces in Lorentzian manifolds, which corresponds to critical points of (1.1) without potentials and the target being a Lorentzian manifold.

This article is organized as follows: In Section 2 we study the basics of the action functional (1.1). In Section 3 we derive an existence result for critical points of (1.1) in the special case of a vanishing scalar potential $V(\phi)$ making use of the method of characteristics. In addition, we point out how this result can be obtained by purely analytical methods. In the last section we prove the existence of a global weak solution to (1.1) for a non-vanishing scalar potential $V(\phi)$.

2. WAVE MAPS WITH SCALAR AND TWO-FORM POTENTIAL

In this section we investigate the basic features of (1.1).

Let (M,h) be two-dimensional Minkowski space with global coordinates (t,x) and the metric of signature $(+,-)$. Moreover, let (N,g) be a Riemannian manifold of dimension $\dim N = n \geq 3$. We will use Latin letters for indices on the target N and Greek letters for indices on the domain M. In addition, we make use of the summation convention, that is, we sum over repeated indices.

Proposition 2.1. *The Euler–Lagrange equation of the functional* (1.1)*is given by*

$$\tau(\phi) = Z(d\phi(\partial_t) \wedge d\phi(\partial_x)) + \nabla V(\phi), \tag{2.1}$$

where $\tau(\phi) := Tr(\nabla^{\phi^* TN} d\phi)$ *denotes the wave map operator of the map* ϕ*. The vector-bundle homomorphism* $Z \in \Gamma(\mathrm{Hom}(\Lambda^2 T^*N, TN))$ *is defined by the equation*

$$\Omega(\eta,\xi_1,\xi_2) = \langle Z(\xi_1 \wedge \xi_2),\eta \rangle , \tag{2.2}$$

where $\Omega = dB$ *is a three-form on* N *and* $\eta, \xi_1, \xi_2 \in TN$*. Moreover* ∂_t, ∂_x *is the global orthonormal basis of* $TR^{1,1} \simeq R^{1,1}$*.*

Proof. This follows as in the Riemannian case, see [2, Proposition 2.1]. □

We call solutions of (2.1)*wave maps with scalar and two-form potential.*

In terms of global coordinates (t,x) on two-dimensional Minkowski space and local coordinates y^i, $I = 1,\ldots,n$ on the target N the Euler–Lagrange equation (2.1) acquires the form

$$\frac{\partial^2 \phi^i}{\partial t^2} - \frac{\partial^2 \phi^i}{\partial x^2} + \Gamma^i_{jk}\left(\frac{\partial \phi^j}{\partial t}\frac{\partial \phi^k}{\partial t} - \frac{\partial \phi^j}{\partial x}\frac{\partial \phi^k}{\partial x}\right) - Z^i(\partial_{y^k} \wedge \partial_{y^j})\frac{\partial \phi^k}{\partial t}\frac{\partial \phi^j}{\partial x} - g^{ij}\frac{\partial V(\phi)}{\partial y^j} = 0 \tag{2.3}$$

Here, Γ^i_{jk} are the Christoffel symbols of the Levi-Cevita connection on N.

Remark 2.2. The standard wave map equation can also be studied from higher-dimensional Minkowski space. However, the two-form potential in (1.1) only gives a well-defined action functional in two dimensions such that the study of (1.1) is restricted to a two-dimensional domain.

Remark 2.3. The vector-bundle homomorphism Z defined in (2.2) can be interpreted as arising from a metric connection with totally antisymmetric torsion. In this case one has

$$\nabla^{Tor}_X Y = \nabla^{LC}_X Y + A(X,Y)$$

where ∇^{LC} denotes the Levi-Cevita connection, X, Y are vector fields and $A(X,Y)$ is a skew-adjoint endomorphism. The endomorphism $A(X,Y)$ satisfies

$$\langle A(X,Y), Z\rangle = \Omega(X,Y,Z)$$

with $\Omega \in \Gamma(\Lambda^3 T^*N)$ similar to (2.2). For more details on metric connections with torsion see [1] and references therein.

By varying the action functional with respect to the domain metric we obtain the energy–momentum tensor:

$$T_{\alpha\beta} := \frac{1}{2}h_{\alpha\beta}|d\phi|^2 - \langle d\phi(e_\alpha), d\phi(e_\beta)\rangle + V(\phi)h_{\alpha\beta}$$

By a direct calculation it follows that the energy–momentum tensor $T_{\alpha\beta}$ is divergence free whenever ϕ is a wave map with scalar and two-form potential, see [2, Lemma 3.3] for more details.

For many of the calculations in this article it turns out to be useful to exploit the geometry of two-dimensional Minkowski space by introducing lightcone coordinates:

$$\xi = \frac{x+t}{2}, \qquad \eta = \frac{x-t}{2}, \qquad \frac{\partial}{\partial x} = \frac{1}{2}\left(\frac{\partial}{\partial \xi} + \frac{\partial}{\partial \eta}\right), \qquad \frac{\partial}{\partial t} = \frac{1}{2}\left(\frac{\partial}{\partial \xi} - \frac{\partial}{\partial \eta}\right)$$

In these coordinates (2.3) acquires the form

$$\frac{\partial^2\phi^i}{\partial\xi\partial\eta} + \left(\Gamma^i_{jk} + \frac{1}{2}Z^i(\partial_{y^k} \wedge \partial_{y^j})\right)\frac{\partial\phi^j}{\partial\xi}\frac{\partial\phi^k}{\partial\eta} + g^{ij}\frac{\partial V(\phi)}{\partial y^j} = 0, \qquad i = 1,\dots,n \tag{2.4}$$

At several places it becomes necessary to make use of the Nash embedding theorem to isometrically embed N into some $\mathbb{R}^q$.

Lemma 2.4. *Assume that* $N \subset \mathbb{R}^q$. *Then the Euler–Lagrange equation* (2.1)*acquires the form*

$$\frac{\partial^2 \phi^i}{\partial \xi \partial \eta} + (\Gamma^i_{jk} + \frac{1}{2} Z^i(\partial_{y^k} \wedge \partial_{y^j}))\frac{\partial \phi^j}{\partial \xi}\frac{\partial \phi^k}{\partial \eta} + g^{ij}\frac{\partial V(\phi)}{\partial y^j} = 0, \qquad i = 1, \dots, n \tag{2.5}$$

where $u\colon \mathbb{R}^{1,1} \to \mathbb{R}^q$ *and* $\mathrm{I\!I}$*denotes the second fundamental form of* N *in* $\mathbb{R}^q$. *Moreover,* $\tilde{Z}$ *and* ΔV *denote the extensions of* Z *and* ∇V *to the ambient space* $\mathbb{R}^q$. *In terms of lightcone coordinates on* M *the Euler–Lagrange equation* (2.1)*is given by*

$$\frac{\partial^2 u}{\partial \xi \partial \eta} = \mathrm{I\!I}(\frac{\partial u}{\partial \eta}, \frac{\partial u}{\partial \xi}) - \frac{1}{2}\tilde{Z}(du(\partial_\xi) \wedge du(\partial_\eta)) - \widetilde{\nabla V}(u) \tag{2.6}$$

Proof. For a proof see [2, Lemma 3.8] and references therein. □

In the following we will omit the tildes at Z and ∇V when extending them to the ambient space.

Besides the energy–momentum tensor there is another conserved quantity:

Proposition 2.5. *Let* u: $\mathbb{R}^{1,1} \to \mathbb{R}^q$ *be a solution of* (2.5). *Then the energy*

$$E(t) = \int_{\mathbb{R}} (|\frac{\partial u}{\partial t}|^2 + |\frac{\partial u}{\partial x}|^2 - 2V(u))dx \tag{2.7}$$

is constant with respect to t.

Proof. Using the extrinsic version of the equation for wave maps with scalar and two-form potential (2.5) we calculate

$$\begin{aligned}
\frac{d}{dt}\int_{\mathbb{R}} (|\frac{\partial u}{\partial t}|^2 + |\frac{\partial u}{\partial x}|^2)dx &= 2\int_{\mathbb{R}} \langle \frac{\partial u}{\partial t}, \frac{\partial^2 u}{\partial t^2} - \frac{\partial^2 u}{\partial x^2}\rangle dx \\
&= 2\int_{\mathbb{R}} (\underbrace{\langle \frac{\partial u}{\partial t}, \mathrm{I\!I}(du, du)\rangle}_{=0 \text{ since } \mathrm{I\!I} \perp \partial_u} + \langle \frac{\partial u}{\partial t}, Z(\frac{\partial u}{\partial t} \wedge \frac{\partial u}{\partial x})\rangle + \langle \frac{\partial u}{\partial t}, \nabla V(u)\rangle)dx \\
&= 2\int_{\mathbb{R}} (\underbrace{\Omega(\frac{\partial u}{\partial t}, \frac{\partial u}{\partial t}, \frac{\partial u}{\partial x})}_{=0} + \frac{\partial}{\partial t}V(u))dx,
\end{aligned}$$

where we interchanged the derivatives with respect to t and x and used integration by parts in the first step. □

3. WAVE MAPS WITH TWO-FORM POTENTIAL

In this section we consider the simpler case of a vanishing scalar potential, that is $V=0$. The action functional under consideration then reduces to

$$S(\phi) = \int_{\mathbb{R}^{1,1}} (\frac{1}{2}|d\phi|^2 + \phi^* B) d\mu$$

and its critical points are given by

$$\tau(\phi) = Z(d\phi(\partial_t) \wedge d\phi(\partial_x)). \tag{3.1}$$

We call solutions of (3.1)*wave maps with two-form potential.*

Note that solutions of (3.1) are invariant under scaling in the sense that if $\phi(t,x)$ is a solution of (3.1), then $\phi_\lambda(t,x)$ defined by

$$\phi_\lambda(t,x) := \phi(\lambda t, \lambda x)$$

is also a solution of (3.1), where λ is a positive real-valued parameter.

Remark 3.1. In the case that the domain is a Riemannian surface, the target a three-dimensional Riemannian manifold and ϕ an isometric embedding, the equation (3.1) is known as *prescribed curvature equation.* Since we are considering a domain with an indefinite metric here, it does not seem to be possible to give an analogue geometric interpretation of solutions of (3.1).

By Proposition 2.7, using the extrinsic version of (3.1), the energy

$$E(t) := \frac{1}{2} \int_{\mathbb{R}} (|\frac{\partial \phi}{\partial t}|^2 + |\frac{\partial \phi}{\partial x}|^2) dx$$

is constant with respect to t whenever ϕ is a solution of (3.1).

Locally, (3.1) looks like

$$\frac{\partial^2\phi^i}{\partial t^2} - \frac{\partial^2\phi^i}{\partial x^2} + \Gamma^i_{jk}\left(\frac{\partial\phi^j}{\partial t}\frac{\partial\phi^k}{\partial t} - \frac{\partial\phi^j}{\partial x}\frac{\partial\phi^k}{\partial x}\right) - Z^i(\partial_{y^k} \wedge \partial_{y^j})\frac{\partial\phi^k}{\partial t}\frac{\partial\phi^j}{\partial x} = 0, \qquad i = 1,\dots,n \tag{3.2}$$

and in terms of lightcone coordinates (ξ,η) equation (3.2) acquires the form

$$\frac{\partial^2\phi^i}{\partial\xi\partial\eta} + (\Gamma^i_{jk} + \frac{1}{2}Z^i(\partial_{y^k} \wedge \partial_{y^j}))\frac{\partial\phi^j}{\partial\xi}\frac{\partial\phi^k}{\partial\eta} = 0, \qquad i = 1,\dots,n \tag{3.3}$$

We will establish the following existence result:

Theorem 3.2. *Let (N,g) be a complete Riemannian manifold. Then the Cauchy problem for wave maps with two-form potential, that is*

$$\tau(\phi) = Z(d\phi(\partial_t) \wedge d\phi(\partial_x)) \tag{3.4}$$

with the smooth initial data

$$\phi(0,x) = \phi_0(x), \qquad \frac{\partial\phi}{\partial t}(0,x) = \phi_1(x)$$

has a smooth unique solution.

To prove Theorem 3.2 we adapt the proof obtained for (standard) wave maps from two-dimensional Minkowski space [6].

We will divide the proof of Theorem 3.2 into several steps.

Lemma 3.3. *The Cauchy problem* (3.4)*is equivalent to the system*

$$\begin{aligned} &\frac{\partial u^i}{\partial\eta} + (\Gamma^i_{jk} + \frac{1}{2}Z^i(\partial_{y^k} \wedge \partial_{y^j}))u^j v^k = 0, \qquad \frac{\partial y^i}{\partial\xi} = u^i, \qquad i = 1,\dots,n, \\ &\frac{\partial v^i}{\partial\xi} + (\Gamma^i_{jk} + \frac{1}{2}Z^i(\partial_{y^k} \wedge \partial_{y^j}))v^j u^k = 0, \qquad \frac{\partial z^i}{\partial\eta} = v^i, \qquad i = 1,\dots,n \end{aligned} \tag{3.5}$$

for functions u,v,y,z: $\mathrm{R}^{1,1} \to \mathrm{R}^n$ with the initial conditions

$$y^i(0,x) = z^i(0,x) = \phi^i_0(x), \qquad u^i(0,x) = \frac{\partial\phi^i_0}{\partial x}(x) + \phi^i_1(x), \qquad v^i(0,x) = -\frac{\partial\phi^i_0}{\partial x}(x) + \phi^i_1(x) \tag{3.6}$$

Proof. This follows by a direct calculation using lightcone coordinates. □

In the following let U be a coordinate system in the target N with coordinates y^i, $I = 1 \dots n$, and $\tilde{U}$ a subset of U fixed by the requirement

$$\sum_{i=1}^{n} (y^i)^2 \leq 1$$

Let a be a positive constant such that

$$g(y)(\lambda,\lambda) \geq a^2|\lambda|^2$$

holds for all $y \in \tilde{U}$ and all $\lambda \in R^n$.

We set

$$M := \sup_{|x| \leq L} \{\sqrt{g(\phi_0)(dy(\partial_\xi), dy(\partial_\xi))}, \sqrt{g(\phi_0)(dy(\partial_\eta), dy(\partial_\eta))}\}$$

for a large number L.

As a next step we prove the existence of a solution to the system (3.5) on the characteristic triangle Λ_k, which we define by

$$-\frac{1}{2}k \leq -\eta \leq \xi \leq \frac{1}{2}k$$

with

$$k := \min\{L, \frac{a}{2M\sqrt{n}}\}$$

Proposition 3.4. *There exists a smooth solution to the Cauchy problem (3.4)on the characteristic triangle Λ_k.*

Proof. To establish the claim we will make use of the system (3.5). Let u_0^i, v_0^i be smooth functions satisfying the initial condition (3.6)

$$\frac{\partial y_0^i}{\partial \xi} = u_0^i, \qquad \frac{\partial z_0^i}{\partial \eta} = v_0^i, \qquad y_0^i(0,x) = z_0^i(0,x) = \phi_0^i(x)$$

and the restrictions

$$\sum_{i=1}^{n}(y_0^i)^2 \leq 1, \qquad \sum_{i=1}^{n}(z_0^i)^2 \leq 1 \tag{3.7}$$

Suppose we have constructed $y_{m-1}^i, z_{m-1}^i, u_{m-1}^i, v_{m-1}^i$ that satisfy the initial condition (3.6). Moreover, suppose that y_{m-1}^i and z_{m-1}^i satisfy the constraint (3.7). We define $y_m^i, z_m^i, u_m^i, v_m^i$ by the equations

$$\begin{aligned} &\frac{\partial u_m^i}{\partial \eta} + (\Gamma_{jk}^i(z_{m-1}) + \frac{1}{2}Z^i(z_{m-1})(\partial_{y^k} \wedge \partial_{y^j}))u_m^j v_{m-1}^k = 0, \qquad \frac{\partial y_m^i}{\partial \xi} = u_m^i, \\ &\frac{\partial v_m^i}{\partial \xi} + (\Gamma_{jk}^i(y_{m-1}) + \frac{1}{2}Z^i(y_{m-1})(\partial_{y^k} \wedge \partial_{y^j}))v_m^j u_{m-1}^k = 0, \qquad \frac{\partial z_m^i}{\partial \eta} = v_m^i. \end{aligned} \tag{3.8}$$

Note that the equations (3.8) are linear. Hence, their solutions exist in the whole characteristic triangle Λ_k.

In addition, the equations (3.8) also have a nice geometric interpretation. We can think of these equations describing two parallel vector fields u_m^i, v_m^i along the curves $z^i = z_{m-1}^i(\xi_0, \eta)$ and $y^i = y_{m-1}^i(\xi, \eta_0)$ for a metric connection with totally antisymmetric torsion.

Since parallel translation is an isometry (this is not affected by the presence of torsion), we have

$$g(z_{m-1})(u_m, u_m) = g(z_{m-1})(u_m, u_m)\big|_{t=0}, \qquad g(z_{m-1})(v_m, v_m) = g(z_{m-1})(v_m, v_m)\big|_{t=0}.$$

Thus, we obtain the estimates

$$|u_m^i| \leq \frac{M}{a}, \qquad |v_m^i| \leq \frac{M}{a}, \qquad i = 1, \ldots, n$$

Consequently, we also find

$$\sum_{i=1}^{n}(y_m^i)^2 \leq 4n\frac{M^2}{a^2}k^2 \leq 1, \qquad \sum_{i=1}^{n}(z_m^i)^2 \leq 1$$

We can deduce that y_m^i, z_m^i remain in the set $\bar{U}$ and that u_m, v_m are bounded. Consequently, the sequences $\{u_m\}$, $\{v_m\}$, $\{z_m\}$, $\{y_m\}$ and the sequences of their partial derivatives converge uniformly on the characteristic triangle Λ_k. Hence, the limit

$$y(t,x) = \lim_{m\to\infty} y_m(t,x)$$

is a solution to the Cauchy problem (3.4) on Λ_k. □

Lemma 3.5 Global existence. *The solution constructed on the characteristic triangle Λ_k can be extended to the whole Minkowski space.*

Proof. To prove that there exists a global smooth solution of (3.4) it suffices to know that there exists a smooth solution on Λ_k. This can be seen as follows: We know that the Cauchy problem (3.4) is solvable up to $t = k$. We perform the following rescaling of the initial data

$$\phi_0(k,x) = \phi(\sigma k, x), \qquad \phi_1(k,x) = \frac{\partial \phi}{\partial t}(\sigma k, x), \qquad -L + \sigma k \leq x \leq L - \sigma k$$

where $\sigma \geq 1$ is a fixed positive integer, defined by $t = \sigma k$. We infer that the Cauchy problem (3.4) is solvable up to $t = (\sigma + 1)k$. By iteration of this procedure we conclude that there exists a global smooth solution of (3.4). □

In the following we will need the extrinsic version of (3.1), which is given by

$$\frac{\partial^2 u}{\partial t^2} - \frac{\partial^2 u}{\partial x^2} = \mathrm{II}(du, du) + Z(du(\partial_t) \wedge du(\partial_x)) \tag{3.9}$$

where $u\colon R^{1,1} \to R^q$.

Lemma 3.6 Uniqueness. *The global solution to the Cauchy problem* (3.4)*constructed in the last Lemma is unique.*

Proof. Suppose we have two solutions u, v of the extrinsic equation for wave maps with two-form potential (3.9) that have the same initial data, that is, $u = v$ and $du = dv$ at $t = 0$. We set $w := u - v$. Making use of the extrinsic version for wave maps with two-form potential we find by a direct calculation

$$\begin{aligned}\frac{d}{dt}\int_{\mathbb{R}} \left(\left|\frac{\partial w}{\partial t}\right|^2 + \left|\frac{\partial w}{\partial x}\right|^2\right)dx &= \int_{\mathbb{R}} \langle \frac{\partial w}{\partial t}, \mathrm{II}(u)(du, du) - \mathrm{II}(v)(dv, dv)\rangle dx \\ &+ \int_{\mathbb{R}} \langle \frac{\partial w}{\partial t}, Z(u)(du(\partial_t) \wedge du(\partial_x)) - Z(v)(dv(\partial_t) \wedge dv(\partial_x))\rangle dx\end{aligned}$$

We have a pointwise bound on du and dv from (3.8), which allows us to estimate

$$|\langle \mathrm{II}(u)(du,du) - \mathrm{II}(v)(dv,dv), \frac{\partial w}{\partial t}\rangle| \leq C(|dw|^2 + |dw||w|),$$

$$|\langle Z(u)(du(\partial_t) \wedge du(\partial_x)) - Z(v)(dv(\partial_t) \wedge dv(\partial_x)), \frac{\partial w}{\partial t}\rangle| \leq C(|dw|^2 + |dw||w|).$$

Thus, we find

$$\frac{d}{dt}\int_{\mathbb{R}} (w^2 + |dw|^2)dx \leq C\int_{\mathbb{R}} (w^2 + |dw|^2)dx$$

By integrating the differential equation and using that $w = dw = 0$ at $t = 0$ □ we obtain the claim.

This finishes the proof of Theorem 3.2.

Remark 3.7. The action functional for the full bosonic string presented in the introduction contains an additional scalar potential. If we try to extend the strategy of proof used in this section to the latter it turns out that we do not succeed due to the presence of the scalar potential.

On the one hand we lose the scale invariance of the Euler–Lagrange equations and secondly, we cannot rewrite the Euler–Lagrange equation into a system which has a geometric interpretation as parallel transport.

The proof of Theorem 3.2 presented above is of a geometric nature. In the following we want to sketch briefly how Theorem 3.2 can also be proven making use of analytical methods, which is inspired from [12]. These methods have the advantage that they can also be applied for non-smooth initial data.

For a solution u of (3.9) we define the energy

$$e(t,x) := \frac{1}{2}|\frac{\partial u}{\partial t}|^2 + \frac{1}{2}|\frac{\partial u}{\partial x}|^2$$

If we test (3.9) with $\frac{\partial u}{\partial t}$ we find

$$\begin{aligned}
0 &= \langle \mathrm{II}(du,du), \frac{\partial u}{\partial t}\rangle + \langle Z(\frac{\partial u}{\partial t} \wedge \frac{\partial u}{\partial x}), \frac{\partial u}{\partial t}\rangle \\
&= \langle \frac{\partial^2 u}{\partial t^2}, \frac{\partial u}{\partial t}\rangle - \langle \frac{\partial^2 u}{\partial x^2}, \frac{\partial u}{\partial t}\rangle \\
&= \frac{1}{2}\frac{\partial}{\partial t}|\frac{\partial u}{\partial t}|^2 + \frac{1}{2}\frac{\partial}{\partial t}|\frac{\partial u}{\partial x}|^2 - \frac{\partial}{\partial x}\langle \frac{\partial u}{\partial x}, \frac{\partial u}{\partial t}\rangle
\end{aligned}$$

which yields the identity

$$\frac{\partial}{\partial t}e(t,x) = \frac{\partial}{\partial x}\langle\frac{\partial u}{\partial x}, \frac{\partial u}{\partial t}\rangle. \tag{3.10}$$

Interchanging the roles of *t* and *x* a similar calculation leads to

$$\frac{\partial}{\partial x}e(t,x) = \frac{\partial}{\partial t}\langle\frac{\partial u}{\partial x}, \frac{\partial u}{\partial t}\rangle \tag{3.11}$$

Differentiating (3.10) with respect to *t* and (3.11) with respect to *x* yields the identity

$$(\frac{\partial^2}{\partial t^2} - \frac{\partial^2}{\partial x^2})e(t,x) = 0 \tag{3.12}$$

which means that the energy e(t,x) satisfies the linear wave equation in two-dimensional Minkowski space. Consequently, we get the following energy equality

$$\tilde{E}(t) := \frac{1}{2}\int_{\mathbb{R}} (|\frac{\partial}{\partial t}e(t,x)|^2 + |\frac{\partial}{\partial x}e(t,x)|^2)dx = \tilde{E}(0)$$

Hence, by the Sobolev embedding theorem, the energy $e(t,x)$ is uniformly bounded in terms of initial data. Depending on the regularity of the initial data we get a global solution of the corresponding regularity of the Cauchy problem (3.4).

4. A GLOBAL WEAK SOLUTION FOR WAVE MAPS WITH SCALAR AND TWO-FORM POTENTIAL

In this section we derive an existence result for wave maps with scalar and two-form potential. In order to take into account the additional scalar potential we have to make use of a more general approach than in the last section.

We use the notation ∂_t and ∂_x to represent partial derivatives with respect to t and x. We set $\Box := \partial_t^2 - \partial_x^2$. Recall the extrinsic version of the equation for wave maps with scalar and two-form potential

$$\Box u = \mathrm{II}(du, du) + Z(\partial_t u \wedge \partial_x u) + \nabla V(u) \tag{4.1}$$

for a function $u : \mathrm{R}^{1,1} \to \mathrm{N} \subset \mathrm{R}^q$. In this section we study the well-posedness of the Cauchy problem with initial data

$$(u, \partial_t u)|_{t=0} = (u_0, u_1),$$

where $u_0(x) \in N \subset \mathrm{R}^q$ and $u_1(x) \in T_{u0(x)N} \subset \mathrm{R}^q$. It turns out that we have to consider a bigger class of initial data compared to the last section.

Remark 4.1. For an arbitrary scalar potential $V(u)$ solutions of (4.1) will not be invariant under scaling. However, in the special case that we perform the rescaling

$$u(t,x) \to u_\lambda(t,x) := \lambda^\beta u(\lambda t, \lambda t), \qquad \lambda > 0, \qquad \beta \in \mathbb{R}$$

and the potential $V(u)$ is such that it satisfies

$$\nabla V(u) \to \lambda^{2+\beta} \nabla V(u_\lambda),$$

then solutions of (4.1) are scale invariant.

4.1. Existence of a global weak solution

By controlling the H^2-norm of u we will derive the following existence result:

Theorem 4.2. *Let (N,g) be a compact Riemannian manifold. Suppose that the scalar potential V is of class $C^2(N,R)$. For any initial data of the regularity*

$$(u_0, u_1) \in H^2 \times H^1(R, TN)$$

the Cauchy problem (4.1) admits a unique solution of class H^2 for all times t.

First of all, we note that for a continuous potential $V(u)$ and due to the assumption that N is compact, we have the following estimate

$$\int_{\mathbb{R}} \left(\left|\frac{\partial u}{\partial t}\right|^2 + \left|\frac{\partial u}{\partial x}\right|^2\right)dx = \int_{\mathbb{R}} |du|^2 dx = E(0) + 2\int_{\mathbb{R}} V(u)dx \leq C$$

For a general treatment of semilinear wave equations of the form (4.1) we refer to [13, Chapter 5]. Moreover, by [13, Remark 5.3] there exists a unique local (in time) solution to the Cauchy problem (4.1).

We set

$$E_2(t) := \frac{1}{2}\int_{\mathbb{R}} (|\partial_x^2 u|^2 + 2|\partial_x\partial_t u|^2 + |\partial_t^2 u|^2)dx$$

To extend the local solution to a global one we establish the following

Proposition 4.3. *Let* $u\colon [0,T)\times R \to N \subset \mathrm{R}^q$ *be a weak solution of* (4.1). *Then the following inequality holds*

$$\frac{dE_2(t)}{dt} \leq C\Big(E_2(t)\int_{\mathbb{R}} |du|^2 dx + E_2(t)^{\frac{1}{2}}\Big(\int_{\mathbb{R}} |du|^2 dx\Big)^{\frac{1}{2}}\Big) \tag{4.2}$$

where the constant C depends on Z, ∇Z, ∇V and Hess *V.*

Proof. We calculate

$$\frac{dE_2(t)}{dt} = \int_{\mathbb{R}} (\langle \partial_x(\Box u), \partial_x\partial_t u\rangle + \langle \partial_t(\Box u), \partial_t^2 u\rangle)dx$$

Using (4.1) we get

$$\langle \partial_x(\Box u), \partial_x\partial_t u\rangle = \langle \partial_x(\mathrm{I\!I}(du, du)), \partial_x\partial_t u\rangle + \langle \partial_x(Z(\partial_t u \wedge \partial_x u)), \partial_x\partial_t u\rangle + \langle \partial_x(\nabla V(u)), \partial_x\partial_t u\rangle$$

and similarly for the derivatives with respect to *t*. First of all, we note that

$$\begin{aligned}\langle \partial_x(\mathrm{I\!I}(du, du)), \partial_x\partial_t u\rangle &= \langle (\partial_{du(\partial_x)}\mathrm{I\!I})(du, du), \partial_x\partial_t u\rangle + 2\langle \mathrm{I\!I}(d\partial_x u, du), \partial_x\partial_t u\rangle \\ &= \langle (\partial_{du(\partial_x)}\mathrm{I\!I})(du, du), \partial_x\partial_t u\rangle - 2\langle (\partial_{du(\partial_x)}\mathrm{I\!I})(d\partial_x u, du), \partial_t u\rangle\end{aligned}$$

where we used that $\partial_t \perp \mathrm{I\!I}$. Hence, we get the estimate

$$|\langle \partial_x(\mathrm{I\!I}(du, du)), \partial_x\partial_t u\rangle + \langle \partial_t(\mathrm{I\!I}(du, du)), \partial_t^2 u\rangle| \leq C|du|^3|d^2 u|$$

Moreover, we have

$$\begin{aligned}
&\langle \partial_x(Z(\partial_t u \wedge \partial_x u)), \partial_x\partial_t u\rangle + \langle \partial_t(Z(\partial_t u \wedge \partial_x u)), \partial_t^2 u\rangle \\
&\quad = \langle (\partial_{du(\partial_x)} Z)(\partial_t u \wedge \partial_x u), \partial_x\partial_t u\rangle + \Omega(\partial_t u, \partial_x^2 u, \partial_x\partial_t u) \\
&\quad + \langle (\partial_{du(\partial_t)} Z)(\partial_t u \wedge \partial_x u), \partial_t^2 u\rangle + \Omega(\partial_t u, \partial_x\partial_t u, \partial_t^2 u) \\
&\quad = \langle (\partial_{du(\partial_x)} Z)(\partial_t u \wedge \partial_x u), \partial_x\partial_t u\rangle + \langle (\partial_{du(\partial_t)} Z)(\partial_t u \wedge \partial_x u), \partial_t^2 u\rangle \\
&\quad + \Omega(\partial_t u, \partial_x\partial_t u, \Box u).
\end{aligned}$$

Hence, we may conclude

$$|\langle \partial_x(Z(\partial_t u \wedge \partial_x u)), \partial_x\partial_t u\rangle + \langle \partial_x(Z(\partial_t u \wedge \partial_x u)), \partial_t^2 u\rangle| \leq C(|du|^3|d^2u| + |du||d^2u|)$$

where we used (4.1) again. Finally, we note that

$$\langle \partial_x(\nabla V(u)), \partial_x\partial_t u\rangle = \operatorname{Hess} V(u)(\partial_x u, \partial_x\partial_t u), \qquad \langle \partial_t(\nabla V(u)), \partial_t^2 u\rangle = \operatorname{Hess} V(u)(\partial_t u, \partial_t^2 u)$$

We estimate

$$\begin{aligned}
\big\||du|^3|d^2u|\big\|_{L^1(\mathbb{R})} &\leq \big\||du|^3\big\|_{L^2(\mathbb{R})}\big\||d^2u|\big\|_{L^2(\mathbb{R})} \\
&\leq C\big\|d|du|^3\big\|_{L^{\frac{2}{3}}(\mathbb{R})}\big\||d^2u|\big\|_{L^2(\mathbb{R})} \\
&\leq C\big\||du|^2|d^2u|\big\|_{L^{\frac{2}{3}}(\mathbb{R})}\big\||d^2u|\big\|_{L^2(\mathbb{R})} \\
&\leq C\big\||du|\big\|^2_{L^2(\mathbb{R})}\big\||d^2u|\big\|^2_{L^2(\mathbb{R})},
\end{aligned}$$

where we used the Sobolev embedding theorem in one dimension in the second step. Thus, we obtain

$$\frac{dE_2(t)}{dt} \leq C\Big(\int_{\mathbb{R}} |du|^2 dx \int_{\mathbb{R}} |d^2u|^2 dx + \int_{\mathbb{R}} |du||d^2u| dx\Big)$$

which completes the proof. □

Corollary 4.4. *Let* $u\colon [0,T)\times \mathbb{R} \to N \subset \mathbb{R}^q$ *be a weak solution of* (4.1). *Then the following inequality holds*

$$E_2(t) \leq C_1 e^{C_2 T},$$

where the positive constants C_1 and C_2 depend on Z, ∇Z, V, ∇V, Hess *V and the initial data.*

Proof. Using the conserved energy $\int_R |du|^2 dx$ and dividing by $E_2(t)^{\frac{1}{2}}$ equation (4.2) becomes an inequality of the form

$$\frac{d}{dt} f(t) \leq c_1 f(t) + c_2$$

where c_1 and c_2 are positive constants. The claim follows by application of the Gronwall Lemma. □

Having gained control over the spatial H^2-norm of u we are able to show the global existence of a solution to (4.1).

Corollary 4.5. *Let* $u\colon [0,T)\times \mathbb{R}\to N\subset \mathbb{R}^q$ *be a weak solution of* (4.1). *The solution exists for all* $t\in[0,\infty)$.

Proof. Since we have a bound on the spatial H^2-norm of u we can extend the solution for all $t\in[0,\infty)$. □

Finally, we address the issue of uniqueness.

Proposition 4.6 Uniqueness. *Suppose* $u,v\colon \mathbb{R}^{1,1}\to N\subset \mathbb{R}^q$ *are two* H^2 *solutions of* (4.1). *If they coincide at* $t = 0$, *then they coincide for all times.*

Proof. We set $w := u - v$. By assumption both u and v are solutions of (4.1). By the last Proposition we know that the H^2-norm of both u and v is finite. Making use of the Sobolev embedding $W^{2,2}(\mathbb{R}) \subset C^{1,\frac{1}{2}}(\mathbb{R})$ we obtain a pointwise bound on $|du|$ and $|dv|$. We will use the notation u_x to denote the partial derivative of u with respect to x and similarly for the derivative with respect to t. Performing a direct calculation we obtain

$$\begin{aligned}\frac{d}{dt}\int_{\mathbb{R}} |dw|^2 dx &= \int_{\mathbb{R}} (\langle w_t, \mathrm{I\!I}(u)(du,du) - \mathrm{I\!I}(v)(dv,dv)\rangle \\ &+ \langle w_t, Z(u)(u_t\wedge u_x) - Z(v)(v_t\wedge v_x)\rangle + \langle w_t, \nabla V(u) - \nabla V(v)\rangle)dx \\ &:= A_1 + A_2 + A_3.\end{aligned}$$

We rewrite the A_1 term as follows

$$\mathrm{I\!I}(u)(du,du) - \mathrm{I\!I}(v)(dv,dv) = \mathrm{I\!I}(u)(du,du) - \mathrm{I\!I}(v)(du,du) + \mathrm{I\!I}(v)(du+dv,dw)$$

This yields

$$\langle w_t, (\mathrm{II}(u) - \mathrm{II}(v))(du, du)\rangle \leq C|du|^2|w||dw|$$

Using the orthogonality $\partial_t \perp \mathrm{II}$ we find

$$\langle w_t, \mathrm{II}(v)(du, dw)\rangle = \langle u_t, \mathrm{II}(v)(du, dw)\rangle = \langle u_t, (\mathrm{II}(v) - \mathrm{II}(u))(du, dw)\rangle \leq C|du|^2|w||dw|$$

The same argument also holds for the term involving dv . Consequently, the A_1-term can be bounded as

$$A_1 \leq C\int_{\mathbb{R}} |du|^2|w||dw|dx \leq C\int_{\mathbb{R}} |w||dw|dx$$

where we used the pointwise bound on $|du|$. To control the A_2-term we rewrite

$$\langle w_t, Z(u)(u_t \wedge u_x) - Z(v)(v_t \wedge v_x)\rangle = \langle w_t, (Z(u) - Z(v))(u_t \wedge u_x)\rangle + \langle w_t, Z(v)(w_t \wedge u_x)\rangle$$
$$+ \langle w_t, Z(v)(v_t \wedge w_x)\rangle$$

such that we can estimate

$$B \leq C\int_{\mathbb{R}} (|w||dw||du|^2 + |dw|^2(|du| + |dv|))dx$$

By assumption ∇V is continuous, hence we may estimate

$$|\langle w_t, \nabla V(u) - \nabla V(v)\rangle| \leq C|dw||w|.$$

Applying Young's inequality several times we are led to

$$\frac{d}{dt}\int_{\mathbb{R}} (|w|^2 + |dw|^2)dx \leq C\int_{\mathbb{R}} (|w|^2 + |dw|^2)dx$$

which can be integrated as

$$\int_{\mathbb{R}} (|u-v|^2 + |du-dv|^2)dx\Big|_{t=T} \le e^{CT} \int_{\mathbb{R}} (|u(0)-v(0)|^2 + |du(0)-dv(0)|^2)dx$$

By assumption we have $u(0) = v(0)$ and $du(0) = dv(0)$. Consequently, we have $u(t) = v(t)$ for all $t \in [0,\infty)$. □

This finishes the proof of Theorem 4.2.

Remark 4.7. The wave map equation has been studied for less regular initial data, see [15] and [8]. However, a sophisticated analysis performed in these references shows that the Cauchy problem is ill-posed then.

4.2. A priori estimates

In this subsection we derive an a priori estimate for weak solutions of (4.1), which heavily makes use of the special structure of the right hand side of (4.1). Our result is inspired by a similar estimate for wave maps from [9].

We define the following quantity

$$F_{\pm}(u) := (\partial_t u \pm \partial_x u) \qquad (4.3),$$

which leads to the following

Proposition 4.8. *Let* $u\colon [0,T) \times \mathrm{R} \to N$ *be a weak solution of* (4.1). *Then the following energy inequality holds*

$$\int_0^T \int_{\mathbb{R}} (F_-^2(u)F_+^2(u) - 4V^2(u))dxdt \le 2E(0)^2 + 2\int_{x\ge y}\int (V(u)(|\partial_x u|^2 + |\partial_t u|^2))dxdy\Big|_{t=0} \qquad (4.4)$$

Proof. We rewrite the extrinsic version of the Euler–Lagrange equation (4.1) as

$$(\partial_t + \partial_x)(\partial_t - \partial_x)u = \mathrm{I\!I}(du, du) + Z(\partial_t u \wedge \partial_x u) + \nabla V(u)$$

Taking the scalar product with $(\partial_t \pm \partial_x)u$ we find

$$(\partial_t \pm \partial_x)|(\partial_t \mp \partial_x)u|^2 = 2(\partial_t \mp \partial_x)V(u).$$

Making use of (4.3) we can rewrite this as a conservation law in the form

$$\partial_t(F_{\pm}^2(u) - 2V(u)) = \pm\partial_x(F_{\pm}^2(u) + 2V(u))$$

Note that

$$\int_{\mathbb{R}} F_{\pm}^2(u)dx = \int_{\mathbb{R}} (|\partial_x u|^2 + |\partial_t u|^2 \pm 2\langle\partial_x u, \partial_t u\rangle)dx \leq 2E(0) + 2\int_{\mathbb{R}} V(u)dx$$

We define the quantity

$$Z(t) := \int\int_{x\geq y} (F_-^2(u) - 2V(u))(F_+^2(u) - 2V(u))dxdy$$

which satisfies the bound

$$Z(t) \leq \int_{\mathbb{R}} (F_-^2(u) - 2V(u))dx \int_{\mathbb{R}} (F_+^2(u) - 2V(u))dx \leq 4E(0)^2$$

We compute

$$\begin{aligned}
\frac{d}{dt}Z(t) &= \int\int_{x\geq y} (\partial_t(F_-^2(u) - 2V(u))(F_+^2(u) - 2V(u)) \\
&\quad + (F_-^2(u) - 2V(u))\partial_t(F_+^2(u) - 2V(u)))dxdy \\
&= -\int_{-\infty}^{\infty} \Big(\int_y^{\infty} \partial_x(F_-^2(u) + 2V)dx\Big)(F_+^2(u) - 2V(u))dy \\
&\quad + \int_{-\infty}^{\infty} \Big(\int_{\infty}^{x} \partial_y(F_+^2(u) + 2V(u))dy\Big)(F_-^2(u) - 2V(u))dx \\
&= 2\int_{\mathbb{R}} (F_-^2(u)F_+^2(u) - 4V^2(u))dx.
\end{aligned}$$

The result follows by integration with respect to t. □

Remark 4.9. The last Proposition gives a H^2 bound of u in spacetime in terms of the potential $V(u)$ and the initial data.

Remark 4.10. If we perform the same calculation that led to (3.12) taking into account the scalar potential, we find

$$\Box e(t,x) = (\frac{\partial^2}{\partial x^2} + \frac{\partial^2}{\partial t^2})V(u) \tag{4.5}$$

where

$$e(t,x) := \frac{1}{2}|\frac{\partial u}{\partial t}|^2 + \frac{1}{2}|\frac{\partial u}{\partial x}|^2$$

Making use of Duhamel's Principle we can write down a formal solution to this equation, which would give a pointwise bound on $e(t,x)$. However, it seems that these bounds cannot be used to improve Theorem 4.2.

ACKNOWLEDGMENTS

The author gratefully acknowledges the support of the Austrian Science Fund (FWF) through the START-Project Y963-N35 of Michael Eichmair.

REFERENCES

1. Volker Branding **Dirac-harmonic maps with torsion** Commun. Contemp. Math., 18 (4) (2016), Article 1550064
2. Volker Branding **The heat flow for the full bosonic string** Ann. Global Anal. Geom., 50 (4) (2016), pp. 347–365
3. Yvonne Choquet-Bruhat **Applications harmoniques hyperboliques** C. R. Acad. Sci. Paris Sér. I Math., 303 (4) (1986), pp. 109–113
4. Thomas Deck **A geometric Cauchy problem for timelike minimal surfaces** Ann. Global Anal. Geom., 12 (4) (1994), pp. 305–312
5. J. Ginibre, G. Velo **The Cauchy problem for the O (N) , CP (N − 1) , and $G_C(N, p)$ models** Ann. Physics, 142 (2) (1982), pp. 393–415
6. Chao Hao Gu **On the Cauchy problem for harmonic maps defined on two-dimensional Minkowski space** Comm. Pure Appl. Math., 33 (6) (1980), pp. 727–737
7. Frédéric Hélein, John C. Wood **Harmonic maps** Handbook of Global Analysis, Elsevier Sci. B.V., Amsterdam (2008), pp. 417–491 1213
8. Markus Keel, Terence Tao **Local and global well-posedness of wave maps on $\mathbf{R}^{1+1}$ for rough data** Int. Math. Res. Not. IMRN (21) (1998), pp. 1117–1156
9. S. Klainerman, M. Machedon **Smoothing estimates for null forms and applications** Duke Math. J., 81 (1) (1995), pp. 99–133 a celebration of John F. Nash, Jr
10. Olaf Müller **The Cauchy problem of Lorentzian minimal surfaces in globally hyperbolic manifolds** Ann. Global Anal. Geom., 32 (1) (2007), pp. 67–85

11. Joseph Polchinski **String Theory, vol. I: An Introduction to the Bosonic String** Cambridge Monogr. Math. Phys.Cambridge University Press, Cambridge (2005) reprint of the 2003 edition
12. Jalal Shatah **Weak solutions and development of singularities of the SU(2) σ-model** Comm. Pure Appl. Math., 41 (4) (1988), pp. 459–469
13. Jalal Shatah, Michael Struwe **Geometric Wave Equations** Courant Lect. Notes Math., vol. 2, New York University, Courant Institute of Mathematical Sciences/American Mathematical Society, New York/Providence, RI (1998)
14. Michael Struwe **Wave maps with and without symmetries** Evolution Equations, Clay Math. Proc., vol. 17, , Amer. Math. Soc., Providence, RI (2013), pp. 483–510
15. Terence Tao **Ill-posedness for one-dimensional wave maps at the critical regularity** Amer. J. Math., 122 (3) (2000), pp. 451–463
16. Daniel Tataru **The wave maps equation** Bull. Amer. Math. Soc. (N.S.), 41 (2) (2004), pp. 185–204

CHAPTER 8

Preventions and Controls on Congenital Transmissions of Zika: Mathematical Analysis

Nita H. Shah[1*], Zalak A. Patel[2], Bijal M. Yeolekar[1]

[1]*Department of Mathematics, Gujarat University, Ahmedabad, Gujarat, India*

[2]*L. D. College of Engineering, Ahmedabad, Gujarat, India*

ABSTRACT

Vector-borne diseases threat lives of millions of people in many countries of the world. Zika is one of the vector-borne diseases which initially spread by the bite of an infected Aedes species mosquito (*Ae. aegypti and Ae. albopictus)* and then it transmits vertically from a pregnant woman to her fetus or from an infected human to their sexual partners. The congenital transmission of Zika virus (ZIKV) results in new born with microcephaly and other neurological abnormalities. The control of infected mosquitos is the best efficient way to control spread of ZIKV. Spraying insecticide is the safest and easiest way to control mosquitos, but sometimes it is cost worthy for long period of spraying. Controlled prevention from the vector bites can also help to control disease spread. To control congenital transmission and sexual transmission of ZIKV, preventions should be taken to reduce/stop pregnancy rate and safe heterosexual transmission among adults. Also, there is no specific treatment available for Zika disease. Treatment is aimed at relieving symptoms with rest, fluids and medications. Controlled combinations of rest, fluids and medications will help to recover early. As costs are incorporated with spraying, preventions and treatment, our aim is to minimise the total cost associated by controlling spraying, preventions and treatment. To fulfil this purpose a mathematical model is developed with disease dynamics in nine compartments namely Susceptible human child, Susceptible human male, Susceptible human female, Infected human child, Infected human male, Infected human female, Recovered human, Susceptible vector and Infected vector including vertical transmission of Zika disease. Numerical simulations have been carried out to optimise controls, and basic reproduction number and stability are calculated.

Keywords: Zika, Vertical Transmission, Dynamical System, Control, Stability

1. INTRODUCTION

Zika virus (ZIKV) is a *Flavivirus.* It is initially transmitted to humans by the bites of infected female mosquitoes from the Aedes genus. The pathogen responsible for spread of Zika disease is known as Zika virus (ZIKV).In past two years remarkable changes has been seen in the epidemiology of Zika virus (ZIKV). The transmission of ZIKV has been first reported from continental America and the Caribbean. Also, recent reports indicate an increase in detected cases of congenital malformations and neurological complications associated with ZIKV infection. To treat, prevent, or diagnose ZIKV infection there is no specific treatment, vaccine, or fast diagnostic test available at this time.

Dick et al. [1] , *Rodrigez et al.* [2] and *Macnamara et al.* [3] observed that in the Zika forest of Uganda the Zika virus was initially isolated from a rhesus monkey in 1947. In 1954 same virus was isolated from humans in Nigeria. There after only sporadic infected human cases were reported from Africa and Southeast Asia. *Duffy et al.* [4] observed that first largest outbreak of ZIKV infection was reported in 2007 on Yap Island of the North Pacific. *Musso et al.* [5] observed that, in French Polynesia during October 2013, 28000 ZIKV infection cases were reported. Pan American Health Organization [6] noted that the recent outbreak began in April 2015 from Brazil, has covered many southern, central American countries and the Caribbean with spread of disease, and more than 140,000 suspected and confirmed cases are reported by the end of February 2016.

Foy et al. [7] reported that infection of ZIKV spread from an infected male to a female during their sexual intercourse. Thereafter, in February 2016 cases of sexually transmitted ZIKV were reported from Dallas [8] County, U.S. and France by CDC [9] , *Hills et al.* [10] and *Mansuy et al.* [11] respectively. According to the Toronto Star [12] spread of Zika infection because of sexual activities has been reported in Argentina, Canada, Chile, France, Italy,

New Zealand, Peru, Portugal, and the USA from 2015. *Hills et al.* [10] recorded that the disease has minor impact on sexual activity as ZIKV infections have mild symptoms for two to seven days. As ZIKV is sexually transmissible, CDC [9] issued related guidelines for preventions should be taken for safe sex during a Zika outbreak. Didier et al. [13] noted that ZIKV has been found in semen samples which point out the possibility of transmission of ZIKV through sexual activity.

Gatherer et al. [14] found that like other *flaviviruses* ZIKV could also be transmitted by blood transfusion.

Mlakar *et al.* [15] and Cauchemez *et al.* [16] indicated that ZIKV increases the chances of microcephaly in new-born babies of infected mothers. Cao-Lormeau et al. [17] derived that it causes Guillain-Barre syndrome (GBS) as well. According to Pan American Health Organization, from October 2015 to February 2016 approximately 6000 suspected cases with 139 deaths of microcephaly in new- borns due to ZIKV infection were reported in Brazil, and from December 2015 to February 2016, more than 200 cases and 118 cases with 5 deaths of GBS due to suspected ZIKV infection were recorded in Colombia and in El Salvador respectively. As per estimation of WHO [18] , based on the reported dengue data from 2015, about four million people in the Americas may be infected by ZIKV in 2016.*Oliveira et al.* [19] observed that in February 2015, abnormal brain development in the foetuses of pregnant and ZIKV infected women were found.

Dalia et al. [20] suggested for prevent pregnancy and effective strategies against the vector will control disease spread. *Ebenezer et al.* [21] suggested that to control disease the best strategy is to combine all preventive, treatment and Insecticides controls simultaneously. Daozhou et al. [22] noted that to control spread of ZIKV preventive steps are necessary during sexual activities in ZIKV outbreak area.

Together all these facts lead to the increase of the people susceptible to the disease. As there is no vaccination available for the ZIKV disease, it is consider as a severe problem. Mathematical models are essential tools to study the dynamics of the spread of infectious diseases like ZIKV. Basic reproduction number provides information about how infection will be sustained. In this study, a new model with optimal control on spraying insecticides, preventions and treatment is examined. Pontryagin's maximum principle, established by Pontrayagin *et al.* [23] is used to determine the optimal control. Result proves that optimal control gives significant reduction in ZIKV spread.

The paper is organised as follow. In Section 2, mathematical model by a system of ordinary differential equations with notations, assumptions and the flow of populations between compartments are described. For autonomous model having fixed rates of controls basic reproduction number of the whole system (human-mosquito combined) is calculated at disease free equilibrium and endemic equilibrium points. In Section 3, Stability of model has been discussed. In Section 4, cost control function for controls on spraying, preventions and treatment is formed and validated for the system of equations obtained in Section 2. In Section 5, numerical simulations are carried out, for both autonomous and control model. In Section 6, conclusions suggest that how to control disease spread with minimal cost.

2. MATHEMATICAL MODEL

The mathematical model is developed with following notations.

A system of non-linear differential equations is formulated to investigate spread of congenital transmission of ZIKV with control spraying, preventions and treatment. Dynamics of human population N_H(t) and Vector population N_v(t) are developed. Human population is classified amongst eight compartments viz. number of susceptible human child S_{HC} (t), number of susceptible human female $S_{HF}(t)$, number of susceptible human male $S_{HM}(t)$, number of infected human child $I_{HC}(t)$, number of infected human female $I_{HF}(t)$, number of infected human male $I_{HM}(t)$, number of recovered human adult R_H(t), while, vector population N_V(t)is divided in to two compartments viz. number of susceptible vectors $S_V(t)$, number of infected vectors $I_V(t)$. The population dynamics of above compartments is shown in Figure 1.

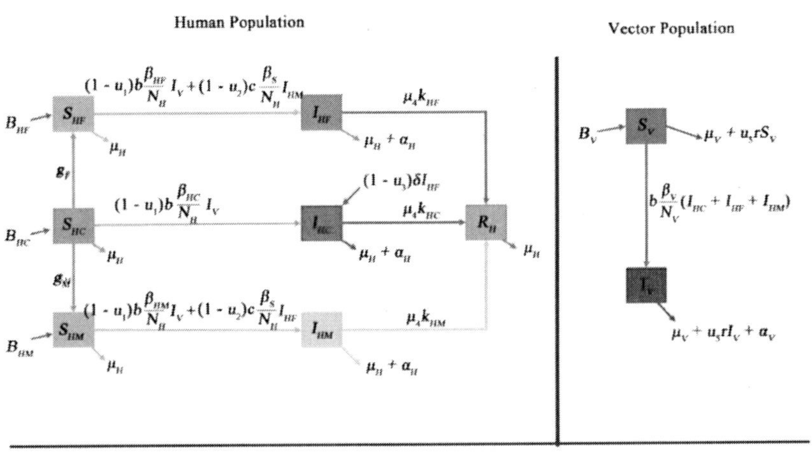

Figure 1. Dynamics of individuals in different compartments.

To investigate effects of congenital transmission of disease and its control by spraying on mosquitoes, preventions and treatment, human population is distributed in human child and human male (adult) and human female (adult) classes.

For model formulation increase in total population of vectors (mosquitoes) is considered, as subpopulation of vectors (mosquitoes) survive from the spraying, get matures and reproduces.

To prepare model, following possibilities of disease spread are considered:

1) Vertical Transmission of infection to new-borns from infected mothers;
2) Horizontal transmission to human child/adult (male and female) from infected vectors (mosquitoes) when they bite to human child/adult (male and female);
3) Heterosexual transmissions amongst human adults (male and female);
4) Horizontal transmission to vectors (mosquitoes) from infected human child/adult (male and female) when vectors (mosquitoes) bite to human child/ adult (male and female).

2.1. Disease Dynamics amongst Human Population

After birth from an infected human female, infected human children will join infected human child class at rate $(1-u_3)\delta I_{HF}$. Susceptible human children will either join infected human child class at rate $(1-u_1)b\frac{\beta_{HC}}{N_H}I_V$ due to bite of an infected vector (mosquito) or susceptible human male (adult) or female (adult) classes at growth rates *gm* and g*F* respectively. Infected human child will join recovered human class at recovery/treatment rate u_4K_{HC}. Disease induced death rate a_{HC} will make very high effect on infected child population as possibility of microcephaly and Guillain-Barre syndrome due to ZIKV transmission occurred during pregnancy.

Susceptible human male (adult) and female (adult) will either join infected human male (adult) and female (adult) class at rates

$$(1-u_1)b\frac{\beta_{HM}}{N_H}I_V+(1-u_2)c\frac{\beta_S}{N_H}I_{HF} \quad \text{and} \quad (1-u_1)b\frac{\beta_{HF}}{N_H}I_V+(1-u_2)c\frac{\beta_S}{N_H}I_{HM}$$

respectively due to vector bite and sexual intercourse with infected human female (adult) and male (adult) respectively. Infected human male (adult) and female (adult) will join recovered human class at recovery/treatment rate u_4K_{HM} and u_4k_{HF} respectively. Disease induced death rate a_H will effect on infected human male and female (adult) population. At each stage, the natural death rate in human child/adult classes ?H is taken into account.

2.2. Disease Dynamics amongst Vector (Mosquito) Population

Vector (mosquito) population is divided in to two compartments susceptible vectors and infected vectors. Susceptible vector gets infection from infected human child/adult, when vectors (mosquitoes) bite them, and joins infected vector class at rate $b\frac{\beta_V}{N_V}(I_{HC}+I_{HF}+I_{HM})$. A portion of susceptible and infected vectors (mosquitoes) will be eliminated due to either natural death rate in vectors (mosquitoes) μ_V or spraying at rate $u_5 r$. Disease induced death rate a_v will also affect the infected vector (mosquito) population.

With above discussion the dynamics of Zika disease can be represented by the system of non-linear differential equations as

$$\frac{dS_{HC}(t)}{dt}=B_{HC}-(1-u_1)b\frac{\beta_{HC}}{N_H}I_V S_{HC}-\mu_H S_{HC}-(g_F+g_M)S_{HC} \tag{1}$$

$$\frac{dS_{HF}(t)}{dt}=B_{HF}-\left((1-u_1)b\frac{\beta_{HF}}{N_H}I_V+(1-u_2)c\frac{\beta_S}{N_H}I_{HM}\right)S_{HF}+g_F S_{HC}-\mu_H S_{HF} \tag{2}$$

$$\frac{dS_{HM}(t)}{dt}=B_{HM}-\left((1-u_1)b\frac{\beta_{HM}}{N_H}I_V+(1-u_2)c\frac{\beta_S}{N_H}I_{HF}\right)S_{HM}+g_M S_{HC}-\mu_H S_{HM} \tag{3}$$

$$\frac{dI_{HC}(t)}{dt}=(1-u_1)b\frac{\beta_{HC}}{N_H}I_V S_{HC}+(1-u_3)\delta I_{HF}-u_4 k_{HC}I_{HC}-(\mu_H+\alpha_{HC})I_{HC} \tag{4}$$

$$\frac{dI_{HF}(t)}{dt}=\left((1-u_1)b\frac{\beta_{HF}}{N_H}I_V+(1-u_2)c\frac{\beta_S}{N_H}I_{HM}\right)S_{HF}-u_4 k_{HF}I_{HF}-(\mu_H+\alpha_H)I_{HF} \tag{5}$$

$$\frac{dI_{HM}(t)}{dt}=\left((1-u_1)b\frac{\beta_{HM}}{N_H}I_V+(1-u_2)c\frac{\beta_S}{N_H}I_{HF}\right)S_{HM}-u_4 k_{HM}I_{HM}-(\mu_H+\alpha_H)I_{HM} \tag{6}$$

$$\frac{\mathrm{d}R_H(t)}{\mathrm{d}t} = u_4\left(k_{HC}I_{HC} + k_{HF}I_{HF} + k_{HM}I_{HM}\right) - \mu_H R_H \tag{7}$$

$$\frac{\mathrm{d}S_V(t)}{\mathrm{d}t} = B_V - \mu_V S_V - b\frac{\beta_V}{N_V}\left(I_{HC} + I_{HF} + I_{HM}\right)S_V - u_5 r S_V \tag{8}$$

$$\frac{\mathrm{d}I_V(t)}{\mathrm{d}t} = b\frac{\beta_V}{N_V}\left(I_{HC} + I_{HF} + I_{HM}\right)S_V - \left(\mu_V + \alpha_V\right)I_V - u_5 r I_V \tag{9}$$

Adding Equations (1) to (7),

$$\begin{aligned}\frac{\mathrm{d}N_H}{\mathrm{d}t} &= B_{HC} + B_{HF} + B_{HM} - \mu_H\left(S_{HC} + S_{HF} + S_{HM} + I_{HC} + I_{HF} + I_{HM} + R_H\right)\\ &\quad + \left(1 - u_3\right)\delta I_{HF} - \alpha_{HC}I_{HC} - \alpha_H\left(I_{HF} + I_{HM}\right)\end{aligned}$$

And, adding Equations (8) and (9),

$$\frac{\mathrm{d}N_V}{\mathrm{d}t} = B_V - \mu_V\left(S_V + I_V\right) - u_5 r\left(S_V + I_V\right) - \alpha_V I_V$$

Hence,

$$\frac{\mathrm{d}N_H}{\mathrm{d}t} \le B_{HC} + B_{HF} + B_{HM} - \mu_H N_H \Rightarrow \lim_{t\to\infty}\mathrm{Sup}N_H \le \frac{B_{HC} + B_{HF} + B_{HM}}{\mu_H}$$

and,

$$\frac{\mathrm{d}N_V}{\mathrm{d}t} \le B_V - \mu_V N_V \Rightarrow \lim_{t\to\infty}\mathrm{Sup}N_V \le \frac{B_V}{\mu_V}.$$

So, the feasible region of the system is

$$\Omega = \begin{cases} \left(S_{HC}, S_{HF}, S_{HM}, I_{HC}, I_{HF}, I_{HM}, R_H, S_V, I_V\right): \\ S_{THC} + S_{UHC} + S_{THA} + S_{UHA} + I_{THC} + I_{UHC} + I_{THA} + I_{UHA} \le \dfrac{B_{HC} + B_{HF} + B_{HM}}{\mu_H}, \\ S_V + I_V \le \dfrac{B_V}{\mu_V}, \\ S_{HC} > 0, S_{HF} > 0, S_{HM} > 0, I_{HC} > 0, I_{HF} \ge 0, I_{HM} \ge 0, R_H \ge 0, S_V > 0, I_V \ge 0 \end{cases}$$

2.3. Computation of Basic Reproduction Number R_0

As at disease free equilibrium, $I_{HC}=0, I_{HF}=0, I_{HM}=0, I_V=0$ and $R_H=0$.

Hence, let

$$X_0 = \left(\frac{B_{HC}}{\mu_H + g_F + g_M}, \frac{B_{HF}}{\mu_H - g_F}, \frac{B_{HM}}{\mu_H - g_M}, 0, 0, 0, 0, \frac{B_V}{\mu_V + u_5 r}, 0\right)$$

be disease free equilibrium point of system.

Using next generation method [24] , let

$$X' = \left(I_{HC}, I_{HF}, I_{HM}, I_V, S_{HC}, S_{HF}, S_{HM}, S_V, R_H\right)'$$

where dash denotes derivative

$$\therefore X' = \frac{dX}{dt} = \Im(x) - v(x)$$

where,

$$\Im(X)=\begin{pmatrix} (1-u_1)b\frac{\beta_{HC}}{N_H}I_V S_{HC} \\ (1-u_1)b\frac{\beta_{HF}}{N_H}I_V S_{HF}+(1-u_2)c\frac{\beta_S}{N_H}I_{HM}S_{HF} \\ (1-u_1)b\frac{\beta_{HM}}{N_H}I_V S_{HM}+(1-u_2)c\frac{\beta_S}{N_H}I_{HF}S_{HM} \\ b\frac{\beta_V}{N_V}(I_{HC}+I_{HF}+I_{HM})S_V \\ 0 \\ 0 \\ 0 \\ 0 \\ 0 \end{pmatrix}$$

and

$$\upsilon(X)=\begin{pmatrix} -(1-u_3)\delta I_{HF}+u_4k_{HC}I_{HC}+\mu_H I_{HC}+\alpha_{HC}I_{HC} \\ u_4k_{HF}I_{HF}+\mu_H I_{HF}+\alpha_H I_{HF} \\ u_4k_{HM}I_{HM}+\mu_H I_{HM}+\alpha_H I_{HM} \\ \mu_V I_V+\alpha_V I_V+u_5 r I_V \\ -B_{HC}+(1-u_1)b\frac{\beta_{HC}}{N_H}I_V S_{HC}+\mu_H S_{HC}+(g_F+g_M)S_{HC} \\ -B_{HF}+(1-u_1)b\frac{\beta_{HF}}{N_H}I_V S_{HF}+(1-u_2)c\frac{\beta_S}{N_H}I_{HM}S_{HF}+\mu_H S_{HF}-g_F S_{HC} \\ -B_{HM}+(1-u_1)b\frac{\beta_{HM}}{N_H}I_V S_{HM}+(1-u_2)c\frac{\beta_S}{N_H}I_{HF}S_{HF}+\mu_H S_{HM}-g_M S_{HC} \\ -B_V+b\frac{\beta_V}{N_V}(I_{HC}+I_{HF}+I_{HM})S_V+u_5 r S_V+\mu_V S_V \\ \alpha_V S_V+\mu_V I_V+u_5 r I_V \end{pmatrix}$$

Using, $F=\left[\frac{\partial \Im_i(X_0)}{\partial X_j}\right]$ and $V=\left[\frac{\partial \upsilon_i(X_0)}{\partial X_j}\right]$ for $i, j = 1,2,3,\cdots,9$

Therefore,

$$F=\begin{pmatrix} 0 & 0 & 0 & b_{14} & 0 & 0 & 0 & 0 & 0 \\ 0 & 0 & b_{23} & b_{24} & 0 & 0 & 0 & 0 & 0 \\ 0 & b_{32} & 0 & b_{34} & 0 & 0 & 0 & 0 & 0 \\ b_{41} & b_{42} & b_{43} & 0 & 0 & 0 & 0 & 0 & 0 \\ 0 & 0 & 0 & 0 & 0 & 0 & 0 & 0 & 0 \\ 0 & 0 & 0 & 0 & 0 & 0 & 0 & 0 & 0 \\ 0 & 0 & 0 & 0 & 0 & 0 & 0 & 0 & 0 \\ 0 & 0 & 0 & 0 & 0 & 0 & 0 & 0 & 0 \\ 0 & 0 & 0 & 0 & 0 & 0 & 0 & 0 & 0 \end{pmatrix}$$

where,

$$b_{14}=\frac{(1-u_1)b\beta_{HC}B_{HC}}{(\mu_H+g_F+g_M)N_H},\qquad b_{23}=\frac{(1-u_2)c\beta_S B_{HF}}{(\mu_H-g_F)N_H},$$

$$b_{24}=\frac{(1-u_1)b\beta_{HF}B_{HF}}{(\mu_H-g_F)N_H},\qquad b_{32}=\frac{(1-u_2)c\beta_S B_{HM}}{(\mu_H-g_M)N_H},$$

$$b_{34}=\frac{(1-u_1)b\beta_{HM}B_{HM}}{(\mu_H-g_M)N_H},\qquad b_{41}=\frac{b\beta_V B_V}{(\mu_V+u_5 r)N_V},$$

$$b_{42}=\frac{b\beta_V B_V}{(\mu_V+u_5 r)N_V},\qquad b_{43}=\frac{b\beta_V B_V}{(\mu_V+u_5 r)N_V}$$

and

$$V=\begin{pmatrix} a_{11} & a_{12} & 0 & 0 & 0 & 0 & 0 & 0 & 0 \\ 0 & a_{22} & 0 & 0 & 0 & 0 & 0 & 0 & 0 \\ 0 & 0 & a_{33} & 0 & 0 & 0 & 0 & 0 & 0 \\ 0 & 0 & 0 & a_{44} & 0 & 0 & 0 & 0 & 0 \\ 0 & 0 & 0 & a_{54} & a_{55} & 0 & 0 & 0 & 0 \\ 0 & 0 & a_{63} & a_{64} & 0 & a_{66} & 0 & 0 & 0 \\ 0 & a_{72} & 0 & a_{74} & 0 & 0 & a_{77} & 0 & 0 \\ a_{81} & a_{82} & a_{83} & 0 & 0 & 0 & 0 & a_{88} & 0 \\ a_{91} & a_{92} & a_{93} & 0 & 0 & 0 & 0 & 0 & a_{99} \end{pmatrix}$$

where,

$$a_{11}=u_4k_{HC}+\mu_H+\alpha_{HC}, \qquad a_{12}=-(1-u_3)\delta,$$

$$a_{22}=u_4k_{HF}+\mu_H+\alpha_H, \qquad a_{33}=u_4k_{HM}+\mu_H+\alpha_H$$

$$a_{33}=u_4k_{HM}+\mu_H+\alpha_H, \qquad a_{44}=\mu_V+\alpha_V+u_5r,$$

$$a_{54}=\frac{(1-u_1)b\beta_{HC}B_{HC}}{(\mu_H+g_F+g_M)N_H}, \qquad a_{55}=\mu_H+g_F+g_M,$$

$$a_{63}=\frac{(1-u_2)c\beta_S B_{HF}}{(\mu_H-g_F)N_H}, \qquad a_{64}=\frac{(1-u_1)b\beta_{HF}B_{HF}}{(\mu_H-g_F)N_H},$$

$$a_{66}=\mu_H-g_F, \qquad a_{72}=\frac{(1-u_2)c\beta_S B_{HM}}{(\mu_H-g_M)N_H},$$

$$a_{74}=\frac{(1-u_1)b\beta_{HM}B_{HM}}{(\mu_H-g_M)N_H}, \qquad a_{77}=\mu_H-g_M,$$

$$a_{81}=\frac{b\beta_V B_V}{(\mu_V+u_5r)N_V}, \qquad a_{82}=\frac{b\beta_V B_V}{(\mu_V+u_5r)N_V},$$

$$a_{83}=\frac{b\beta_V B_V}{(\mu_V+u_5r)N_V}, \qquad a_{88}=\mu_V+u_5r,$$

$$a_{91}=-u_4k_{HC}, \qquad a_{92}=-u_4k_{HF},$$

$$a_{93}=-u_4k_{HM}, \qquad a_{99}=\mu_H$$

The basic reproduction number R_0 is spectral radius of matrix FV^{-1}.

With the parametric values given in Table 1,R_0=0.4757.

3. STABILITY

3.1. Stability at Disease Free Equilibrium

If all eigenvalues of Jacobian matrix of the system of differential Equations (1) to (10) have negative real parts at

$$X_0=\left(\frac{B_{HC}}{\mu_H+g_F+g_M},\frac{B_{HF}}{\mu_H-g_F},\frac{B_{HM}}{\mu_H-g_M},0,0,0,0,\frac{B_V}{\mu_V+u_5r},0\right)$$

then disease free equilibrium point becomes stable. Jacobian matrix of the system at DFE is,

$$J=\begin{pmatrix} c_{11} & 0 & 0 & 0 & 0 & 0 & 0 & 0 & c_{19} \\ 0 & c_{22} & 0 & 0 & 0 & c_{26} & 0 & 0 & c_{29} \\ 0 & 0 & c_{33} & 0 & c_{35} & 0 & 0 & 0 & c_{39} \\ 0 & 0 & 0 & c_{44} & c_{45} & 0 & 0 & 0 & c_{49} \\ 0 & 0 & 0 & 0 & c_{55} & c_{56} & 0 & 0 & c_{59} \\ 0 & 0 & 0 & 0 & c_{65} & c_{66} & 0 & 0 & c_{69} \\ 0 & 0 & 0 & c_{74} & c_{75} & c_{76} & c_{77} & 0 & 0 \\ 0 & 0 & 0 & c_{84} & c_{85} & c_{86} & 0 & c_{88} & 0 \\ 0 & 0 & 0 & c_{94} & c_{95} & c_{96} & 0 & 0 & c_{99} \end{pmatrix}$$

where,

$$c_{11}=-\mu_H-g_F-g_M,\quad c_{19}=-\frac{(1-u_1)b\beta_{HC}B_{HC}}{(\mu_H+g_F+g_M)N_H},$$

$$c_{21}=g_F,\quad c_{22}=-\mu_H,\quad c_{26}=-\frac{(1-u_2)c\beta_S B_{HF}}{(\mu_H-g_F)N_H},$$

$$c_{29}=-\frac{(1-u_1)b\beta_{HF}B_{HF}}{(\mu_H-g_F)N_H},\quad c_{31}=g_M,\quad c_{33}=-\mu_H,$$

$$c_{35}=-\frac{(1-u_2)c\beta_S B_{HM}}{(\mu_H-g_M)N_H},\quad c_{39}=-\frac{(1-u_1)b\beta_{HM}B_{HM}}{(\mu_H-g_M)N_H},$$

$$c_{44}=-u_4k_{HC}-\mu_H-\alpha_{HC},\quad c_{45}=(1-u_3)\delta,$$

$$c_{49}=-c_{19},\quad c_{55}=-u_4k_{HF}-\mu_H-\alpha_H,\quad c_{56}=-c_{26},\quad c_{59}=-c_{29},$$

$$c_{65}=-c_{35},\quad c_{66}=-u_4k_{HM}-\mu_H-\alpha_H,\quad c_{69}=-c_{39},$$

$$c_{74}=u_4k_{HC},\quad c_{75}=u_4k_{HF},\quad c_{76}=u_4k_{HM},\quad c_{77}=-\mu_H,$$

$$c_{84}=-\frac{b\beta_V B_V}{(\mu_V+u_5r)N_V},\quad c_{85}=c_{84},\quad c_{86}=c_{84},\quad c_{88}=-\mu_V-u_5r,$$

$$c_{94}=-c_{84},\quad c_{95}=-c_{84},\quad c_{96}=-c_{84},\quad c_{99}=c_{88}-\alpha_V.$$

With the parametric values given in Table 1, trace(*J*)<0and also det(*J*)>0, if $R_0<1$.

This implies that disease free equilibrium is locally asymptotically stable if $R_0<1$ and unstable, if $R_0>1$.

3.2. Stability at Endemic Equilibrium Point

For the endemic equilibrium, using

$$\frac{dS_{THC}}{dt}=0,\quad \frac{dS_{UHC}}{dt}=0,\quad \frac{dS_{THA}}{dt}=0,$$
$$\frac{dS_{UHA}}{dt}=0,\quad \frac{dI_{THC}}{dt}=0,$$
$$\frac{dI_{UHC}}{dt}=0,\quad \frac{dI_{THA}}{dt}=0,$$
$$\frac{dI_{UHA}}{dt}=0,\quad \frac{dS_{V}}{dt}=0,\quad \frac{dI_{V}}{dt}=0$$

These give, $X^*=\left(S^*_{THC},S^*_{UHC},S^*_{THA},S^*_{UHA},I^*_{THC},I^*_{UHC},I^*_{THA},I^*_{UHA},S^*_{V},I^*_{V}\right)$ as an endemic equilibrium point.

$$J^*=\begin{pmatrix} d_{11} & 0 & 0 & 0 & 0 & 0 & 0 & 0 & d_{19} \\ d_{21} & d_{22} & 0 & 0 & 0 & d_{26} & 0 & 0 & d_{29} \\ d_{31} & 0 & d_{33} & 0 & d_{35} & 0 & 0 & 0 & d_{39} \\ d_{41} & 0 & 0 & d_{44} & d_{45} & 0 & 0 & 0 & d_{49} \\ 0 & d_{52} & 0 & 0 & d_{55} & d_{56} & 0 & 0 & d_{59} \\ 0 & 0 & d_{63} & 0 & d_{65} & d_{66} & 0 & 0 & d_{69} \\ 0 & 0 & 0 & d_{74} & d_{75} & d_{76} & d_{77} & 0 & 0 \\ 0 & 0 & 0 & d_{84} & d_{85} & d_{86} & 0 & d_{88} & 0 \\ 0 & 0 & 0 & d_{94} & d_{95} & d_{96} & 0 & d_{98} & d_{99} \end{pmatrix}$$

where,

$$d_{11} = -\frac{(1-u_1)b\beta_{HC}}{N_H} I_V^* - \mu_H - g_F - g_M,$$

$$d_{19} = -\frac{(1-u_1)b\beta_{HC}}{N_H} S_{HC}^*, \quad d_{21} = g_F,$$

$$d_{22} = -\frac{(1-u_1)b\beta_{HF}}{N_H} I_V^* - \frac{(1-u_2)c\beta_S}{N_H} I_{HM}^* - \mu_H,$$

$$d_{26} = -\frac{(1-u_2)c\beta_S}{N_H} S_{HF}^*, \quad d_{29} = -\frac{(1-u_1)b\beta_{HF}}{N_H} S_{HF}^*,$$

$$d_{31} = g_M, \quad d_{33} = -\frac{(1-u_1)b\beta_{HM}}{N_H} I_V^* - \frac{(1-u_2)c\beta_S}{N_H} I_{HF}^* - \mu_H,$$

$$d_{35} = -\frac{(1-u_2)c\beta_S}{N_H} S_{HM}^*, \quad d_{39} = -\frac{(1-u_1)b\beta_{HM}}{N_H} I_V^*,$$

$$d_{41} = -d_{39}, \quad d_{44} = -u_4 k_{HC} - \mu_H - \alpha_{HC}, \quad d_{45} = (1-u_3)\delta,$$

$$d_{49} = -d_{19}, \quad d_{52} = \frac{(1-u_1)b\beta_{HF}}{N_H} I_V^* + \frac{(1-u_2)c\beta_S}{N_H} I_{HM}^*,$$

$$d_{55} = -u_4 k_{HF} - \mu_H - \alpha_H, \quad d_{56} = -d_{26}, \quad d_{59} = -d_{29},$$

$$d_{63} = \frac{(1-u_1)b\beta_{HM}}{N_H} I_V^* + \frac{(1-u_2)c\beta_S}{N_H} I_{HF}^*, \quad d_{65} = -d_{35},$$

$$d_{66} = -u_4 k_{HM} - \mu_H - \alpha_H, \quad d_{69} = -d_{39}, \quad d_{74} = u_4 k_{HC},$$

$$d_{75} = u_4 k_{HF}, \quad d_{76} = u_4 k_{HM}, \quad d_{77} = -\mu_H, \quad d_{84} = -\frac{b\beta_V}{N_V} S_V^*,$$

$$d_{85} = d_{84}, \quad d_{86} = d_{84}, \quad d_{88} = -\mu_V - \frac{b\beta_V}{N_V}(I_{Hc}^* + I_{HF}^* + I_{HM}^*) - u_5 r,$$

$$d_{94} = -d_{84}, \quad d_{95} = -d_{84}, \quad d_{96} = -d_{84},$$

$$d_{98} = \frac{b\beta_V}{N_V}(I_{Hc}^* + I_{HF}^* + I_{HM}^*), \quad d_{99} = -\mu_V - \alpha_V - u_5 r.$$

With the parametric values given in Table 1, trace (J^*) <0 and also det(J^*)<0, if R_0<1. This implies that endemic equilibrium is locally asymptotically stable if R_0<1 and unstable, if R_0>1.

Table 1. Description of notation and parametric values.

Notation	Description	Parametric values
$N_H(t)$	Human Population at time t	1600
$N_V(t)$	Vector Population at time t	20000
$S_{HC}(t)$	Number of susceptible Human Child at time t	20
$S_{HF}(t)$	Number of susceptible Human female at time t	40
$S_{HM}(t)$	Number of susceptible Human male at time t	30
$I_{HC}(t)$	Number of Infected Human Child at time t	10
$I_{HF}(t)$	Number of Infected Human female at time t	20
$I_{HM}(t)$	Number of Infected Human male at time t	12
$R_H(t)$	Number of Recovered Human at time t	80
$S_V(t)$	Number of susceptible vector at time t	9000
$I_V(t)$	Number of Infected vector at time t	5000
B_{HC}	New recruitments in susceptible human child class	40
B_{HF}	New recruitments in susceptible human female class	80
$S_V(t)$	Number of susceptible vector at time t	9000
$I_V(t)$	Number of Infected vector at time t	5000
B_{HC}	New recruitments in susceptible human child class	40
B_{HF}	New recruitments in susceptible human female class	80
B_{HM}	New recruitments in susceptible human male class	50
B_V	New recruitments in susceptible vector class	100
b	Vector biting rate	50
k_{HC}	Recovery/Treatment Rate in human child (Rate at which infected human child get recovered and moves towards recovered human class)	0.15

k_{HF}	Recovery/Treatment Rate in human female (Rate at which infected human female get recovered and moves towards recovered human class)	0.35
k_{HM}	Recovery/Treatment Rate in human male (Rate at which infected human male get recovered and moves towards recovered human class)	0.35
β_{HC}	Disease transmission rate in human child from vector bite	0.5
β_{HF}	Disease transmission rate in human female from vector bite	0.25
β_{HM}	Disease transmission rate in human male from vector bite	0.25
β_S	Disease transmission rate in human from sexual contacts	0.2
c	Sexual contact rate in human	25
β_V	Disease transmission rate in vector	0.4
g_F	Growth rate of human from child to female	0.35
g_M	Growth rate of human from child to male	0.35
r	Rate of Spraying to kill vectors(insecticides)	0.9
δ	Birth rate coefficient of infected new borns	0.15
μ_H	Mortality rate for human	0.14
μ_V	Mortality rate for vectors	0.5
α_H	Disease induced death rate amongst human in Infection classes	0.7
α_V	Disease induced death rate amongst vectors	0.5
u_1	Prevention control from vector bites	Are controls and taken initially zero, will be optimized to control disease spread.
u_2	Prevention control from sexual activities amongst infected humans	
u_3	Prevention control for infected human female from conceiving child	
u_4	Treatment control amongst human	
u_5	Spraying (insecticide) control amongst vector	

4. OPTIMAL CONTROL

For Zika disease an optimal control model is formulated, to derive optimal prevention from mosquito bite u_1, optimal prevention to stop pregnancy u_2 and sexual transmission u_3, optimal treatment u_4 and optimal spraying u_5 with minimal implementation cost, in order to minimise the number of infected individuals for model described by Equations (1) to (9) in the time interval [0,T] with the feasible region same as given by ? in Section 2.

Considering the cost-functional as,

$$J(u_1,u_2,u_3,u_4,u_5)$$
$$=\int_0^T \Big(w_{21}(S_{HC})^2 + w_{22}(S_{HF})^2 + w_{23}(S_{HM})^2 + w_{24}(I_{HC})^2$$
$$+ w_{25}(I_{HF})^2 + w_{26}(I_{HM})^2 + w_{27}(R_H)^2 + w_{28}(S_V)^2$$
$$+ w_{29}(I_V)^2 + w_{11}u_1^2 + w_{12}u_2^2 + w_{13}u_3^2 + w_{14}u_4^2 + w_{15}u_5^2 \Big) dt$$

where, $X=(S_{HC},S_{HF},I_{HC},I_{HF},I_{HM},R_H,S_V,I_V)$, $u=(u_1,u_2,u_3,u_4,u_5)$ and W_{ij} are weights to regularise the optimal control. In order minimise the disease spread using, $W_{ij}>0$, for $i=2, j=1,2,...9$. Also to minimise the cost associated with strategies applied for preventions, treatment and spraying in a way that spread of infection can be controlled by choosing weights $w_{km}>0$, for $k=1, m=1,2,...5$. So, the optimal control task reads as min $J(u_1,u_2,u_3,u_4,u_5)$ such that $P(x,u)=0$. Where $P(x,u)=0$ denotes the system of equations de-

fined in (1) to (9), i.e.$\dot{x}=M(t,x,u), x(0)=x_0$. The optimal control u^* can be obtained from $J=(u^*_1,u^*_2,u^*_3,u^*_4,u^*_5)=\min J(u_1,u_2,u_3,u_4,u_5)$ such that $P(x,u)=0$.

The control set is

$$\Gamma = \left\{ \begin{array}{l} (u_1,u_2,u_3,u_4,u_5)/u_i(t) \text{ is piecewise continous on } [0,T], \\ a_i \le u_i(t) \le b_i, i=1,2,3,4,5, a_i, b_i \text{ are constants in } [0,1] \end{array} \right\}$$

Using Lagrangian techniques for a problem along with Hamiltonian, the adjoint variable is needed to construct for the optimal control problem given by (1) to (9).

Introducing the Lagrangian to derive the optimality conditions,

$$L(x,u,\lambda)=\left(w_{21}\left(S_{HC}\right)^2+w_{22}\left(S_{HF}\right)^2+w_{23}\left(S_{HM}\right)^2+w_{24}\left(I_{HC}\right)^2\right.$$
$$+w_{25}\left(I_{HF}\right)^2+w_{26}\left(I_{HM}\right)^2+w_{27}\left(R_H\right)^2+w_{28}\left(S_V\right)^2+w_{29}\left(I_V\right)^2$$
$$\left.+w_{11}u_1^2+w_{12}u_2^2+w_{13}u_3^2+w_{14}u_4^2+w_{15}u_5^2\right)$$

To obtain the minimal value of the Lagrangian, defining the Hamiltonian H for the control problem as,

$$H(x,u,\lambda)=w_{21}\left(S_{HC}\right)^2+w_{22}\left(S_{HF}\right)^2+w_{23}\left(S_{HM}\right)^2+w_{24}\left(I_{HC}\right)^2$$
$$+w_{25}\left(I_{HF}\right)^2+w_{26}\left(I_{HM}\right)^2+w_{27}\left(R_H\right)^2+w_{28}\left(S_V\right)^2$$
$$+w_{29}\left(I_V\right)^2+w_{11}u_1^2+w_{12}u_2^2+w_{13}u_3^2+w_{14}u_4^2+w_{15}u_5^2$$
$$+\lambda_1\left(B_{HC}-(1-u_1)b\frac{\beta_{HC}}{N_H}I_VS_{HC}-\mu_HS_{HC}-\left(g_F+g_M\right)S_{HC}\right)$$
$$+\lambda_2\left(B_{HF}-\left((1-u_1)b\frac{\beta_{HF}}{N_H}I_V+(1-u_2)c\frac{\beta_S}{N_H}I_{HM}\right)S_{HF}+g_FS_{HC}-\mu_HS_{HF}\right)$$
$$+\lambda_3\left(B_{HM}-\left((1-u_1)b\frac{\beta_{HM}}{N_H}I_V+(1-u_2)c\frac{\beta_S}{N_H}I_{HF}\right)S_{HM}+g_MS_{HC}-\mu_HS_{HM}\right)$$
$$+\lambda_4\left((1-u_1)b\frac{\beta_{HC}}{N_H}I_VS_{HC}+(1-u_3)\delta I_{HF}-u_4k_{HC}I_{HC}-\left(\mu_H+\alpha_{HC}\right)I_{HC}\right)$$
$$+\lambda_5\left(\left((1-u_1)b\frac{\beta_{HF}}{N_H}I_V+(1-u_2)c\frac{\beta_S}{N_H}I_{HM}\right)S_{HF}-u_4k_{HF}I_{HF}-\left(\mu_H+\alpha_H\right)I_{HF}\right)$$
$$+\lambda_6\left(\left((1-u_1)b\frac{\beta_{HM}}{N_H}I_V+(1-u_2)c\frac{\beta_S}{N_H}I_{HF}\right)S_{HM}-u_4k_{HM}I_{HM}-\left(\mu_H+\alpha_H\right)I_{HM}\right)$$
$$+\lambda_7\left(u_4\left(k_{HC}I_{HC}+k_{HF}I_{HF}+k_{HM}I_{HM}\right)-\mu_HR_H\right)$$
$$+\lambda_8\left(B_V-\mu_VS_V-b\frac{\beta_V}{N_V}\left(I_{HC}+I_{HF}+I_{HM}\right)S_V-u_5rS_V\right)$$
$$+\lambda_9\left(b\frac{\beta_V}{N_V}\left(I_{HC}+I_{HF}+I_{HM}\right)S_V-\left(\mu_V+\alpha_V\right)I_V-u_5rI_V\right)$$

To get optimality Pontrayagin's maximum (minimum) principle, for the model is defined as follow.

If=$(u^*_1,u^*_2,u^*_3,u^*_4,u^*_5)$ is optimal solution of an optimal control problem then there exists a nontrivial vector function $\lambda(t)=(\lambda_1(t),\lambda_2(t),\lambda_3(t),\lambda_4(t),\lambda_5(t),\lambda_6(t),\lambda_7(t),\lambda_8(t),\lambda_9(t))$ satisfying following equations:

1) The state equation $\frac{dx}{dt}=\frac{\partial H\left(t,u_1^*,u_2^*,u_3^*,u_4^*,u_5^*,\lambda(t)\right)}{\partial\lambda}$

2) The optimality condition $0=\frac{\partial H\left(t,u_1^*,u_2^*,u_3^*,u_4^*,u_5^*,\lambda(t)\right)}{\partial u}$

3) The adjoint equation $\frac{\mathrm{d}\lambda}{\mathrm{d}t}=\frac{\partial H\left(t,u_1^*,u_2^*,u_3^*,u_4^*,u_5^*,\lambda(t)\right)}{\partial x}$.

Using equations stated above for the Hamiltonian defined by (13), using state equations as given by (1) to (9), adjoint equations as

$$\dot{\lambda}_1=-2w_{21}S_{HC}+(\lambda_1-\lambda_4)\left((1-u_1)b\frac{\beta_{HC}}{N_H}I_V\right)+\lambda_1(\mu_H+g_F+g_M)$$

$$\dot{\lambda}_2=-2w_{22}S_{HF}+(\lambda_2-\lambda_5)\left((1-u_1)b\frac{\beta_{HF}}{N_H}I_V+(1-u_2)c\frac{\beta_S}{N_H}I_{HM}\right)+\lambda_2(\mu_H-g_F)$$

$$\dot{\lambda}_3=-2w_{23}S_{HM}+(\lambda_3-\lambda_6)\left((1-u_1)b\frac{\beta_{HM}}{N_H}I_V+(1-u_2)c\frac{\beta_S}{N_H}I_{HF}\right)+\lambda_3(\mu_H-g_M)$$

$$\dot{\lambda}_4=-2w_{24}I_{HC}+(\lambda_8-\lambda_9)\left(\frac{b\beta_V}{N_V}S_V\right)+(\lambda_4-\lambda_7)u_4k_{HC}+\lambda_4(\mu_H+\alpha_{HC})$$

$$\dot{\lambda}_5=-2w_{25}I_{HF}+(\lambda_3-\lambda_6)\left((1-u_2)c\frac{\beta_S}{N_H}S_{HM}\right)+(\lambda_8-\lambda_9)\left(\frac{b\beta_V}{N_V}S_V\right)$$
$$+(\lambda_5-\lambda_7)u_4k_{HF}+\lambda_5(\mu_H+\alpha_H)-\lambda_5(1-u_3)\delta$$

$$\dot{\lambda}_6=-2w_{26}I_{HF}+(\lambda_2-\lambda_5)\left((1-u_2)c\frac{\beta_S}{N_H}S_{HF}\right)+(\lambda_8-\lambda_9)\left(\frac{b\beta_V}{N_V}S_V\right)$$
$$+(\lambda_6-\lambda_7)u_4k_{HM}+\lambda_6(\mu_H+\alpha_H)$$

$$\dot{\lambda}_7=-2w_{27}R_H+\mu_H\lambda_7$$

$$\dot{\lambda}_8=-2w_{28}S_V+(\lambda_8-\lambda_9)\left(\frac{b\beta_V}{N_V}(I_{HC}+I_{HF}+I_{HM})\right)+\lambda_8(\mu_V+u_5r)$$

$$\dot{\lambda}_9=-2w_{29}I_V+(\lambda_1-\lambda_4)\left((1-u_1)b\frac{\beta_{HC}}{N_H}S_{HC}\right)+(\lambda_2-\lambda_5)\left((1-u_1)b\frac{\beta_{HF}}{N_H}S_{HF}\right)$$
$$+(\lambda_3-\lambda_6)\left((1-u_1)b\frac{\beta_{HM}}{N_H}S_{HM}\right)+\lambda_9(\mu_V+\alpha_V+u_5r)$$

and optimality conditions are as

$$2w_{11}u_1 + (\lambda_1 - \lambda_4)\left(b\frac{\beta_{HC}}{N_H}I_V S_{HC}\right) + (\lambda_2 - \lambda_5)\left(b1\frac{\beta_{HF}}{N_H}I_V S_{HF}\right)$$
$$+(\lambda_3 - \lambda_6)\left(b\frac{\beta_{HM}}{N_H}I_V S_{HM}\right) = 0$$

$$2w_{12}u_2 + (\lambda_2 - \lambda_5)\frac{c\beta_S}{N_H}S_{HF}I_{HM} + (\lambda_3 - \lambda_6)\frac{c\beta_S}{N_H}S_{HM}I_{HF} = 0$$

$$2w_{13}u_3 - \lambda_4\delta I_{HF} = 0$$

$$2w_{14}u_4 + (\lambda_7 - \lambda_4)k_{HC}I_{HC} + (\lambda_7 - \lambda_5)k_{HF}I_{HF} + (\lambda_7 - \lambda_6)k_{HM}I_{HM} = 0$$

$$2w_{15}u_5 - \lambda_8 r S_V - \lambda_9 r I_V = 0$$

Solving optimality conditions for optimal control and the property of control space u give,

$$u_1^*(t) = \max\left\{\min\left\{1, \frac{1}{2w_{11}}\left[(\lambda_4 - \lambda_1)\left(b\frac{\beta_{HC}}{N_H}I_V S_{HC}\right) + (\lambda_5 - \lambda_2)\left(b1\frac{\beta_{HF}}{N_H}I_V S_{HF}\right) + (\lambda_6 - \lambda_3)\left(b\frac{\beta_{HM}}{N_H}I_V S_{HM}\right)\right]\right\}, 0\right\} \quad (10)$$

$$u_2^*(t) = \max\left\{\min\left\{1, \frac{1}{2w_{12}}(\lambda_5 - \lambda_2)\frac{c\beta_S}{N_H}S_{HF}I_{HM} + (\lambda_6 - \lambda_3)\frac{c\beta_S}{N_H}S_{HM}I_{HF}\right\}, 0\right\} \quad (11)$$

$$u_3^*(t) = \max\left\{\min\left\{1, \frac{1}{2w_{13}}\lambda_4\delta I_{HF}\right\}, 0\right\} \quad (12)$$

$$u_4^*(t) = \max\left\{\min\left\{1, \frac{1}{2w_{14}}(\lambda_4 - \lambda_7)k_{HC}I_{HC} + (\lambda_5 - \lambda_7)k_{HF}I_{HF} + (\lambda_6 - \lambda_7)k_{HM}I_{HM}\right\}, 0\right\} \quad (13)$$

$$u_5^*(t) = \max\left\{\min\left\{1, \frac{1}{2w_{15}}(\lambda_8 r S_V + \lambda_9 r I_V)\right\}, 0\right\} \quad (14)$$

5. NUMERICAL SIMULATIONS AND OBSERVATIONS

For parametric values given in Table 1, the basic reproduction number R_0=0.4757, trace(J)=-4.88<0, det(J)=0.0014>0, trace(J*)=-4.88<0 and det(J*)=0.0041>0 Above values verifies that non-linear system given by Equations (1) to (9) is asymptotically stable at disease free equilibrium point as well as endemic equilibrium point.

In Figure 2, global stability of human population for R_0<1 is shown. It can be seen that in the beginning susceptible human population increases but after 15 days it moves toward stability. Also, infected human population decreases initially but no prevention from vector bite, treatment, sexual activity, pregnancy, and no spraying increase infected population after 10 days. High disease induced death rate controls the infected human population.

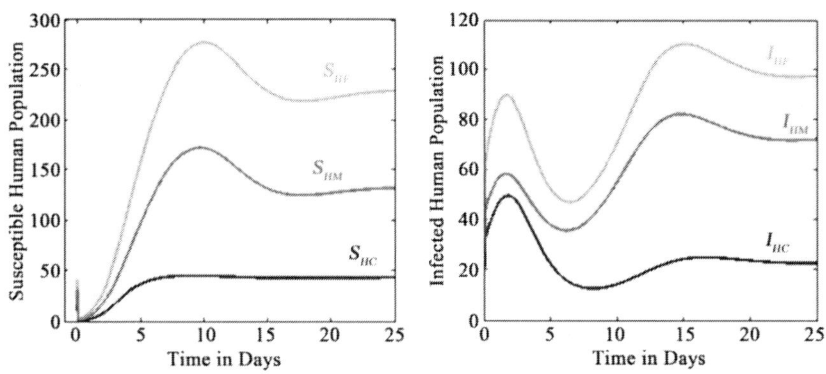

Figure 2. Global stability of human population.

In Figure 3, the effects of increase in prevention from mosquito bites on various infected human classes are analysed. It can be observed that prevention from vector bites in human population is an essential tool to control disease spread.

From Figure 4, it can be analysed that how the infected human population can be controlled by giving proper treatment on time. As there is no specific vaccination or treatment available for Zika disease cure, the treatment is considered as cure of symptoms of disease.

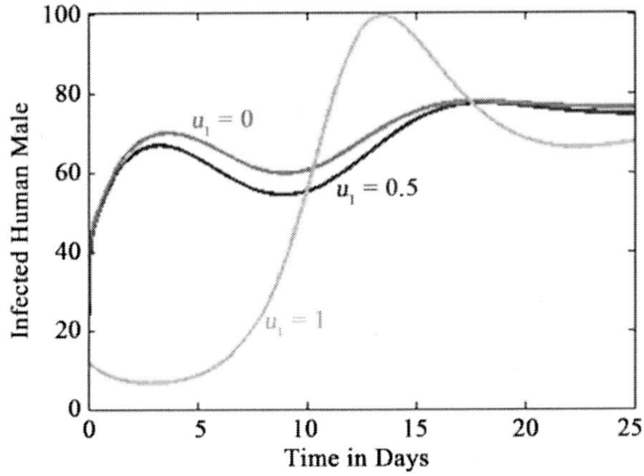

Figure 3. Effects of increase in prevention from mosquito bites on infected population.

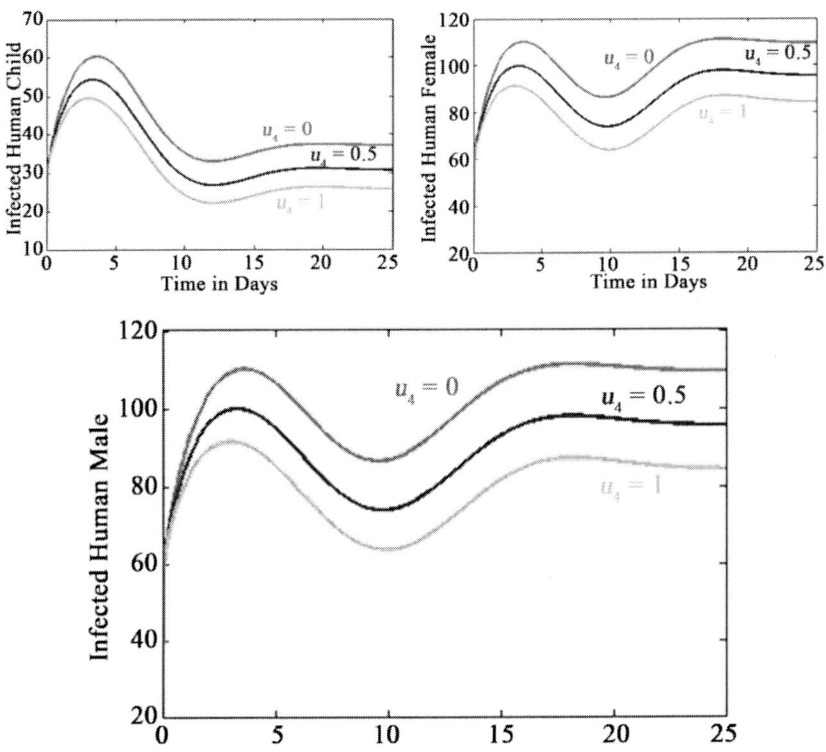

Figure 4. Effects of increase in treatment on infected population.

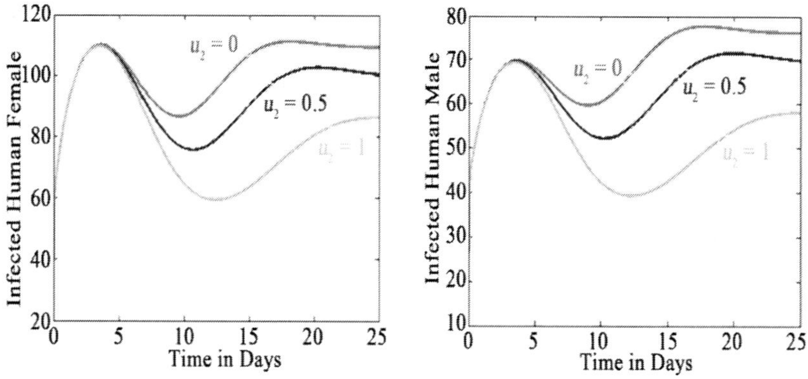

Figure 5. Effects of increase in prevention from sexual contacts on infected population.

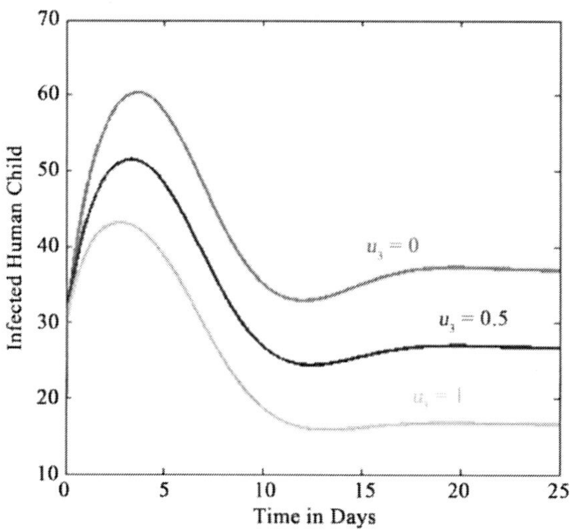

Figure 6. Effects of increase in prevention from pregnancy on infected population.

Figure 5 indicates that proper preventive steps to stop sexual transmission of disease taken at the time of disease out breaks will help to control disease spread.

From Figure 6, it can be observed that the preventions to stop pregnancy amongst infected female and effected area by ZIKV, will reduce the new child births having microcephaly and hence infected child population.

Figure 7 indicates that how the effective spraying can control infected mosquito population and hence finally spraying on mosquito will control disease spread indirectly in human population.

Thus to investigate effects of congenital transmission of ZIKV and its control by spraying on mosquitoes, preventions and treatment on human population, it is required to make the total cost associated with above controls minimise using optimal policy for all controls. (Figure 8)

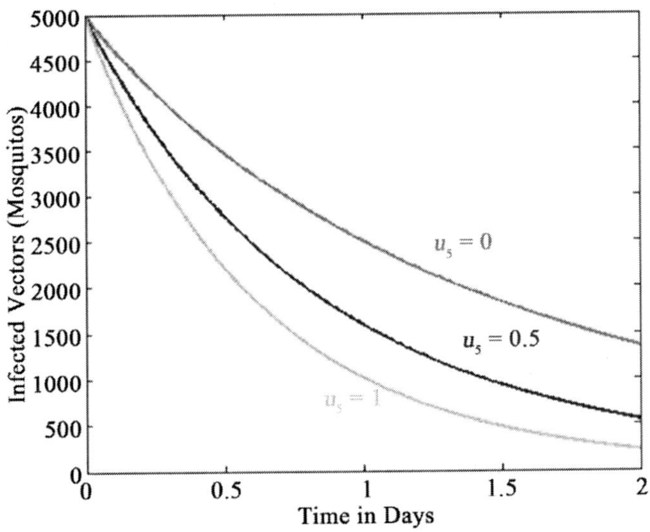

Figure 7. Effects of increase in spraying (insecticide) on infected mosquito population.

To minimise the total effective cost to control disease spray, the policy is to be designed in a way that during first week of disease outbreaks, 26% preventions on treatment and 40% prevention to stop pregnancy, 15% prevention on sexual activities amongst human, 8% spraying is required to have for 1% of prevention from vector bite.

6. CONCLUSIONS

In this paper, the spread of ZIKV considered initially as vector-borne infection and then after its spread amongst sexual partners and from mother to child is analysed. The control over spraying on vectors, prevention from vector bite, prevention during sexual activity, prevention to stop pregnancy and prevention for treatment with time incorporated. It is observed that all controls disease effectively in each compartment.

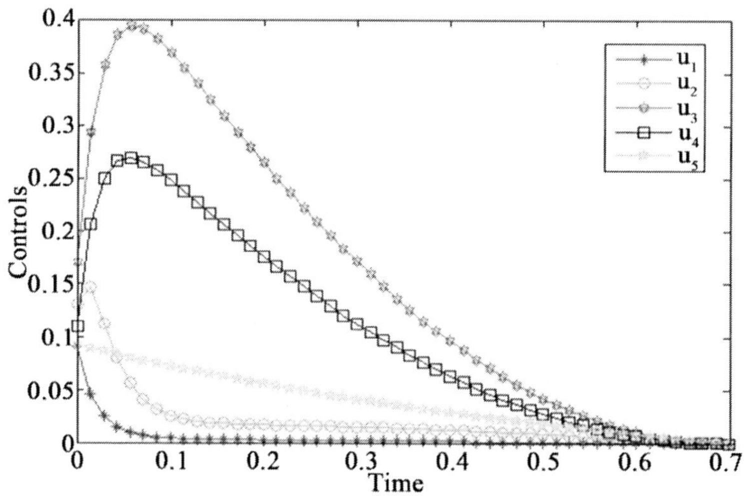

Figure 8. Controls.

In future, the model can be studied with addition of transmission of ZIKV by blood transmission in human population.

ACKNOWLEDGEMENTS

The authors thank DST-FIST file # MS1-097 for support to the department of Mathematics.

REFERENCES

1. Dick, G.W., Kitchen, S.F. and Haddow, A.J. (1952) Zika Virus (I). Isolations and Serological Specificity. Transaction of the Royal Society of Tropical Medicine and Hygiene, 46, 509-520. https://doi.org/10.1016/0035-9203(52)90042-4
2. Rodrigez, A.J. (2015) Zika: The New Arbovirus Threat for Latin America. The Journal of Infection in Developing Countries, 9, 684-685
3. Macnamara, F.N. (1954) Zika Virus: A Report on Three Cases of Human Infection during an Epidemic of Jaundice in Nigeria. Transaction of the Royal Society of Tropical Medicine and Hygiene, 48, 139-145. https://doi.org/10.1016/0035-9203(54)90006-1

4. Duffy, M.R., et al. (2009) Zika Virus Outbreak on Yap Island, Federated States of Micronesia. The New England Journal of Medicine, 360, 2536-2543. https://doi.org/10.1056/NEJMoa0805715
5. Musso, D., Nilles, E.J. and Cao-Lormeau, V.M. (2014) Rapid Spread of Emerging Zika Virus in the Pacific Area. Clinical Microbiology and Infection, 20, 595-596. https://doi.org/ 10. 1111/1469-0691.12707
6. Pan American Health Organization (PAHO), Zika Virus Infection. http://www.paho. org/hq/index.php?option=com [Citation Time(s):1]
7. Foy, B.D., et al. (2011) Probable Non-Vector-Borne Transmission of Zika Virus, Colorado, USA. Emerging Infectious Diseases, 17, 880-882. https://doi.org/ 10.3201/ eid1705.101939
8. Dallas County Health and Human Services (DCHHS), DCHHS Reports First Zika Virus Case in Dallas County Acquired Through Sexual Transmission. 2 February 2016. https: //www.dallascounty.org/department/hhs/documents/DCHHS_Zika
9. Centers for Disease Control and Prevention (CDC). Update: Interim Guidelines for Prevention of Sexual Transmission of Zika Virus—United States, 2016. 23 February 2016. http://emergency.cdc.gov/han/han00388.asp
10. Hills, S.L., et al. (2016) Transmission of Zika Virus through Sexual Contact with Travellers to Areas of Ongoing Transmission—Continental United States. Morbidity and Mortality Weekly Report, 65, 215-216. https://doi.org/10.15585/mmwr.mm6508e2
11. Mansuy, J.M., et al. (2016) Zika Virus: High Infectious Viral Load in Semen, a New Sexually Transmitted Pathogen? The Lancet Infectious Diseases, 16, 405. https://doi.org/10. 1016/S1473-3099(16)00138-9
12. Toronto Star, Canada's First Case of Sexually-Transmitted Zika Virus Confirmed in Ontario. 25 April 2016. https://www.thestar.com/news/gta/2016/04/25/first-canadian-case-of-sexually-transmitted-zika-virus-detected-in-ontario.html
13. Musso, D., Roche, C., Robin, E., Nhan, T., Teissier, A., Cao-Lormeau, V., et al. (2015) Potential Sexual Transmission of Zika Virus. Emerging Infectious Diseases, 21, 359-361. www.cdc.gov/eid
14. Gatherer, D. and Kohl, A. (2015) Zika Virus: A Previously Slow Pandemic Spreads Rapidly through the Americas. Journal of General Virology, 97, 269-273. https://doi.org/ 10.1099/jgv.0.000381
15. Mlakar, J., et al. (2016) Zika Virus Associated with Microcephaly. The New England Journal of Medicine, 374, 951-958. https://doi.org/10.1056/NEJMoa1600651
16. Cauchemez, S., Besnard, M., Bompard, P., Dub, T., Guillemette-Artur, P., Eyrolle-Guignot, D., Salje, H., Van Kerkhove, M.D., Abadie, V., Garel, C., Fontanet, A. and Mallet, H.P. (2016) Association between Zika Virus and Microcephaly in French Polynesia, 2013-15: A Retrospective Study. The Lancet, 387, 2125-2132. https://doi.org/10.1016/S0140-6736(16)00651-6
17. Cao-Lormeau, V.M., Blake, A., Mons, S., Lastère, S., Roche, C., Vanhomwegen, J., Dub, T., Baudouin, L., Teissier, A., Larre, P., Vial, A.L., Decam, C., Choumet, V., Halstead, S.K., Willison, H.J., Musset, L., Manuguerra, J.C., Despres, P., Fournier, E., Mallet, H.P., Musso, D., Fontanet, A., Neil, J. and Ghawché, F. (2016) Guillain-Barré Syndrome Outbreak Associated with Zika Virus Infection in French Polynesia: A Case Control Study. The Lancet, 387, 1531-1539. https://doi.org/10.1016/S0140-6736(16)00562-6
18. WHO Report. http://www.who.int/mediacentre/news/statements/2016/1st-emergency-committee-zika/en/
19. Oliveira Melo, A.S., Malinger, G., Ximenes, R., Szejnfeld, P.O., Alves, S.S. and Bispo de Filippis, A.M. (2016) Zika Virus Intrauterine Infection Causes Fetal Brain Abnormality and Microcephaly: Tip of the Iceberg? Ultrasound in Obstetrics & Gynecology, 47, 6-7. https://doi.org/10.1002/uog.15831
20. Pizza, D.M.M., Loaiza, A.M., Montoya, J.F.A., Manrique, O.A., Sossa, V.A., Munoz, C.A.A., Raigosa, S., Contreras, H.M., Contreras, I.P., Perea, M.E.C. and Enríquez, M.A. (2016) A Model for the Risk of Microcephaly Induced by the Zika Virus (ZIKV). Open Journal of Modelling and Simulation, 4, 109-117. https://doi.org/10.4236/ojmsi.2016.43010

21. Bonyah, E. and Okosun, K.O. (2016) Mathematical Modeling of Zika Virus. Asian Pacific Journal of Tropical Disease, 6, 673-679
22. Gao, D., Lou, Y., He, D., Porco, T.C., Kuang, Y., Chowell, G. and Ruan, S. (2016) Prevention and Control of Zika as a Mosquito-Borne and Sexually Transmitted Disease: A Mathematical Modeling Analysis. Scientific Reports, 6, Article No. 28070
23. Pontryagin, L.S., Boltyanskii, V.G., Gamkrelidze, R.V. and Mishchenko, E.F. (1986) The Mathematical Theory of Optimal Process. Gordon and Breach Science Publishers, New York, 4
24. Van den Driessche, P. and Watmough, J. (2002) Reproduction Numbers and Sub-Threshold Endemic Equilibria for Compartmental Models of Disease Transmission. Mathematical Biosciences, 180, 29-48

CHAPTER 9

Abstract Swiss cheese space and classicalisation of Swiss cheeses

J.F.Feinstein, S.Morley[1], H.Yang[2]

School of Mathematical Sciences, University of Nottingham, University Park, Nottingham, NG7 2RD, UK

ABSTRACT

Swiss cheese sets are compact subsets of the complex plane obtained by deleting a sequence of open disks from a closed disk. Such sets have provided numerous counterexamples in the theory of uniform algebras. In this paper, we introduce a topological space whose elements are what we call "abstract Swiss cheeses". Working within this topological space, we show how to prove the existence of "classical" Swiss cheese sets (as discussed in [6]) with various desired properties. We first give a new proof of the Feinstein–Heath classicalisation theorem [6]. We then consider when it is possible to "classicalise" a Swiss cheese while leaving disks which lie outside a given region unchanged. We also consider sets obtained by deleting a sequence of open disks from a closed annulus, and we obtain an analogue of the Feinstein–Heath theorem for these sets. We then discuss regularity for certain uniform algebras. We conclude with an application of these techniques to obtain a classical Swiss cheese set which has the same properties as a non-classical example of O'Farrell [5].

Keywords: Swiss cheeses, Rational approximation, Uniform algebras, Bounded point derivations, Regularity of $\boldsymbol{R(X)}$

1. INTRODUCTION

Throughout, we use the term *compact plane set* to mean a non-empty, compact subset of the complex plane. Let X be a compact plane set. Then $C(X)$ denotes the set of all continuous, complex-valued functions on X, and $R(X)$ denotes the set of those functions $f \in C(X)$ which can be uniformly approximated on X by rational functions with no poles on X. Both $R(X)$ and $C(X)$ are uniform algebras on X. We refer the reader to [1,2,8,13] for further definitions and background concerning uniform algebras and Banach algebras.

A Swiss cheese set is a compact subset of C obtained by deleting a sequence of open disks from a closed disk. Such sets have been used as examples in the theory of uniform algebras and rational approximation. Swiss cheese sets were introduced by Roth [12], where she gave the first known example of a compact plane set X such that $R(X) \neq C(X)$ but X has empty interior. Since then there have been numerous applications of Swiss cheese sets in the literature.

One notable example of a Swiss cheese construction is due to McKissick [11]. He gave an example of a Swiss cheese set X such that $R(X)$ is regular but $R(X) \neq C(X)$. (We will define regularity in Section 7.) The sequence of open disks used to construct this Swiss cheese set may touch or overlap, which means that the set X might have undesirable topological properties. To improve the topological properties of the resulting Swiss cheese set, while preserving the properties of the uniform algebra, a process that we call *classicalisation* was developed [6].

We may consider a pair consisting of a closed disk and a collection of open disks in the plane, from which we obtain the desired Swiss cheese set (see Definition 2.1 below). We call such a pair a *Swiss cheese* and say it is *classical* if the collection of open disks and the complement of the closed disk have pairwise disjoint closures and the sum of the radii of all open disks is finite. Note that, in the literature, the term 'Swiss cheese' traditionally refers to what we call a Swiss cheese set. Feinstein and Heath [6] considered Swiss cheeses in which the sum of the radii of the open disks is strictly less than the radius of the larger, closed disk. They proved, using Zorn's lemma, that for such a Swiss cheese, the associated Swiss cheese set contains a Swiss cheese set associated to a classical Swiss cheese. Later, Mason [10] gave a proof of this theorem using transfinite induction.

Classical Swiss cheese sets have many desirable topological properties. For example, Dales and Feinstein [3] proved that given two points x,y in a classical Swiss cheese set there is a rectifiable path connecting x,y and such that the length of this path is no more than $\pi|x-y|$; in fact, the constant π can be replaced by $\pi/2$ here. After this observation it is easy to see that a classical Swiss cheese set is path connected (and hence connected), locally path connected (and hence locally connected), and uniformly regular, as defined in [3]. Also as a consequence of connectedness, we see that a classical Swiss cheese set cannot have any isolated points. In [6] it was noted that every classical Swiss cheese set with empty interior is homeomorphic to the Sierpiński carpet as a consequence of a theorem of Whyburn [14].

Browder [1] notes that if X is a classical Swiss cheese set then $R(X)$ is essential (see also [6]). In particular, $R(X) \neq C(X)$, as originally proved by Roth [12]. It follows from the Hartogs–Rosenthal theorem that X must have positive area. A direct proof that every classical Swiss cheese set has positive area is due to Allard, as outlined in [1, pp. 163–164].

Where existing examples of Swiss cheese sets in the literature are not classical, it is of interest to construct classical Swiss cheese sets which solve the same problems. As part of a general classicalisation scheme, we discuss some new techniques for constructing such classical Swiss cheese sets.

In this paper we consider what we call *abstract Swiss cheeses*, which are sequences of pairs consisting of a complex number and a non-negative real number. Each pair in this sequence corresponds to a centre and radius of a disk in the plane. We give the set of all abstract Swiss cheeses a natural topology and use this topology to give a new proof of the Feinstein–Heath theorem. We show that, under some conditions, we can classicalise Swiss cheese sets while only changing open disks which lie in certain regions. We prove an analogue of the Feinstein–Heath theorem for annuli. We give some results regarding regularity of $R(X)$ for unions of compact plane sets, which will be used in the final section. Finally, we give an example of the application of a combination of these results to construct an example of a classical Swiss cheese set X such that $R(X)$ is regular and admits a non-degenerate bounded point derivation of infinite order (as defined in Section 8), which improves an example of O'Farrell [5]. This fits into our general classicalisation scheme.

2. SWISS CHEESES AND ABSTRACT SWISS CHEESE SPACE

We denote the set of all non-negative real numbers by $\mathbb{R}^+$, the set of positive integers by $\mathbb{N}$ and the set of all non-negative integers by $\mathbb{N}_0$. Let $a \in \mathbb{C}$ and let $r > 0$. We denote the open disk of radius r and centre a by $B(a,r)$ and the corresponding closed disk by $\overline{B}(a,r)$. We also set $\overline{B}(a,0)=\{a\}$ and $B(a,0)=\emptyset$. We say a disk with radius zero is *degenerate*. For a non-degenerate open or closed disk D in the plane, let $r(D)$ denote the radius of D; for a degenerate disk D we define r(D)=0. The following is the definition of a Swiss cheese used in [6].

DEFINITION 2.1

Let $\Delta \subseteq C$ be a non-degenerate open disk and let D be a countable collection of non-degenerate, open disks in the plane. Then the ordered pair $E=(\overline{\Delta},D)$ is a *Swiss cheese*. We also define the following.

a) The *Swiss cheese set* X_E *associated with the Swiss cheese E* is defined by

$$X_E = \overline{\Delta} \setminus \bigcup_{D \in \mathcal{D}} D \quad (1)$$

b) The *discrepancy* $\delta(E)$ of E is defined by

$$\delta(E) = r(\overline{\Delta}) - \sum_{D \in \mathcal{D}} r(D)$$

c) The Swiss cheese E is *semiclassical* if $\delta(E) > -\infty$, for each $D \in D$ we have $D \subseteq \overline{\Delta}$, and for each $D' \in D$ with $D \neq D'$ we have $D \cap D' = \emptyset$. In this case we say the Swiss cheese set associated to E is *semiclassical.*

d) The Swiss cheese E is *classical* if δ(E) $> -\infty$, for each $D \in D$ we have $\overline{D} \subseteq \Delta$, and for each $D' \in D$ with $D \neq D'$ we have $\overline{D} \cap \overline{D}' = \emptyset$. In this case we say the Swiss cheese set associated to E is *classical.*

e) The Swiss cheese E is *finite* if the collection D is finite and infinite otherwise.

The condition $\delta(E) > -\infty$ is equivalent to the sum of the radii of the open disks being finite.

We note that without some condition on the disks in D we can obtain every compact plane set as a Swiss cheese set with this definition.

Throughout this paper, we will work in what we call *abstract Swiss cheese space F*, where $F=(C\times R^{+})^{N0}$ with the product topology.

DEFINITION 2.2

Let $A=((a_n,r_n))_{n=0}^{\infty} \in F$. We call A an *abstract Swiss cheese*, and we define the following.

a) The *significant index set of A* is $S_A := \{ n \in N : r_n > 0\}$. We say that A is *finite* if S_A is a finite set, otherwise A is *infinite.*

b) The *associated Swiss cheese set* X_A is defined by

$$X_A = \bar{\mathrm{B}}(a_0, r_0) \setminus \left(\bigcup_{n=1}^{\infty} \mathrm{B}(a_n, r_n) \right) \tag{2}$$

c) We say that A is *semiclassical* if $\sum_{n=1}^{\infty} r_n < \infty,\ r_0 > 0$ and for all $k \in S_A$ the following hold:

- $\mathrm{B}(a_k, r_k) \subseteq \mathrm{B}(a_0, r_0)$;
- whenever $\ell \in S_A$ has $\ell \neq k$, we have $\mathrm{B}(a_k,r_k) \cap \mathrm{B}(a_\ell,r_\ell)=\emptyset$.

d) We say that A is *classical* if $\sum_{n=1}^{\infty} r_n < \infty,\ r_0 > 0$ and for all $k \in S_A$ the following hold:

- $\overline{\mathrm{B}}(a_k,r_k) \subseteq \mathrm{B}(a_0,r_0)$;
- whenever $\ell \in S_A$ with $\ell \neq k$, we have $\overline{\mathrm{B}}(a_k,r_k) \cap \overline{\mathrm{B}}(a_\ell,r_\ell)=\emptyset$.

For $\alpha \geq 1$ we define the *discrepancy function of order* α, $\delta_\alpha: F \to [-\infty,\infty)$ by

$$\delta_\alpha(A) = r_0^\alpha - \sum_{n=1}^{\infty} r_n^\alpha \qquad (A = ((a_n, r_n))_{n=0}^{\infty} \in \mathcal{F}) . \tag{3}$$

Note that in (2) we could instead write

$$X_A := \bar{\mathrm{B}}(a_0, r_0) \setminus \left(\bigcup_{n \in S_A} \mathrm{B}(a_n, r_n) \right)$$

If A is semiclassical or classical, then $\pi\delta_2(A)$ is the area of the Swiss cheese set X_A. We will usually write $A=((a_n,r_n))$ for an abstract Swiss cheese. All sequences, unless otherwise specified, will be indexed by $\mathbb{N}_0$.

We also define the following functions on F.

DEFINITION 2.3

The *radius sum function* is the map $\rho: F \to [0,\infty]$ defined by

$$\rho(A) = \sum_{n=1}^{\infty} r_n \qquad (A = ((a_n, r_n)) \in \mathcal{F}) .$$

The *centre bound function* is the map $\mu: F \to [0,\infty]$ defined by

$$\mu(A) = \sup_{n \in \mathbb{N}} |a_n| \qquad (A = ((a_n, r_n)) \in \mathcal{F})$$

Let $E \subseteq$ C. For an abstract Swiss Cheese $A = ((a_n, r_n))$ we define $H_A(E)$ to be the set of those $n \in S_A$ such that $\bar{B}(a_n, r_n) \cap E \neq \emptyset$. The *local radius sum function on E* is the function $\rho E: F \to [0,\infty]$ defined by

$$\rho_E(A) = \sum_{n \in H_A(E)} r_n \qquad (A = ((a_n, r_n)) \in \mathcal{F})$$

It is easy to see that ρ and μ are both lower semicontinuous from F to $[0,\infty]$. (For ρ, this is an easy consequence of Fatou's lemma for series.)

We now explain the connection between Swiss cheeses, as in Definition 2.1, and abstract Swiss cheeses. We construct a many-to-one surjection of a subset of F onto the collection of all Swiss cheeses as in Definition 2.1. Let $A=((a_n,r_n))$ be an abstract Swiss cheese with $r_0 > 0$. Then we can obtain an associated Swiss cheese E_A by setting

$$E_A := (\bar{\mathrm{B}}(a_0, r_0), \{\mathrm{B}(a_n, r_n) : n \in S_A\}).$$

The associated Swiss cheese sets of A and E_A are equal, and $\delta(E_A) \geq \delta 1(A)$. Moreover, if A is finite then E_A is finite; if A is semiclassical then E_A is semiclassical; and if A is classical then E_A is classical. Conversely, if E is a finite Swiss cheese then there is a finite abstract Swiss cheese A such that $E_A=E$.

Let $E=(\bar{\Delta},\mathrm{D})$ be a Swiss cheese. If E is (semi)classical then there is an abstract Swiss cheese A with E_A=E such that A is (semi)classical. Moreover, when the sum of the radii of open disks in D is finite, we can find an abstract Swiss cheese $A=((a_n,r_n))$ with $\rho(A)<\infty$ and $E=E_A$ such that the sequence $(r_n)_{n=1}^{\infty}$ is non-increasing.

We denote the collection of all abstract Swiss cheeses $A=((a_n,r_n))$ with $\rho(A)<\infty$ and $(r_n)_{n=1}^{\infty}$ non-increasing by N. In addition, for each $M > 0$ and $R > 0$, we denote the set of all those abstract Swiss cheeses $A=((a_n,r_n)) \in N$ such that $\mu(A) \leq M$ and $\rho(A) \leq R$ by $N(M,R)$. Let $M,R > 0$. Although $N(M,R)$ is not itself compact, a closed subset S of $N(M,R)$ is compact if and only if the 0-th coordinate projection maps S to a bounded subset of C×R$^+$. Note that, for $A=((a_n,r_n)) \in N(M,R)$ we have $r_n \leq R/n$ for all $n \in$ N.

LEMMA 2.4

Let M, R > 0. For $\alpha \geq 1$, the function δ_α:F→[−∞,∞)is upper semicontinuous. For α >1, the function $\delta_\alpha|_{N(M,R)}$:N(M,R)→R is continuous.

PROOF

As for the lower semicontinuity of ρ, it is an easy consequence of Fatou's lemma for series that $\delta_\alpha:F\to[-\infty,\infty)$ is an upper semicontinuous function for each $\alpha \geq 1$.

Fix $\alpha > 1$. For each $m \in N_0$ let $A^{(m)}=((a_n^{(m)}, r_n^{(m)})) \in N(M,R)$ and suppose $A^{(m)}\to A \in N(M,R)$ as $m\to\infty$. We have $|r_n^{(m)}|^\alpha \leq R^\alpha/n^\alpha$ for all $n \in$ N. Since $\sum_{n=1}^{\infty} R^\alpha/n^\alpha < \infty$, by the dominated convergence theorem, we have

$$\delta_\alpha(A) = r_0^\alpha - \sum_{n=1}^{\infty} r_n^\alpha = \lim_{m\to\infty}\left((r_0^{(m)})^\alpha - \sum_{n=1}^{\infty}(r_n^{(m)})^\alpha\right) = \lim_{m\to\infty}\delta_\alpha(A^{(m)})$$

So δ_α is continuous from $N(M,R)$ to R□ .

We remark that there are examples showing that δ_1 is only upper semicontinuous, but not continuous.

DEFINITION 2.5

Let $A=((a_n,r_n))$ be an abstract Swiss cheese.

(a) Let $a \in$ C and $r > 0$ and let $m \in N_0$. We say an abstract Swiss cheese $B=((b_n,s_n))$ is obtained from A by *inserting* a disk B(a,r) at index m if, for $0\leq n< m$, we have $b_n=a_n$, $s_n=r_n$; for $n>m$ we have $b_n=a_{n-1}$, $s_n=a_{n-1}$, and $b_m=a$, $s_m=r$.

(b) Let $m \in N_0$. We say an abstract Swiss cheese $B=((b_n,s_n))$ is obtained from A by *deleting* the disk at index m if, for $0\leq n < m$, we have $b_n=a_n$, $s_n=r_n$ and for all $n \geq m$ we have $b_n=a_{n+1}$, $s_n=r_{n+1}$.

(c) Suppose $A \in N$. Let $a \in$ C and $r > 0$ and $k,\ell \in N$ with $k \neq \ell$. We say an abstract Swiss cheese $B=((b_n,s_n))$ is obtained from A by *replacing* the disks B(a_k,r_k),B(a_ℓ,r_ℓ) by B(a,r) if B is obtained by deleting the disks at indices k,ℓ and inserting the disk B(a,r) at the first index in N such that the sequence $(s_n)_{n=1}^{\infty}$ is non-increasing.

Note that, if $A \in N$, then the abstract Swiss cheese B obtained by deleting or replacing disks, as defined in Definition 2.5, is also in N.

3. SOME GEOMETRIC RESULTS

Throughout, we shall require the following elementary geometric lemmas. The first is probably well-known, and the proof is elementary.

LEMMA 3.1

Let $z, w \in C$ and $r, s \in R^+$, then $\overline{B}(z,r) \subseteq \overline{B}(w,s)$ if and only if $|z-w| \leq s-r$. If $r>0$, then $B(z,r) \subseteq C \setminus \overline{B}(w,s)$ if and only if $|w-z| \geq s+r$.

The following two elementary lemmas are essentially those used in [6,10], but including some additional information distilled from the original proofs. These lemmas are summarised in Fig. 1. In the first lemma, we allow for the line segment to be degenerate.

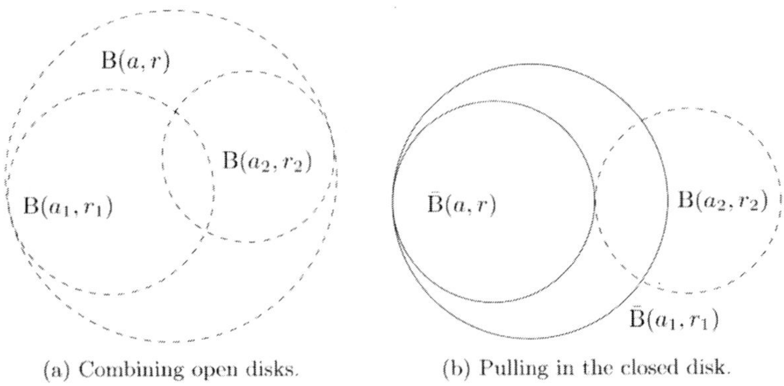

(a) Combining open disks. (b) Pulling in the closed disk.

Figure. 1. Elementary lemmas for combining and pulling in disks.

LEMMA 3.2

Let $a_1, a_2 \in C$ and $r_1, r_2 > 0$. Then there exists a unique pair $(a,r) \in C \times R^+$ with r minimal such that $B(a_1,r_1) \cup B(a_2,r_2) \subseteq B(a,r)$. Moreover, a lies on the line segment joining a_1 and a_2. Suppose further that $\overline{B}(a_1,r_1) \cap \overline{B}(a_2,r_2) \neq \emptyset$. Then $r \leq r_1+r_2$, and equality holds if and only if $B(a_1,r_1) \cap B(a_2,r_2) = \emptyset$.

LEMMA 3.3

Let $a_1, a_2 \in C$ and $r_1 > r_2 > 0$ with $\overline{B}(a_2,r_2) \nsubseteq B(a_1,r_1)$. Then there exists a unique pair $(a,r) \in C \times R^+$ with $\overline{B}(a,r) \subseteq \overline{B}(a_1,r_1)$ and $B(a_2,r_2) \cap \overline{B}(a,r) = \emptyset$ such that r is maximal. Moreover, $r \geq r_1 - r_2$ and equality holds if and only if $B(a_2,r_2) \subseteq B(a_1,r_1)$.

The cases in which equality holds in Lemmas 3.2 and 3.3 are illustrated in Fig. 2.

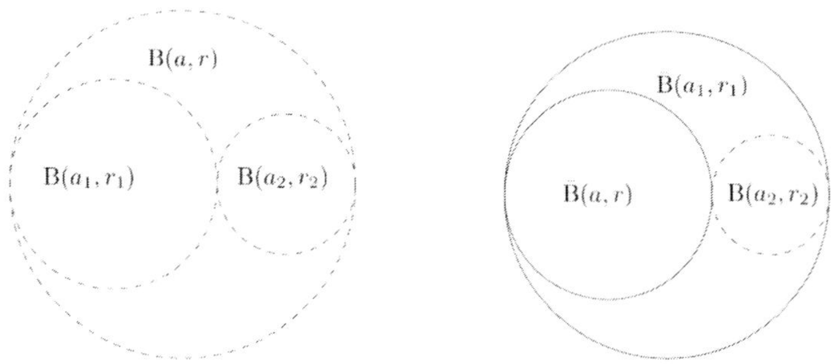

(a) Case where equality holds in Lemma 3.2. (b) Case where equality holds in Lemma 3.3.

Figure. 2. Extreme cases in the combining and pulling in lemmas.

4. CLASSICALISATION OF SWISS CHEESES

We aim to give a topological proof of the Feinstein–Heath classicalisation theorem (Theorem 4.1), as described in the introduction, stated below in the language of abstract Swiss cheeses.

THEOREM 4.1

Let $A=((a_n,r_n))$ *be an abstract Swiss cheese with* $\delta_1(A) > 0$*. Then there exists a classical, abstract Swiss cheese* $\mathrm{B} \in F$ *such that* $X_B \subseteq X_A$ *and* $\delta_1(B) \geq \delta_1(A)$.

We will see below that it is enough to prove this theorem for abstract Swiss cheeses where some redundancy has been eliminated, as the general case then follows. We first introduce the following terminology.

DEFINITION 4.2

Let $A=((a_n,r_n))$ be an abstract Swiss cheese. Then A is *redundancy-free* if, for all $k \in S_A$, we have $\mathrm{B}(a_k,r_k) \cap \overline{\mathrm{B}}(a_0,r_0) \neq \emptyset$, and for all $\ell \in S_A$ with $k \neq \ell$ we have $\mathrm{B}(a_k,r_k) \nsubseteq \mathrm{B}(a_\ell,r_\ell)$.

An elementary argument, which we leave to the reader, shows that it is easy to eliminate redundancy from abstract Swiss cheeses with finite radius sum, as in the following lemma.

LEMMA 4.3

Let $A = ((a_n, r_n)) \in F$ with $\rho(A) < \infty$. Then there exists a redundancy-free abstract Swiss cheese $=((b_n, s_n)) \in N$ with $X_B = X_A, \mu(B) < \infty$ and $\overline{B}(b_0, s_0) = \overline{B}(a_0, r_0)$ such that $\rho_E(B) \leq \rho_E(A)$ for each subset $E \subseteq \mathbb{C}$. In particular, $\rho(B) \leq \rho(A)$.

Note that, since $\overline{B}(b_0, s_0) = \overline{B}(a_0, r_0)$ and $\rho(B) \leq \rho(A)$ in the above lemma we have $\delta_1(B) \geq \delta_1(A)$, as we claimed before. It is clear, by Lemma 4.3, that to prove Theorem 4.1 it is enough to consider A such that $\delta_1(A) > 0$ and A is redundancy-free.

We now define a relation on F which will help us to construct a compact subset of *F*. Then we prove the existence of classical abstract Swiss cheeses with desired properties in this compact subset.

DEFINITION 4.4

Let $A = ((a_n, r_n))$ and $B = ((b_n, s_n))$ be abstract Swiss cheeses. We say *B* is *partially above A* if $\overline{B}(b_0, s_0) \subseteq \overline{B}(a_0, r_0)$, and, for each $n \in \mathbb{N}$, either $B(a_n, r_n) \subseteq \mathbb{C} \setminus \overline{B}(b_0, s_0)$, or there exists $m \in N$ such that $B(a_n, r_n) \subseteq B(b_m, s_m)$, or both.

It is clear that *A* is partially above itself and that if *B* is partially above *A*, then $X_B \subseteq X_A$.

Fix a redundancy-free abstract Swiss cheese $A = ((a_n, r_n)) \in N$ with $\delta_1(A) > 0$. Note that $\rho(A) < \infty$ and, since *A* is redundancy-free, $\mu(A) < \infty$. We set $R = \rho(A)$ and $M = \mu(A)$.

Let $S(A)$ be the collection of all $B = ((b_n, s_n)) \in N(M, R)$ such that *B* is partially above *A*. Recall that, since $B \in N(M, R)$, we have $s_n \leq R/n$ for all $n \in S_B$ so that

$$n \leq \frac{R}{r_n} \qquad (n \in S_B) \tag{4}$$

By our conditions on *A* it is clear that $A \in S(A)$. We now prove that $S(A)$ is compact.

LEMMA 4.5

The set $S(A)$ is a compact subset of F.

PROOF

As noted earlier, it is enough to prove that $S(A)$ is closed in $N(M, R)$ and that the 0-th coordinate projection is bounded on $S(A)$. The latter is clear from the definition of $S(A)$, so we prove that $S(A)$ is closed in $N(M, R)$.

For each $m \in N_0$, let $A^{(m)} = ((a_n^{(m)}, r_n^{(m)}))_{n=0}^{\infty}$ be an abstract Swiss cheese in $S(A)$, and suppose the sequence $(A^{(m)})$ converges to $B=((b_n,s_n))\in N(M,R)$. It remains to show that B is partially above A.

It is easy to see (by Lemma 3.1, for example) that $\overline{B}(b0,s0) \subseteq \overline{B}(a_0,r_0)$. Fix $k \in N$. We show that either $B(a_k,r_k) \subseteq C\setminus \overline{B}(b_0,s_0)$ or there exists $\ell \in S_B$ with $B(a_k,r_k)\subseteq B(b_\ell,s_\ell)$. If $r_k=0$ then $B(a_k,r_k)=\emptyset$ and the result is trivial, so we may assume that $k \in S_A$. First assume that there exists $n_0 \in N_0$ such that, for all $m \geq n_0$ we have $B(a_k,r_k)\subseteq C\setminus \overline{B}(a_0^{(m)},r_0^{(m)})$. Then we have $|a_k-a_0^{(m)}|\geq r_k+r_0^{(m)}$ for all $m \geq n_0$ by Lemma 3.1. Letting $m\to\infty$, we obtain $|a_k-a_0|\geq r_k+r_0$, and so, by Lemma 3.1 again, $B(a_k,r_k)\subseteq C\setminus \overline{B}(b_0,s_0)$.

Otherwise for each $n_0 \in N_0$, there exist $m \geq n_0$ and $\ell\, m \in N$ such that

$$B(a_k, r_k) \subseteq B(a_{\ell_m}^{(m)}, r_{\ell_m}^{(m)}) \tag{5}$$

By passing to a subsequence of $A^{(m)}$ if necessary, we can assume (5) holds for all $m \in N_0$. For each m, since $r_{\ell_m}^{(m)} \geq r_k$,by (4) we have $\ell_m\leq R/_{rk}$. Thus there must be a $p \in N$ that appears infinitely many times in the sequence $(\ell_m)_m$. Passing to a subsequence again if necessary, we may assume $\ell_m=p$ for all m. Since $A^{(m)}\to B$ as $m\to\infty$ and $B(a_k,r_k)\subseteq B(a_p^{(m)},r_p^{(m)})$, it is again easy to show, using Lemma 3.1, that $B(a_k,r_k)\subseteq B(b_p,s_p)$. Thus B is partially above A and we have proved that $S(A)$ is closed.

Since δ_1 is upper semicontinuous and $S(A)$ is compact and non-empty, δ_1 attains a maximum value on $S(A)$ and this value is at least $\delta_1(A)>0$. Let

$$\mathcal{S}_1 := \{A' \in \mathcal{S}(A) : \delta_1(A') = \sup_{B\in\mathcal{S}(A)} \delta_1(B)\},$$

which is also compact and non-empty.

LEMMA 4.6

Let $B=((b_n,s_n)) \in S(A)$.

(a)*Suppose that $k, \ell\in S_B$ with $k \neq \ell$ such that $\overline{B}(b_k,s_k)\cap \overline{B}(b_\ell,s_\ell) \neq\emptyset$. If we have $B(b_k,s_k)\cap B(b_\ell,s_\ell) \neq \emptyset$ then there exists $B'\in S(A)$ such that $\delta_1(B')>\delta_1(B)$. Otherwise, there exists $B'\in S(A)$ such that $\delta_1(B')=\delta_1(B)$ and $\delta_2(B')<\delta_2(B)$.*

(b)*Suppose that $k \in S_B$ with $s_k<s_0$ such that $\overline{B}(b_k,s_k)\not\subseteq B(b_0,s_0)$. If we have $B(b_k,s_k)\not\subseteq B(b_0,s_0)$ then there exists $B' \in S(A)$ such that $\delta_1(B')>\delta_1(B)$. Otherwise, there exists $B'\in S(A)$ with $\delta_1(B')=\delta_1(B)$ and $\delta_2(B')< \delta_2(B)$.*

PROOF

(a) Let $B(b,s)$ be the open disk obtained by applying Lemma 3.2 to the disks $B(b_k,s_k)$ and $B(b_\ell,s_\ell)$. Let $B'=((b_n',s_n'))$ be obtained by replacing the disks $B(b_k,s_k)$ and $B(b_\ell,s_\ell)$ by $B(b,s)$.

If $B(b_k,s_k)\cap B(b_\ell,s_\ell) \neq \emptyset$ then we have $s < s_k+s_\ell$ and so $\delta_1(B')>\delta_1(B)$. Otherwise, we have $s = s_k+s_\ell$ and hence $s^2 > s_k^2 + s_\ell^2$. In this case, we have $\delta_1(B')=\delta_1(B)$ and $\delta_2(B')<\delta_2(B)$.

We now show that $B' \in S(A)$. Clearly $B' \in N$ by our definition of replacing disks in an abstract Swiss cheese. Since b lies on the line segment connecting b_k and b_ℓ, it follows that $\mu(B')\leq\mu(B)$ and since $s \leq s_k+s_\ell$ we have $\rho(B')\leq\rho(B)$. Thus $B'\in N(M,R)$. It remains to show that B' is partially above A.

We have $\bar{B}(b_0', s_0') = \bar{B}(b_0, s_0)$ so that $\bar{B}(b_0', s_0') \subseteq B(a_0, r_0)$. Fix $p \in \mathbb{N}$. Since B is partially above A, we have $B(a_p,r_p)\subseteq B(b_m,s_m)$ for some $m \in S_B$ or $B(a_p,r_p)\subseteq \mathbb{C}\setminus \bar{B}(b_0,s_0)$. If $B(a_p,r_p)\subseteq \mathbb{C}\setminus \bar{B}(b_0,s_0)$ then we also have $B(a_p, r_p) \subseteq \mathbb{C} \setminus \bar{B}(b_0', s_0')$. Otherwise, let $m \in S_B$ with $B(a_p,r_p)\subseteq B(b_m,s_m)$. If $m = k$ or $m = \ell$, then, with q as the index where $B(b,s)$ was inserted, we have $B(a_p, r_p) \subseteq B(b_q', s_q')$. If $m \neq k, \ell$, then there exists $q \in S_{B'}$ such that $B(b_q', s_q') = B(b_m, s_m)$. Thus $B(a_p, r_p) \subseteq B(b_q', s_q')$. Hence B' is partially above A, and so $B'\in S(A)$ as required.

(b) Let $\bar{B}(b,s)$ be the closed disk obtained by applying Lemma 3.3 to the disks $B(b_0,s_0)$ and $B(b_k,s_k)$. Let $B' = ((b_n', s_n'))$ be the abstract Swiss cheese obtained by deleting the disks at indices 0 and k and inserting the disk $\bar{B}(b,s)$ at index 0.

If $B(b_k,s_k) \nsubseteq B(b_0,s_0)$ then we have $s > s_0-s_k$ so that $\delta_1(B')>\delta_1(B)$. Otherwise, we have $s_0=s+s_k$ and $s_0^2 > s^2 + s_k^2$ so that $\delta_1(B')=\delta_1(B)$ and $\delta_2(B')<\delta_2(B)$.

The proof that $B'\in S(A)$ is similar to the roof in part (a).

We are now ready to prove the main results of this section.

THEOREM 4.7

All abstract Swiss cheeses in S_1 are semiclassical.

PROOF

Let $B=((b_n,s_n))\in S_1$. Suppose for contradiction that B is not a semiclassical abstract Swiss cheese. Consider first the case where there are distinct $k,\ell \in S_B$ with $B(b_k,s_k)\cap B(b_\ell,s_\ell)\neq\emptyset$. By Lemma 4.6(a) there exists $B'\in S(A)$ with $\delta_1(B')>\delta_1(B)$, which is a contradiction.

The remaining case is where there is a $k \in S_B$ with $\mathrm{B}(b_k,s_k) \not\subseteq \mathrm{B}(b_0,s_0)$. We have $\delta_1(B) \geq \delta_1(A) > 0$ so that $s_k < s_0$. By Lemma 4.6(b) there exists $B' \in S(A)$ with $\delta_1(B') > \delta_1(B)$, which is a contradiction.

Since S_1 is compact and non-empty, δ_2 attains both maximum and minimum values on S_1. Let

$$S_2 := \{A' \in S_1 : \delta_2(A') = \inf_{B \in S_1} \delta_2(B)\},$$

which is again non-empty and compact. Since all the abstract Swiss cheeses in S_1 are semiclassical, $\pi\delta_2(B)$ is the area of X_B for all $B \in S_1$, and hence for all $B \in S_2$. So the abstract Swiss cheeses in S_2 are obtained by finding those $B \in S_1$ for which the area of X_B is minimal on S_1.

THEOREM 4.8

All abstract Swiss cheeses in S_2 are classical.

PROOF

Let $B=((b_n,s_n)) \in S_2$. Suppose for contradiction that B is not classical. If there are distinct $k,\ell \in S_B$ with $\overline{\mathrm{B}}(b_k,s_k) \cap \overline{\mathrm{B}}(b_\ell,s_\ell) \neq \emptyset$ then, by Lemma 4.6(a), there exists $B' \in S(A)$ such that either $\delta_1(B') > \delta_1(B)$ or $\delta_1(B') = \delta_1(B)$ and $\delta_2(B') < \delta_2(B)$. In either case we obtain a contradiction since $B \in S_2$.

Otherwise there exists $k \in S_B$ with $\overline{\mathrm{B}}(b_k,s_k) \not\subseteq \mathrm{B}(b_0,s_0)$. Note that $s_k < s_0$ since $\delta_1(B) > 0$. By Lemma 4.6(b) there exists $B' \in S(A)$ such that either $\delta_1(B') > \delta_1(B)$ or $\delta_1(B') = \delta_1(B)$ and $\delta_2(B') < \delta_2(B)$. In either case we obtain a contradiction since $B \in S_2$.

In the next theorem, we show that if X_A has empty interior then we do not have to minimise δ_2 on S_1 to find classical abstract Swiss cheeses.

THEOREM 4.9

If int $X_A = \emptyset$ *then each abstract Swiss cheese in S_1 is classical.*

PROOF

Let $B=((b_n,s_n)) \in S_1$. Then, by Theorem 4.7, B is semiclassical. Suppose for contradiction that B is not classical. Then there are two cases summarised in Fig. 3. First suppose there exist distinct $k,\ell \in S_B$ with $\overline{\mathrm{B}}(b_k,s_k) \cap \overline{\mathrm{B}}(b_\ell,s_\ell) \neq \emptyset$. Then by Lemma 3.2, since $B(b_k,s_k) \cap \mathrm{B}(b_\ell,s_\ell) = \emptyset$, there exists an open disk $\mathrm{B}(a,r) \supseteq \mathrm{B}(b_k,s_k) \cup \mathrm{B}(b_\ell,s_\ell)$ with $r = s_k + s_\ell$. By replacing the disks $\mathrm{B}(b_k,s_k)$ and $\mathrm{B}(b_\ell,s_\ell)$ with $\mathrm{B}(a,r)$ we obtain a new abstract Swiss cheese $B' = ((b'_n, s'_n))$ such that $B' \in S_1$ (following the proof of Lemma 4.6). Let p be the index at which the disk $\mathrm{B}(a,r)$ was inserted. Since X_B has empty interior, there exists $m \in S_B$ with $m \neq p$ such that $\mathrm{B}(a,r) \cap \mathrm{B}(b_m,s_m) \neq \emptyset$. Let $q \in S_{B'}$ be such that

$B(b'_q,s'_q)=B(b_m,s_m)$. Note that $p \neq q$. Applying Lemma 4.6(a) to $p,q \in S_{B'}$ and B', we obtain an abstract Swiss cheese $B'' \in S(A)$ which has $\delta_1(B'')>\delta_1(B')$. But this is a contradiction.

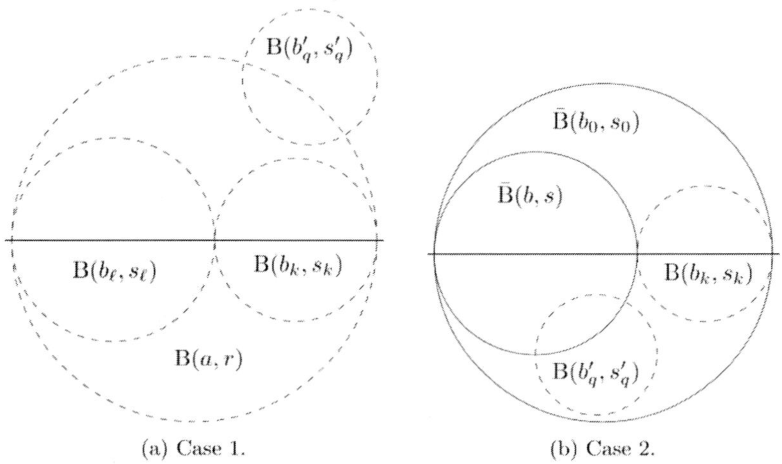

Figure. 3. The two cases in the proof of Theorem 4.9.

Now suppose there exists $k \in S_B$ with $\overline{B}(b_k,s_k) \not\subseteq B(b_0,s_0)$. Let $\overline{B}(b,s)$ be the closed disk obtained by applying Lemma 3.3 to the disks $\overline{B}(b_0,s_0)$ and $B(b_k,s_k)$. Since B is semiclassical, we have $s=s_0-s_k$ (as in Fig. 2b). By deleting the disks at indices 0 and k and inserting $B(b,s)$ at index 0, we obtain a new abstract Swiss cheese $B'=((b'_n,s'_n)) \in S_1$ such that $\delta_1(B')=\delta_1(B)$ (again following the proof of Lemma 4.6). Since X_B has empty interior, there exists $q \in S_{B'}$ such that $B(b_q,s_q) \not\subseteq \overline{B}(b,s)$. Applying Lemma 4.6(b) to q and B', we obtain an abstract Swiss cheese $B'' \in S(A)$ which has $\delta_1(B'')>\delta_1(B')$. But this is a contradiction.

Let B be an abstract Swiss cheese satisfying $\delta_1(B)>0$, so that B satisfies the conditions of Theorem 4.1. Then we can apply Lemma 4.3 to obtain a redundancy-free abstract Swiss cheese $A \in N$ with $X_A = X_B$ and such that $\delta_1(A) \geq \delta_1(B)$.

We can then apply the above constructions to A. Each abstract Swiss cheese A' from the corresponding non-empty set S_2 is classical by Theorem 4.8 and has $X_{A'} \subseteq X_A=X_B$ and $\delta_1(A') \geq \delta_1(A) \geq \delta_1(B)$. So we obtain the Feinstein–Heath classicalisation theorem as a corollary of Theorem 4.8.

5. CONTROLLED CLASSICALISATION

In this section we discuss some situations in which it is possible to make a Swiss cheese classical without changing certain disks. This process we call "controlled classicalisation".

Recall that, for $E \subseteq \mathbb{C}$ and an abstract Swiss cheese $A=((a_n,r_n))$, the set $H_A(E)$ is the set of all $n \in S_A$ such that $\bar{B}(a_n,r_n) \cap E \neq \emptyset$.

LEMMA 5.1

Let U be a non-empty open subset of $\mathbb{C}$. For each $m \in \mathbb{N}_0$, $A^{(m)} = ((a_n^{(m)}, r_n^{(m)})) \in \mathcal{F}$ let and suppose that $A^{(m)} \to A=((a_n,r_n)) \in F$ as $m \to \infty$. Then $\rho_U(A) \le \liminf_{m\to\infty} \rho_U(A^{(m)})$.

PROOF

Since U is open and $A^{(m)} \to A$ as $m \to \infty$, for each $k \in H_A(U)$ there exists $m_0 \in \mathbb{N}_0$ such that, for all $m \ge m_0$, we have $k \in S_{A^{(m)}}$ and

$$\bar{B}(a_k^{(m)}, r_k^{(m)}) \cap U \neq \emptyset.$$

Let χ_m denote the characteristic function of $H_{A^{(m)}}(U) \cap H_A(U)$. Then χ_m converges pointwise to the function $\chi := \chi_{H_A(U)}$ as $m \to \infty$. Since $r_k^{(m)} \to r_k$ as $m \to \infty$ for each k, by Fatou's lemma for series, we have

$$\rho_U(A) = \sum_{n=1}^{\infty} \chi(n) r_n \le \liminf_{m\to\infty} \sum_{n=1}^{\infty} \chi_m(n) r_n^{(m)} \le \liminf_{m\to\infty} \rho_U(A^{(m)}),$$

as required.

For the rest of this section $A =((a_n,r_n)) \in N$ will be a fixed redundancy-free abstract Swiss cheese. Note that both $\rho(A)$ and $\mu(A)$ are finite and $r_n \le \rho(A)/n$ for all $n \in \mathbb{N}$. We define the (classical) *error set of A* to be

$$E(A) := \bigcup_{\substack{m,n\in S_A \\ m\neq n}} \left(\bar{B}(a_m, r_m) \cap \bar{B}(a_n, r_n) \right) \cup \bigcup_{n\in S_A} ((\mathbb{C} \setminus B(a_0, r_0)) \cap \bar{B}(a_n, r_n)).$$

Note that if $E(A) \subseteq \mathrm{B}(a0,r_0)$ then $\overline{\mathrm{B}}(a_n,r_n) \subseteq \mathrm{B}(a_0,r_0)$ for all $n \in S_A$. We aim to prove that, under suitable conditions, we can classicalise A while leaving many of the open disks unchanged.

As in Section 4, we seek to construct a compact subset of F on which the function δ_1 can be maximised and then the function δ_2 minimised to give a suitable classical abstract Swiss cheese.

In the rest of this paper, we will frequently need to consider indexed collections of pairs of sets of the following form. Let $I \subseteq \mathrm{N}$ be non-empty. Let $C=((K_n,U_n))_{n\in I}$, where each K_n is a compact plane set and each U_n is an open set with $K_n \subseteq U_n$. We call such an indexed collection a *controlling collection of pairs.* In the special case where I has only one member, we say C is a *controlling pair* and write $C=(K,U)$.

DEFINITION 5.2

Let $C=((K_n,U_n))_{n\in I}$ be a controlling collection of pairs. Define

$$V(C) := \bigcup_{n\in I} U_n, \quad F(C) := \bigcup_{n\in I} K_n \; .$$

Let $L_A(C)$ denote the set of all $B=((b_n,s_n)) \in N\,(\mu(A),\rho(A))$ such that:

a) for each $(K,U) \in C$ we have $\rho_U(B) \leq \rho_U(A)$;

b) $\overline{\mathrm{B}}(b_0,s_0)= \overline{\mathrm{B}}(a_0,r_0)$;

c) for all $k \in S_A$ with $\overline{\mathrm{B}}\,(a_k,r_k) \cap V\,(C)=\emptyset$ there exists $\ell \in S_B$ such that $\mathrm{B}(b_\ell,s_\ell)=\mathrm{B}(a_k,r_k)$;

d) for each $n \in I$ and for all $k \in S_A$ with $\overline{\mathrm{B}}(a_k,r_k) \cap U_n \neq \emptyset$:

- there exists $\ell \in S_B$ with $\mathrm{B}(b_\ell,s_\ell)=\mathrm{B}(a_k,r_k)$; or
- there exists $\ell \in H_B(K_n)$ with $\mathrm{B}(a_k,r_k) \subseteq \mathrm{B}\,(b_\ell,s_\ell)$.

Note that $A \in L_A(C)$, and if $B \in L_A(C)$ then B is partially above A. Thus if $B \in L_A(C)$ then $X_B \subseteq X_A$. The properties (a)–(d) reflect the properties we desire for the final abstract Swiss cheese. We will use the open sets U to bound the error set $E(A)$. Under some technical assumptions, conditions (c) and (d) ensure that abstract Swiss cheeses maximising δ_1 in $L_A(C)$ have the property that any open disk which lies outside $V(C)$ is the same as an open disk from A.

We first require some preliminary lemmas. The following lemma is probably well-known and can be proved using a Hausdorff metric argument, but we include an elementary proof for the convenience of the reader.

LEMMA 5.3

Let K be a compact plane set. Let (z_n) be a sequence in C, and let (t_n) be a sequence in $\mathbb{R}^+$. Suppose that $\overline{B}(z_n,t_n) \cap K \neq \emptyset$ for all n, and that $z_n \to z$ and $t_n \to t$ as $n \to \infty$. Then $\overline{B}(z,t) \cap K \neq \emptyset$.

PROOF

For each $n \in N_0$ there exists a point $w_n \in \overline{B}(z_n,t_n) \cap K$. Now since (w_n) is a sequence in K there is a convergent subsequence (w_{nk}) converging to a point $w \in K$. For each $k \in N_0$, we have $w_{nk} \in \overline{B}(z_{nk},t_{nk})$ so that $|w_{nk}-z_{nk}| \leq t_{nk}$. Hence, taking the limit as $k \to \infty$, we have $|w-z| \leq t$ so that $w \in \overline{B}(z,t) \cap K$ as required.

We now prove that the space $L_A(C)$ is a compact subspace of F for an arbitrary countable collection C of pairs (K,U) where K is a compact plane set and U an open neighbourhood of K.

LEMMA 5.4

Let $C:=((K_n,U_n))_{n \in I}$ be a controlling collection of pairs. Then the set $L_A(C) \subseteq F$ is compact.

PROOF

It is sufficient to show that $L_A(C)$ is closed in $N(\mu(A),\rho(A))$, since the 0-th coordinate projection is clearly bounded on $L_A(C)$. For each $m \in N_0$, let $A^{(m)} = ((a_n^{(m)}, r_n^{(m)}))_{n=0}^{\infty} \in \mathcal{L}_A(C)$. Let $B=((b_n,s_n))$ and suppose that $A^{(m)} \to B \in N(\mu(A),\rho(A))$ as $m \to \infty$; we need to show that $B \in L_A(C)$.

By Lemma 5.1 we see that B also satisfies (a), and it is immediate that (b) is also satisfied.

It remains to prove (c) and (d) hold for B. Fix $k \in S_A$. Suppose that $\overline{B}(a_k,r_k) \cap V(C) = \emptyset$. Since, for each $m \in N_0$, we have $A^{(m)} \in L_A(C)$ it follows that for each m there exists an integer ℓ_m such that $B(a_k, r_k) = B(a_{\ell_m}^{(m)}, r_{\ell_m}^{(m)})$. Now since $r_k^{(m)} = r_k$ for each m we have $1 \leq \ell_m \leq \rho(A)/r_k$ for all m. But then there must exist an integer $1 \leq p \leq \rho(A)/r_k$ such that $\ell_k = p$ infinitely often so we can find a subsequence $(A^{(m_j)})_j$ such that $\ell_{m_j} = p$ for all j. Since $B(a_k, r_k) = B(a_k^{(m_j)}, r_k^{(m_j)})$ for all j and $A^{(m_j)} \to B$ as $j \to \infty$, it follows that $B(a_k,r_k) = B(b_p,s_p)$. This proves that (c) holds for B.

Now suppose that $\overline{B}(a_k,r_k) \cap U \neq \emptyset$ for some $(K,U) \in C$. As above, for each $m \in N_0$ there exists an integer ℓ_m such that $B(a_k, r_k) \subseteq B(a_{\ell_m}^{(m)}, r_{\ell_m}^{(m)})$ and $r_{\ell_m}^{(m)} \geq r_k$. We choose ℓ_m as follows: if in $A^{(m)}$ there is an open disk $B(a,r) = B(a_k,r_k)$ then we pick ℓ_m to be the index of that open disk, otherwise we choose ℓ_m to be the index of an open disk $B(a,r)$ that properly contains

$B(a_k,r_k)$ and $\overline{B}(a,r) \cap F(C) \neq \emptyset$. Hence we have $1 \leq \ell_m \leq \rho(A)/r_k$ for all m and so there exists an integer $1 \leq p \leq \rho(A)/r_k$ such that $\ell_m = p$ infinitely often. By considering a subsequence we can assume that $\ell_m = p$ for all m. If $B(a_p^{(m)}, r_p^{(m)}) = B(a_k, r_k)$ holds for infinitely many m then there is a subsequence $(A^{(m_j)})_j$ such that $B(a_k, r_k) = B(a_p^{(m_j)}, r_p^{(m_j)})$ for all j. Since $A^{(m_j)} \to B$ as $j \to \infty$ it follows that $B(a_k,r_k) = B(b_p,s_p)$. If $B(a_k, r_k) = B(a_p^{(m)}, r_p^{(m)})$ for only finitely many m then we must have

$$B(a_k, r_k) \subseteq B(a_p^{(m)}, r_p^{(m)}) \quad \text{and} \quad \bar{B}(a_p^{(m)}, r_p^{(m)}) \cap K \neq \emptyset$$

for infinitely many m. Then there exists a subsequence $(A^{(m_j)})_j$ such that $B(a_k, r_k) \subseteq B(a_p^{(m_j)}, r_p^{(m_j)})$ and $\bar{B}(a_p^{(m_j)}, r_p^{(m_j)}) \cap K \neq \emptyset$ for all j. But then $B(a_k,r_k) \subseteq B(b_p,s_p)$ and, by Lemma 5.3, $\overline{B}(b_p,s_p) \cap K \neq \emptyset$. This proves that (d) holds for B.

Thus we have proved that $B \in L_A(C)$ and hence $L_A(C)$ is compact.

We are interested in those abstract Swiss cheeses B in a space $L_A(C)$ on which the discrepancy function δ_1 is maximised. These abstract Swiss cheeses have some desirable properties. Let $L^*_A(C)$ denote the subset of $L_A(C)$ of all abstract Swiss cheeses where $\delta_1$achieves its maximum. Since $L_A(C)$ is non-empty and compact, $L^*_A(C)$ is non-empty and compact. Recall that $A \in N$ is assumed to be redundancy-free.

LEMMA 5.5

Let $C := ((K_n,U_n))_{n \in I}$ be a controlling collection of pairs. Let $B = ((b_n,s_n)) \in L^*_A(C)$. Then B has the following properties.

(a)*For all $k,\ell \in S_B$ with $k \neq \ell$, we have $B(b_k,s_k) \neq B(b_\ell,s_\ell)$.*

(b) *For each $k \in S_B$, there exists $\ell \in S_A$ such that $B(a_\ell,r_\ell) \subseteq B(b_k,s_k)$. Moreover, if $\overline{B}(b_k,s_k) \cap F(C) = \emptyset$ then this $\ell \in S_A$ is unique, and we have $B(b_k,s_k) = B(a_\ell,r_\ell)$.*

(c) *Let E be a fixed subset of C. Let $H_1 := H_B(E) \setminus H_B(V(C))$and let $H_2 := H_A(E) \setminus H_A(V(C))$. There exists a bijection $\sigma: H_1 \to H_2$ satisfying the following condition: for each $k \in H_1$ and $\ell \in H_2$, we have $\sigma(k) = \ell$ if and only if $B(b_k,s_k) = B(a_\ell,s_\ell)$. In particular,*

$$\sum_{n \in H_1} s_n = \sum_{n \in H_2} r_n .$$

PROOF

(a) If $k,\ell \in S_B$ with $k \neq \ell$ such that $B(b_k,s_k) = B(b_\ell,s_\ell)$ then we can obtain an abstract Swiss cheese B' by deleting the disk at index ℓ which has $\delta_1(B') > \delta_1(B)$. It is easy to see that $B' \in L_A(C)$, which is a contradiction.

(b) Let $k \in S_B$. Assume, for contradiction, there does not exist $\ell \in S_A$ such that $\mathrm{B}(a_\ell,r_\ell) \subseteq \mathrm{B}(b_k,s_k)$. Then we can delete the disk at index k from B to obtain an abstract Swiss cheese B' with $\delta_1(B')>\delta_1(B)$. It is clear that $B' \in L_A(C)$, which contradicts the maximality of $\delta1(B)$. Thus there exists $\ell \in S_A$ such that $\mathrm{B}(a_\ell,r_\ell)\subseteq \mathrm{B}(b_k,s_k)$.

Now suppose, in addition, that $\overline{\mathrm{B}}(b_k,s_k) \cap F(C)=\emptyset$. We show that the $\ell \in S_A$ found above with $\mathrm{B}(a_\ell,r_\ell)\subseteq \mathrm{B}(b_k,s_k)$ is unique and that we have $\mathrm{B}(a_\ell,r_\ell)=\mathrm{B}(b_k,s_k)$. Assume, for contradiction, that $\mathrm{B}(a_\ell,r_\ell)\neq\mathrm{B}(b_k,s_k)$. Then, since A is redundancy-free, we must have $\mathrm{B}(a_m,r_m)\neq \mathrm{B}(b_k,s_k)$ for all $m \in S_A$. We claim that the abstract Swiss cheese B' obtained by deleting the disk at index k from B has $B' \in L_A(C)$; this will lead to a contradiction.

Clearly $B' \in N(\mu(A),\rho(A))$ and it is also clear that B' satisfies conditions (a) and (b) of Definition 5.2(a). Since $\mathrm{B}(a_m,r_m)\neq \mathrm{B}(b_k,s_k)$ for all $m \in S_A$, it follows that 5.2(c) remains true for B'. Similarly, since $\overline{\mathrm{B}}(b_k,s_k)\cap F(C)= \emptyset$, 5.2(d) remains true for B'. This proves our claim.

But now $\delta_1(B')>\delta_1(B)$, which contradicts the maximality of $\delta_1(\mathrm{B})$. Thus $\mathrm{B}(a_\ell,r_\ell)=\mathrm{B}(b_k,s_k)$. The uniqueness of ℓ follows from the fact that A is redundancy-free.

(c) Note that if, for some $k \in S_B$ and $\ell \in S_A$, $\mathrm{B}(b_k,s_k)=\mathrm{B}(a_\ell,r_\ell)$ then $k \in H_1$ if and only if $\ell \in H_2$. Combining this with (b), for each $k \in H_1$ there exists a unique $\ell \in H_2$ such that $\mathrm{B}(b_k,s_k)=\mathrm{B}(a_\ell,r_\ell)$. Thus we may define $\sigma(k)=\ell$ for such k,ℓ. We must show that σ is a bijection. By (a), σ is injective. Let $\ell \in H$. By 5.2(c), there exists $k \in S_B$ with $\mathrm{B}(b_k,s_k)= \mathrm{B}(a_\ell,r_\ell)$. By the remark above, $k \in H_1$, and so $\sigma(k)=\ell$. This proves that σ is surjective. It is now immediate that $\sum_{n\in H_1} s_n = \sum_{n\in H_2} r_n$. This completes the proof.

We shall see that by imposing some additional technical conditions on C we can obtain a controlled classicalisation theorem. Recall that if $E \subseteq \mathbb{C}$ is non-empty and $z \in \mathbb{C}$ then we define the distance of z to E by $\mathrm{dist}(z,E):=\inf\{|z-x|:x \in E\}$. For a non-empty compact set $K \subseteq C$ and positive real number M we define $U(K,M):=\{z \in C:\mathrm{dist}(z,K)<M\}$.

LEMMA 5.6

Let $I \subseteq \mathbb{N}$ be non-empty. Let $(K_n)_{n\in I}$ be a collection of compact plane sets and let $(M_n)_{n\in I}$ be a collection of positive real numbers. For each $n \in I$, let $U_n:=U(K_n,M_n)$. Suppose that $\rho\, U_k(A)< M_k/2$ and $U_k \subseteq \mathrm{B}(a_0,r_0)$ for all $k \in I$ and suppose that $U_k \cap U_\ell= \emptyset$ for all distinct $k,\ell \in I$. Let C be the controlling collection$((K_n,U_n))_{n\in I}$. Let $B=((b_n,s_n))\in L_A(C)$ and fix $m \in I$. Suppose there exist $k, \ell \in S_B$ with $k \neq \ell$ such that $\overline{\mathrm{B}}(b_k,s_k) \cap K_m \neq \emptyset$ and $\overline{\mathrm{B}}(b_k,s_k) \cap \overline{\mathrm{B}}(b_\ell,s_\ell) \neq \emptyset$. Then there exists $B' \in L_A(C)$ such that either $\delta_1(B')>\delta_1(B)$ or $\delta_1(B')=\delta_1(B)$ and $\delta_2(B')< \delta_2(B)$.

PROOF

Let B (b,s) be the disk obtained by the application of Lemma 3.2 to the disks $B(b_k,s_k)$ and $B(b_\ell,s_\ell)$. Let $B'=((b'_n,s'_n))$ be an abstract Swiss cheese obtained from B by replacing the disks at indices k,ℓ with the disk B(b,s). Since B $\in L_A(C)$ we have $\rho U_m(B) \leq \rho U_m(A) < M_m/2$, so that $s \leq sk + s\ell < M_m/2$. Since $\bar{B}(b_k,s_k) \cap K_m \neq \emptyset$, we must have $\bar{B}(b,s) \subseteq U_m$ and hence $\bar{B}(b,s) \cap U_n = \emptyset$ for all $n \in I$ with $n \neq m$.

It is clear that either $\delta_1(B') > \delta_1(B)$, when $s < sk + s\ell$, or we have $\delta_1(B') = \delta_1(B)$ and $\delta_2(B') < \delta_2(B)$, when $s = s_k + s_\ell$, so it remains to show that $B' \in L_A(C)$. By construction, and since $\bar{B}(b,s) \subseteq U_m$ and $\bar{B}(b,s) \cap U_n = \emptyset$ for $n \in I$ with $n \neq m$, we have $B' \in$ N$(\mu(A),\rho(A))$ and satisfies (a) and (b) in Definition 5.2.

Fix $j \in S_A$. If $\bar{B}(a_j,r_j) \cap V(C) = \emptyset$, then there exists $p \in S_B$ with $p \neq k,\ell$ and B(b_p,s_p)=B(a_j,r_j). Hence there is a $p' \in S_{B'}$ such that B$(b'_{p'},s'_{p'})$=B(a_j,r_j) and B'satisfies (c) in Definition 5.2.

Suppose that $\bar{B}(a_j,r_j) \cap V(C) \neq \emptyset$. Let $n \in I$ such that $\bar{B}(a_j,r_j) \cap U_n \neq \emptyset$. Since $B \in L_A(C)$, there exists $p \in S_B$ such that B$(a_j,r_j) \subseteq$ B (b_p,s_p), where equality holds unless $\bar{B}(b_p,s_p) \cap K_n \neq \emptyset$. If $p \neq k, \ell$, then there exists $q \in S_{B'}$ such that B$(b_{q'},s_{q'})$=B(b_p,s_p). Thus B$(a_j,r_j) \subseteq$ B $(b_{q'},s_{q'})$ and equality holds if $\bar{B}(b_{q'},s_{q'}) \cap K_n = \emptyset$. If $n \neq m$ then we cannot have $p=k$ or $p=\ell$ since $\bar{B}(b,s) \subseteq U_m$ and $U_n \cap U_m = \emptyset$. If $n = m$ and either $p = k$ or $p=\ell$, then there exists $q \in S_{B'}$ such that $B(b'_q,s'_q)$=B(b,s), so that B$(a_j,r_j) \subseteq$B(b'_q,s'_q) and $\bar{B}(b'_q,s'_q) \cap K_n \neq \emptyset$. Moreover, $\bar{B}(b'_q,s'_q) \cap U_i = \emptyset$ for all $i \in I$ with $i \neq m$. It follows that B' satisfies 5.2(d) and hence $B' \in L_A(C)$ This completes the proof.

Similar geometric reasoning and induction shows that, under the conditions of the lemma, given $n_1,\ldots,n_p \in S_A$ and $m \in I$ such that

$$\bar{B}(a_{n_1}, r_{n_1}) \cap K_m \neq \emptyset \quad \text{and} \quad \bar{B}(a_{n_{j-1}}, r_{n_{j-1}}) \cap \bar{B}(a_{n_j}, r_{n_j}) \neq \emptyset$$

for $j=2,\ldots,p$ we have $\bar{B}(b_{nj},r_{nj}) \subseteq U_m$ for each $j=1,\ldots,p$.

We are now ready to prove the controlled classicalisation theorem.

THEOREM 5.7

Let $I \subseteq$ N be non-empty. Let $(K_n)_{n\in I}$ be a collection of compact plane sets and let $(M_n)_{n\in I}$ be a collection of positive real numbers. For each $n \in I$, let $U_n := U(K_n, M_n)$. Suppose that $U_k \subseteq$ B (a_0, r_0) and $\rho U_k(A) < M_k/2$ for all $k \in I$ and suppose that $U_k \cap U\_ = \emptyset$ for all distinct $k, \ell \in I$. Let C be the controlling collection $((K_n, U_n))_{n\in I}$ and suppose $E(A) \subseteq F$(C). Then there exists $B=((b_n, s_n)) \in L^*_A(C)$ such that $X_B \setminus V(C) = X_A \setminus V(C)$and B is classical.

PROOF

We know that $L^*_A(C)$ is non-empty and compact so δ_2 obtains its minimum on= L^*_A(C). Let $B \in L^*_A$ (C) such that δ_2 is minimised on L^*_A (C) at B. We first show that $\bar{\mathrm{B}}(b_k,s_k) \subseteq \mathrm{B}(b_0,s_0)$ for all $k \in S_B$. Let C be the complement of the disk $\mathrm{B}(a_0,r_0)$=$\mathrm{B}(b_0,s_0)$. Let $k \in S_B$ and assume, for contradiction, that $C \cap \bar{\mathrm{B}}(b_k,s_k) \neq \emptyset$. If there exists $u \in S_A$ such that $\mathrm{B}(a_u,r_u)$=$\mathrm{B}(b_k,s_k)$ then

$$\bar{\mathrm{B}}(b_k, s_k) \subseteq U_m \subseteq \mathrm{B}(a_0, r_0) = \mathrm{B}(b_0, s_0),$$

which is impossible. Otherwise, by Lemma 5.5, there exists $u \in S_A$ with $\mathrm{B}(a_u,r_u) \subseteq \mathrm{B}(b_k,s_k)$. Since $B \in L^*_A$ (C), it follows that there exists $m \in I$ such that $\bar{\mathrm{B}}(b_k,s_k) \cap K_m \neq \emptyset$, and so

$$\bar{\mathrm{B}}(b_k, s_k) \subseteq U_m \subseteq \mathrm{B}(a_0, r_0) = \mathrm{B}(b_0, s_0),$$

which contradicts the fact that $C \cap \bar{\mathrm{B}}(b_k,s_k) \neq \emptyset$.

We must now show that there do not exist distinct $k,\ell \in S_B$ such that $\bar{\mathrm{B}}(b_k,s_k) \cap \bar{\mathrm{B}}(b_\ell,s_\ell) \neq \emptyset$. Suppose, for contradiction, that such a pair exists. If $\bar{\mathrm{B}}(b_k,s_k) \cap F(C) = \emptyset$ and $\bar{\mathrm{B}}(b_\ell,s_\ell) \cap F(C) = \emptyset$ then there exist $u,v \in S_A$ with $\mathrm{B}(a_u,r_u)$=$\mathrm{B}(b_k,s_k)$ and $\mathrm{B}(a_v,r_v)$=$\mathrm{B}(b_\ell,s_\ell)$, which is a contradiction since $E(A) \subseteq F(C)$. Thus at least one of these disks has non-empty intersection with at least one compact set Km.

We may assume, without loss of generality, that $\mathrm{B}(b_k,s_k) \cap K_m \neq \emptyset$ for some $m \in I$. It follows that s_k, $s_\ell < M_m/2$ and $\bar{\mathrm{B}}(b_k,s_k) \subseteq U_m$ and $\bar{\mathrm{B}}(b_\ell,s_\ell) \cap U_m \neq \emptyset$. Let $\mathrm{B}(b,s)$ be the open disk obtained by an application of Lemma 3.2 to the disks $\mathrm{B}(b_k,s_k)$ and $\mathrm{B}(b_\ell,s_\ell)$. Then, by Lemma 5.6, the abstract Swiss cheese $B' \in L_A(C)$ obtained by replacing the disks $\mathrm{B}(b_k,s_k)$ and $\mathrm{B}(b_\ell,s_\ell)$ with $\mathrm{B}(b,s)$ has either $\delta_1(B')>\delta_1(B)$ or $\delta_1(B')=\delta_1(B)$ and $\delta_2(B')<\delta_2(B)$. Both of these cases are impossible since we assumed that δ_1 was maximised on B and δ_2 was minimised on B. It follows that no such pair k,ℓ can exist and hence B is classical.

It remains to show that $X_B \backslash V(C) = X_A \backslash V(C)$. Since $B \in L_A(C)$, we have $X_B \subseteq X_A$, and so $X_B \backslash V(C) \subseteq X_A \backslash V(C)$. Let $U_A := (C \backslash X_A) \cup V(C)$ and $U_B := (C \backslash X_B) \cup V(C)$. Let $z \in U_B$, we show that $z \in U_A$. If z is outside of $\bar{\mathrm{B}}(b_0,s_0)$ then it is also outside of $\bar{\mathrm{B}}$ (a_0,r_0) since the closed balls are the same. If z is in $\bar{\mathrm{B}}$ (b_0,s_0), there exists $k \in S_B$ such that $z \in \mathrm{B}(b_k,s_k)$. Note that $\bar{\mathrm{B}}(b_k,s_k) \cap F(C) = \emptyset$, otherwise $\bar{\mathrm{B}}(b_k,s_k) \subseteq V(C)$. By Lemma 5.5, there exists $\ell \in S_A$ such that $\mathrm{B}(a_\ell,r_\ell)$=$\mathrm{B}(b_k,s_k)$. Thus z ∈

$B(a_\ell,r_\ell)$ and $U_B \subseteq U_A$. It follows that $C\setminus U_B \supseteq C\setminus U_A$ and hence $X_B\setminus V(C)=X_A\setminus V(C)$.

Note here that the classical, abstract Swiss cheese B obtained from this theorem is an element of L^*A (C) and therefore satisfies properties (a)–(d) of Definition 5.2, and the conclusion of Lemma 5.5 holds for B. Note also that, in contrast to the Feinstein–Heath classicalisation theorem, $\delta 1(B)$ may be negative here. We can obtain similar results using transfinite induction.

Taking I to have just one element in Theorem 5.7, we obtain the following corollary, which we use in Section 8.

COROLLARY 5.8

Let K be a compact plane set and let M be a positive real number. Let $U=U(K,M)$ and let C be the controlling pair(K,U). Suppose that $\rho_U(A)<M$ and $E(A) \subseteq K$. Then there exists $B=((b_n,s_n))\in L^*_A(C)$ such that $X_B\setminus U=X_A\setminus U$ and B is classical.

In Section 8 we give an application of controlled classicalisation to construct an example of a classical Swiss cheese set X such that $R(X)$ is regular and admits a non-degenerate bounded point derivation of infinite order, which improves the example constructed by O'Farrell [5]. First we need to discuss annular classicalisation and discuss regularity of $R(X)$.

6. ANNULAR CLASSICALISATION

In this section we give some results about Swiss cheese like sets obtained by deleting open disks from a closed annulus, rather than a closed disk. Let K be a closed annulus in the plane. Then we can write $K=\overline{B}(a_0,r_0)\setminus B(a_1,r_1)$ for some $a_0=a_1\in C$ and $r_0>r_1>0$ real. We say an abstract Swiss cheese $A=((a_n,r_n))$ is *annular* if $a_0=a_1$ and $0<r_1<r_0$ and let K_A denote the annulus $\overline{B}(a_0,r_0)\setminus B(a_1,r_1)$. We shall usually omit 'abstract' from the statement A *is an annular abstract Swiss cheese.*

LEMMA 6.1

Let $a\in C$ and $r_0>r_1>0$ and let $K:=\overline{B}(a,r_0)\setminus B(a,r_1)$. Let $b\in C$ and $0<s<(r_0-r_1)/2$ such that $B(b,s)\cap \overline{C\setminus K}\neq\emptyset$. Then there exist $r'_0,r'_1>0$ such that $K':=\overline{B}(a,r'_0)\setminus B(a,r'_1)\subseteq K$ with $K'\cap B(b,s)=\emptyset$ and $r'_0-r'_1\geq r_0-r_1-2s$.

PROOF

Set $D=B(b,s)$. If $D\subseteq C\setminus K$ then there is nothing to prove so suppose not. Since $s<(r_0-r_1)/2$ there are only two possible cases. We must have either $\overline{D}\cap \overline{B}(a,r_1)\neq\emptyset$ or $D\cap C\setminus B(a,r_0)\neq\emptyset$.

In the first case, where $\overline{D} \cap \overline{B}(a,r_1) \neq \emptyset$, let $r'_0=r_0$ and $r'_1=|b-a|+s$. We have $|b-a| > r_1-s$ and $|b-a| \leq r_1+s$. Hence $r'_1>r_1-s+s=r_1$ and $r'_1\leq r_1+2s<r_1+r0-r_1=r_0$ and

$$r'_0 - r'_1 = r_0 - (|b-a| + s) \geq r_0 - s - r_1 - s = r_0 - r_1 - 2s.$$

Since for each $z \in D$ we have $|b-a|-s<|z-a|<|b-a|+s$ it follows immediately that $D \subseteq C\backslash K$.

In the second case, where $D \cap C\backslash B(a,r_0) \neq \emptyset$, let $r'_0=|b-a|-s$ and $r'_1=r_1$. We have $|b-a|<r_0+s$ and $|b-a|\geq r_0-s$. Hence $r'_0<r_0+s-s=r_0$ and

$$r'_0 > r_0 - s - s > r_0 - (r_0 - r_1) = r_1$$

and so

$$r'_0 - r'_1 = |b-a| - s - r_1 \geq r_0 - r_1 - 2s.$$

Similarly, for all $z \in D$ we have $|b-a|-s<|z-a|<|b-a|+s$ and so $D \subseteq C\backslash K$. This completes the proof.

DEFINITION 6.2

The *annular radius sum function* $\rho_{ann}:F\to[0,\infty]$ is defined by

$$\rho_{\text{ann}}(A) := \sum_{n=2}^{\infty} r_n \qquad (A = ((a_n, r_n)) \in \mathcal{F})$$

and the *annular discrepancy function* $\delta_{ann}:F\to[-\infty,\infty)$ is given by

$$\delta_{\text{ann}}(A) = r_0 - r_1 - 2\rho_{\text{ann}}(A) \qquad (A = ((a_n, r_n)) \in \mathcal{F}).$$

Note that if $\delta_{ann}(B) > 0$ then $r_0 > r_1$. We aim to prove an analogue of the Feinstein–Heath classicalisation theorem (Theorem 4.1) for annular Swiss cheeses by constructing a suitable compact subset of F.

It is easy for the reader to check that the following analogue of Lemma 4.3 holds for annular Swiss cheeses.

LEMMA 6.3

Let A be an annular Swiss cheese with $\rho_{\text{ann}}(A) < \infty$. Then there exists an annular Swiss cheese $B=((b_n,s_n))$ with the following properties: $\rho_{\text{ann}}(B) \leq \rho_{\text{ann}}(A)$, $X_B=X_A$ and $K_B=K_A$; $\mu(B) < \infty$; the sequence (s_n) $n \geq 2$ is non-increasing; for each $j \in S_B\setminus\{1\}$, we have $B(b_j,s_j) \cap K_B \neq \emptyset$ and, for all $k \in S_B\setminus\{1,j\}$, we have $B(b_j,s_j) \nsubseteq B(b_k,s_k)$. Moreover, for each $E \subseteq C$, we have $\rho_E(B) \leq \rho_E(A)$.

Note that, in the previous lemma, the properties $K_B=K_A$ and $\rho_{\text{ann}}(B) \leq \rho_{\text{ann}}(A)$ together imply that $\delta_{\text{ann}}(B) \geq \delta_{\text{ann}}(A)$.

For the rest of this section, let $A=((a_n,r_n))$ be an annular Swiss cheese with $\delta_{\text{ann}}(A)>0$, such that $\mu(A) < \infty$ and $(r_n)_{n=2}^{\infty}$ is non-increasing.

LEMMA 6.4

Let A be the family of all $B=((b_n,s_n)) \in F$ such that

(a) *the sequence (s_n) $n \geq 2$ is non-increasing,*

(b) $\rho_{\text{ann}}(B) \leq \rho_{\text{ann}}(A)$,

(c) $\mu(B) \leq \mu(A)$,

(d) *B is partially above A, and*

(e) *$b_0=b_1=a_0$, and $r_0 \geq {}_0 \geq s_1 \geq r_1$.*

Then A is compact in F, each abstract Swiss cheese $B \in A$ with $\delta_{\text{ann}}(B) > 0$ is annular. Moreover, the function $\delta_{\text{ann}}|_A : A \to R$ is upper semicontinuous and the function $\delta_2|_A : A \to R$ is continuous.

PROOF

It is easy to see that the family A is pointwise bounded by properties (b), (c) and (e) so it remains only to prove that A is closed. For each $m \in N_0$, let $A^{(m)}=((a_n^{(m)},r_n^{(m)}))_{n=0}^{\infty} \in A$ and suppose that $A^{(m)} \to B \in F$ as $m \to \infty$. It is clear that B satisfies (a)–(d) (as in the proof of Lemma 4.5). Since convergence is pointwise, we have $b_0=a_0$ and $b_1=a_1$. Since A was annular, it follows that $b_0=b_1$.

Since each $A^{(m)} \in A$ we have $r_0 \geq r_0^{(m)} \geq r_1^{(m)} \geq r_1$, by taking $m \to \infty$ we have

$r_0 \geq s_0 \geq s_1 \geq r_1$.

Hence A is closed and pointwise bounded and is therefore compact by Tychonoff's theorem.

Let $B=((b_n,s_n)) \in A$ with $\delta_{\text{ann}}(B) > 0$. Then we have $b_0=b_1$ and $\delta_{\text{ann}}(B) > 0$ and this implies that $s_0 > s_1$ and it follows that B is annular.

The proof that δ_{ann} is upper semicontinuous is an immediate consequence of Fatou's lemma for series, similar to the upper semicontinuity of δ_1.

To prove that the restriction of δ_2 to A is continuous note that, for $n \in \mathrm{N}$ with $n \geq 2$, we have $s_n^2 \leq \rho_{\mathrm{ann}}(B)^2/n^2$ for each $B=((b_n,s_n)) \in A$. The result then follows from the dominated convergence theorem as in the proof of Lemma 2.4.

It is clear that $A \in A$ and so A is non-empty. For all $B \in A$ we also have $X_B \subseteq X_A$. We require one additional lemma before we prove the main theorem.

LEMMA 6.5

Let A be as in Lemma 6.4. Let $B=((b_n,s_n)) \in A$ be an annular Swiss cheese such that $\delta_{\mathrm{ann}}(B) \geq \delta_{\mathrm{ann}}(A)$. Suppose there exists $k \in S_B \setminus \{1\}$such tha $\overline{\mathrm{B}}(b_k,s_k) \cap C \setminus K_B \neq \emptyset$. Then there exists $B'=((b_n',s_n')) \in A$ with $\delta_{\mathrm{ann}}(B') \geq \delta_{\mathrm{ann}}(B)$. Moreover, if $\delta_{\mathrm{ann}}(B')=\delta_{\mathrm{ann}}(B)$ then $\delta_2(B')<\delta_2(B)$.

PROOF

We begin by setting $b'_0=b'_1=b_0$. Then, as in Lemma 6.1, we can find $s'_0>s'_1>0$ such that $K'_B:= \overline{\mathrm{B}}(b'_0,s'_0) \setminus \mathrm{B}(b'_1,s'_1) \subseteq K_B$, $K'_B \cap \mathrm{B}(b_k,s_k)= \emptyset$ and

$$s_0' - s_1' \geq s_0 - s_1 - 2s_k.$$

Let $b'_\ell=b_\ell$ and $s'_\ell=s_\ell$ if $2 \leq \ell< k$, $b'_\ell=b_{\ell+1}$ and $s'_\ell=s_{\ell+1}$ if $k<\ell$, we obtain an abstract Swiss cheese $B'=((b'_n,s'_n))$.

From construction we see B' satisfies Properties (a), (c) and (e). We have

$$\delta_{\mathrm{ann}}(B') = s_0' - s_1' - 2\sum_{n=2}^{\infty} s_n' \geq s_0 - s_1 - 2s_k - 2\sum_{n=2}^{\infty} s_n + 2s_k = \delta_{\mathrm{ann}}(B).$$

Since $s'_0 \leq s_0$ and $s'_1 \geq s_1$ we must have $\rho_{\mathrm{ann}}(B') \leq \rho_{\mathrm{ann}}(B) \leq \rho_{\mathrm{ann}}(A)$, so (b) is satisfied.

We now show that B' is partially above A. Fix $j \in S_A$. If $B(a_j,s_j)$ lies in the complement of $\mathrm{B}(b_0,s_0)$, then it lies in the complement of $\mathrm{B}(b'_0,s'_0)$ and if $\mathrm{B}(a_j,s_j) \subseteq \mathrm{B}(b_1,s_1)$ then $\mathrm{B}(a_j,s_j) \subseteq \mathrm{B}(b'_1,s'_1)$. Suppose there exists $m \in S_B$ such that $\mathrm{B}(a_j,s_j) \subseteq \mathrm{B}(b_m,s_m)$. If $m \neq k$ there exists $\ell \in S_{B'}$ such that $\mathrm{B}(b'_\ell,s'_\ell)=\mathrm{B}(b_m,s_m)$, and so $\mathrm{B}(a_j,s_j) \subseteq \mathrm{B}(b'_\ell,s'_\ell)$. If $m=k$ then either $\mathrm{B}(a_j,r_j) \subseteq \mathrm{B}(b'_1,s'_1)$ or $\mathrm{B}(a_j,s_j)$ lies in the complement of $\mathrm{B}(b'_0,s'_0)$. It follows that B′ is partially above A, and satisfies 4 and hence $B' \in A$. Since we have $\delta_{\mathrm{ann}}(B') \geq \delta_{\mathrm{ann}}(A) > 0$, it follows that B' is annular.

It remains to show that if $\delta_{\mathrm{ann}}(B')=\delta_{\mathrm{ann}}(B)$ then $\delta_2(B')<\delta_2(B)$. Assume that $\delta_{\mathrm{ann}}(B')=\delta_{\mathrm{ann}}(B)$. Then either $s_0=s'_0+2s_k$ or $s'_1=s_1+2s_k$. In the first case we have $(s_0')^2 < s_0^2 - 4s_k^2 < s_0^2 - s_k^2$ and in the second case we have $(s_1')^2 > s_1^2 + s_k^2$.

In the first case we have $s_0^2 > (s_0')^2 + s_k^2$, and in the second case we have $(s_1')^2 > s_1^2 + s_k^2$. In either case, we have $\delta_2(B') < \delta_2(B)$. This completes the proof.

Note that, as for arbitrary abstract Swiss cheeses, if B is a semiclassical, annular Swiss cheese then $\pi\delta_2(B)$ is the area of X_B.

THEOREM 6.6

Let A be as in Lemma 6.4. Then there exists a classical, annular Swiss cheese $B=((b_n,s_n))$ in A such that $\delta_{\text{ann}}(B) \geq \delta_{\text{ann}}(A)$ and $X_B \subseteq X_A$. Moreover, we have $r_0-2\rho_{\text{ann}}(A) \leq s_0 \leq r_0$ and $r_1 \leq s_1 \leq r_1+2\rho_{\text{ann}}(A)$.

PROOF

Since δ_{ann} is upper semicontinuous on A and A is compact and non-empty, it follows that δ_{ann} achieves its maximum on A. Let A_1 denote the non-empty, compact subset of A on which δ_{ann} is maximised. Then δ_2, which is continuous on A_1, achieves its minimum. Let A_2 denote the non-empty, compact subset of A_1 on which δ_2 is minimised and let $B=((b_n,s_n)) \in A_2$.

Since $\delta_{\text{ann}}(B) \geq \delta_{\text{ann}}(A) > 0$ it follows that B is annular and $X_B \subseteq X_A$. Suppose, for contradiction, that B is non-classical. There are two possible cases.

First suppose that there are $k, \ell \in S_B \backslash \{1\}$ with $k > \ell$ such that $k, \ell \in S_B$ and $\overline{\mathrm{B}}(b_k,s_k) \cap \overline{\mathrm{B}}(b_\ell,s_\ell) \neq \emptyset$. Then, by Lemma 3.2 there exist $b \in C$ and $s > 0$ such that

$$\mathrm{B}(b_k,s_k) \cup \mathrm{B}(b_\ell,s_\ell) \subseteq \mathrm{B}(b,s)$$

and $s \leq s_k + s_\ell$. Let $B'=((b'_n,s'_n))$ be the abstract Swiss cheese obtained by deleting the disks at indices k,ℓ from B and inserting the disk $\mathrm{B}(b,s)$ at the first index in $\mathrm{N}\backslash\{1\}$ such that $(s'_n)_{n=2}^{\infty}$ is non-increasing. It is easy to see that $B' \in A$ and

$$\rho_{\text{ann}}(B) \geq \rho_{\text{ann}}(B) - s_k - s_\ell + s = \rho_{\text{ann}}(B'), \tag{6}$$

so that $\delta_{\text{ann}}(B') \geq \delta_{\text{ann}}(B)$. By the maximality of $\delta_{\text{ann}}(B)$, equality must hold here and in (6). Thus $s=s_k+s_\ell$ and $s^2 = (s_k+s_\ell)^2 > s_k^2 + s_\ell^2$ so that $\delta_2(B')<\delta_2(B)$. This contradicts the minimality of $\delta_2(B)$. It follows that no such k,ℓ exist.

Now suppose there exists $k \in S_B \backslash \{1\}$ such that $\overline{\mathrm{B}}(b_k,s_k) \cap \overline{\mathbb{C} \setminus K_B} \neq \emptyset$ and $s_k>0$. By Lemma 6.5 there exists an annular Swiss cheese $B' \in A$ with $\delta_{\text{ann}}(B') \geq \delta_{\text{ann}}(B)$ such that, if $\delta_{\text{ann}}(B')=\delta_{\text{ann}}(B)$ then $\delta_2(B')<\delta_2(B)$. This is a contradiction, so no such k can exist. It follows that B is classical.

Since B ∈ A, we have $r_0 \geq s_0 \geq s_1 \geq r_1$. We also have

$$s_0 - s_1 \geq \delta_{\text{ann}}(B) \geq \delta_{\text{ann}}(A) = r_0 - r_1 - 2\rho_{\text{ann}}(A)$$

so that

$$s_0 \geq r_0 - 2\rho_{\text{ann}}(A) - (r_1 - s_1) \geq r_0 - 2\rho_{\text{ann}}(A)$$

and $s_1 \leq r_1 + 2\rho_{\text{ann}}(A) - (r_0 - s_0) \leq r_1 + 2\rho_{\text{ann}}(A)$ This completes the proof.

7. REGULARITY OF *R*(*X*)

Let X be a compact plane set. We say that $R(X)$ is *regular* if, for all closed sets $E \subseteq X$ and points $x \in X \backslash E$, there exists a function $f \in R(X)$ such that $f(x)=1$ and $f(y)=0$ for all $y \in E$. We say that $R(X)$ is *normal* if, for each pair of disjoint closed sets $E, F \subseteq X$, there exists a function $f \in R(X)$ such that $f(x)=0$ for all $x \in E$ and $f(x)=1$ for all $x \in F$. It is standard that $R(X)$ is regular if and only if it is normal (see [2, Proposition 4.1.18]).

In order to avoid ambiguity, we introduce the following notation to clarify in which topological space we are taking the interior. Let X be a compact plane set and $E \subseteq X$. Then $\text{int}_X E$ denotes the interior of E in the topological space X.

DEFINITION 7.1

Let X be a compact plane set, and let $x \in X$. We denote by M_x the ideal of all functions in $R(X)$ which vanish at x. We denote by J_x the ideal of all functions in $R(X)$ which vanish on a neighbourhood of x. We say x is an R-point for $R(X)$ if, for all $y \in X$ with $y \neq x$, we have $J_x \nsubseteq M_y$.

It is standard that $R(X)$ is regular if and only if every point $x \in X$ is an R-point of $R(X)$. The following proposition is a special case of [7, Corollary 4.7].

PROPOSITION 7.2

Let X be a compact plane set such that $R(X)$ is not regular. Let E denote the set of non-R-points for $R(X)$. Then E contains a non-empty perfect subset. In particular, E is uncountable.

Our classicalisation theorems involve finding "good" compact subsets of a given compact plane set. The following proposition, stated in [6], lists some properties of $R(X)$ which are inherited when a subset of X is considered.

PROPOSITION 7.3

Let X and Y be compact plane sets with $Y \subseteq X$. Then:

(a) *if $R(X)=C(X)$ then $R(Y)=C(Y)$;*

(b) *if $R(X)$ does not have any non-zero bounded point derivations then neither does $R(Y)$;*

(c) *if $R(X)$ is regular then so is $R(Y)$.*

In this section, we prove some results about regularity of $R(X)$ which we shall require for the construction in the final section. The following proposition is essentially [8, Corollary II.10.3].

PROPOSITION 7.4

Let X be a compact plane set and let $f \in C(X)$. Suppose that for each $x \in X$ there is a closed neighbourhood N_x of x in X such that $f|_{N_x} \in R(N_x)$. Then $f \in R(X)$.

We shall require the following theorem.

THEOREM 7.5

Let X be a compact plane set and let E be a countable subset of X. Let (X_α) be a family of compact plane sets such that $R(X_\alpha)$ is regular for all α and $\bigcup_\alpha \operatorname{int}_X(X \cap X_\alpha) \supseteq X \backslash E$. Then $R(X)$ is regular.

PROOF

We first show that every point in $X \backslash E$ is an R-point for $R(X)$. Let $x \in X \backslash E$ and $y \in X$ with $x \neq y$. Then there exist α and $r > 0$ such that $\overline{B}(x,r) \cap X \subseteq X_\alpha$. Let $\delta < r/3$ such that $|x-y| > 2\delta$. Let F denote the complement of $X \cap B(x,2\delta)$ in $\overline{B}(x,r) \cap X$. Since $X \cap \overline{B}(x,r) \subseteq X_\alpha$, $R(X \cap \overline{B}(x,r))$ is regular by Proposition 7.3 (and hence normal). Thus there exists a function $g \in R(X \cap \overline{B}(x,r))$ with $g(z)=0$ for all $z \in X \cap \overline{B}(x,\delta)$ and $g(z)=1$ for all $z \in F$. Extend g to a function $f \in C(X)$ by setting $f(z)=g(z)$ for all $z \in (\overline{B}(x,r) \cap X)$ and $f(z)=1$ for all $z \in X \backslash \overline{B}(x,r)$. Clearly f satisfies the conditions of Proposition 7.4, so $f \in R(X)$. By our choice of δ, we have $x \in U$ and $y \in F$ so f vanishes on a neighbourhood of x and $f(y)=1$, so x is an R-point for $R(X)$.

It follows that $R(X)$ has at most countably many non-R-points. So, by Proposition 7.2, $R(X)$ is regular.

Note that we do not assume that $X_\alpha \subseteq X$. However, replacing X_α by $X \cap X_\alpha$ does not alter the result. We obtain the following corollaries.

COROLLARY 7.6

Let X be a compact plane set and $x_0 \in X$. Let (X_α) be a family of compact plane sets such that $\bigcup_\alpha \operatorname{int}_X(X \cap X_\alpha) = X \backslash \{x_0\}$ and $R(X_\alpha)$ is regular for all α. Then $R(X)$ is regular.

COROLLARY 7.7

Let X_1, X_2 be compact plane sets such that $X_1 \cap X_2$ is countable. If $R(X_1)$ and $R(X_2)$ are regular then $R(X_1 \cup X_2)$ is regular.

8. CLASSICALISATION OF AN EXAMPLE OF O'FARRELL

In this section we see an application of the results of Sections 5–7. In [5], O'Farrell modified the construction of McKissick [11] to construct a Swiss cheese set X such that $R(X)$ is regular and admits a non-degenerate bounded point derivation of infinite order (defined below). However, this Swiss cheese set is not necessarily classical.

DEFINITION 8.1

Let X be a compact plane set and let $x \in X$. A *point derivation of order* $n \in \mathbb{N}$ (respectively, ∞) at x (on $R(X)$) is a sequence $d_0, d_1, \ldots$ of linear functionals with $d_0 = \varepsilon_x$, the evaluation character at x, satisfying

$$d_j(fg) = \sum_{k=0}^{j} d_k(f) d_{j-k}(g) \qquad (f, g \in R(X)),$$

for all $j = 1, 2, \ldots, n$ (respectively, $j = 1, 2, \ldots$).

Let $\mathrm{d} = (d_j)_{j=0}^{n}$ be a point derivation of order n at x (where we include the possibility that $n = \infty$ when (d_j) is a point derivation of infinite order). We say that **D** is *bounded* if d_j is a bounded linear functional for each j with $j \leq n$ (respectively, all j). We say that **D** is *non-degenerate* if $d_1 \neq 0$.

We refer the reader to [4] (especially Lemma 2.1 and p. 170) for further details, related results and comments concerning non-degenerate higher point derivations.

Following our general scheme of classicalisation, we construct a classical Swiss cheese set X such that $R(X)$ is regular and admits a non-degenerate bounded point derivation of infinite order at one of the points of X.

The following proposition is an immediate corollary of Proposition 7.3 and the result of McKissick [6, Proposition 1.10] (see also [11] and [5]).

PROPOSITION 8.2

Let $b_0=b_1\in$ C, lets $s_0> s_1 > 0$, and let $\varepsilon > 0$. Then there exists an annular Swiss cheese $A=((a_n,r_n))$with $\rho_{\text{ann}}(A) < \varepsilon, a_j=b_j$ and $r_j=s_j$ for $j =0,1$, and such that $R(XA)$is regular.

We now use a sequence of lemmas to show that we can construct a classical annular Swiss cheese with the same properties as those in Proposition 8.2.

LEMMA 8.3

Let $\lambda_0> \lambda_1>0$ and $\varepsilon,\eta >0$ be given and let $a \in$ C. There exists a classical, annular Swiss cheese $B =((b_n,s_n))$with $b_0=b_1=$ a such that $\rho_{\text{ann}}(B) <\varepsilon ,\lambda_0 \leq s_0 \leq \lambda_0+ \eta$ and $\lambda_1-\eta\leq s_1\leq \lambda 1$such that $R(XB)$is regular.

PROOF

We may assume that $\eta<\lambda_1$ and $\varepsilon \leq \eta/2$. Let $A=((a_n,r_n))$ be an abstract Swiss cheese obtained from Proposition 8.2 with $a_0=a_1=a$, $r_0=\lambda_0+\eta$, $r_1=\lambda_1=\eta$ and $\rho_{\text{ann}}(A)< \varepsilon$ and such that $R(XA)$ is regular. By Lemma 6.3, we may assume that the sequence $(r_n)_n=2\infty$ is non-increasing. Apply Theorem 6.6 to the abstract Swiss cheese A to obtain a classical, annular Swiss cheese $B=((b_n,s_n))$ with

$$b_0 = b_1 = a_0 = a_1 = a, \quad r_0 - 2\varepsilon \leq s_0 \leq r_0, \quad \text{and} \quad r_1 \leq s_1 \leq r_1 + 2\varepsilon,$$

such that $\rho_{\text{ann}}(B)\leq\rho_{\text{ann}}(A)$ and $X_B \subseteq X_A$. By Lemma 7.3, $R(XB)$ is regular. Since $2\varepsilon \leq \eta$, we have $\lambda_0\leq s_0\leq\lambda_0+\eta$ and $\lambda_1-\eta \leq s_0\leq \lambda_1$. This completes the proof. □

By instead taking $r_0=\lambda_0$ and $r_1=\lambda_1$ in the proof of the previous lemma we see that we could also approximate the desired annulus with a smaller annulus, rather than a larger annulus as in Lemma 8.3.

In the next lemma, we see how to obtain a classical, annular Swiss cheese A such that $R(XB)$ is regular with one of the first two radii specified exactly while prescribing tight bounds on the other.

LEMMA 8.4

Let$\lambda_0>\lambda_1>0$ and $\varepsilon,\eta > 0$be given and let $a \in$ C.

(a)*There exists a classical, annular Swiss cheese $B^{(1)} = ((b_n^{(1)}, s_n^{(1)}))$ with $s_0^{(1)} = \lambda_0,\ \lambda_1 - \eta \leq s_1^{(1)} \leq \lambda_1$ and $b_0^{(1)} = b_1^{(1)} = a$ such that $R(X_{B^{(1)}})$ is regular and $\rho_{\text{ann}}(B^{(1)})<\varepsilon$.*

(b)*There exists a classical, annular Swiss cheese $((b_n^{(2)}, s_n^{(2)}))$with $b_0^{(2)} = b_1^{(2)} = a$such that $\rho_{\text{ann}}(B^{(2)}) < \varepsilon,\ \lambda_0 \leq s_0^{(2)} \leq \lambda_0 + \eta \text{ and } s_1^{(2)} = \lambda_1$ and $s_1^{(2)} = \lambda_1$ and such that $R(X_{B^{(2)}})$ is regular.*

PROOF

We prove (b); the proof of (a) is similar but easier. We may assume, without loss of generality, that $a=0$. Let $\gamma \in (0,\lambda_1)$ to be chosen later. Apply Lemma 8.3 to obtain a classical, annular Swiss cheese $A=((a_n,r_n))$ such that $\rho_{\text{ann}}(A)< \gamma/2$, $\lambda_0 \le r_0 \le \lambda_0+\gamma$ and $\lambda_1-\gamma \le r_1 \le \lambda_1$ and such that $R(XA)$ is regular.

For each $n \ge 0$ let $b_n^{(2)}:=\lambda_1 a_n/r_1$ and $s_n^{(2)}:=\lambda_1 r_n/r_1$. Then

$$\lambda_0 \le s_0^{(2)} = \frac{\lambda_1}{r_1} r_0 \le M_\gamma r_0 \le M_\gamma(\lambda_0+\gamma)$$

.

Set $M_\gamma:=\lambda_1/(\lambda_1-\gamma) > 1$. We have $s_1^{(2)} = \lambda_1$ and $s_0^{(2)} \ge \lambda_0$ so it remains to show that $s_0^{(2)} \le \lambda_0 + \eta$ provided that γ is sufficiently small. We have

$$\lambda_0 \le s_0^{(2)} = \frac{\lambda_1}{r_1} r_0 \le M_\gamma r_0 \le M_\gamma(\lambda_0+\gamma)$$

.

Since $M_\gamma \to 1$ as $\gamma \to 0$, if γ is small enough then we have $\lambda_0 \le s_0^{(1)} \le \lambda_0 + \eta$ and $M_\gamma \gamma/2 < \varepsilon$. Clearly $R(X_{B^{(2)}})$ is regular since $R(X_A)$ is regular. This completes the proof of the (b).

Our final lemma shows that we can obtain this type of annular Swiss cheese with s_0 and $s_1$precisely prescribed.

LEMMA 8.5

Let $\lambda_0 > \lambda_1 > 0$ and $\varepsilon > 0$ be given and let $a \in \mathbb{C}$. There exists a classical, annular Swiss cheese $B=((b_n,s_n))$with $b_0=b_1=a, s_0=\lambda_0, s_1=\lambda_1, \rho_{\text{ann}}(B)< \varepsilon$ and such that $R(X_B)$is regular.

PROOF

Let $\kappa=(\lambda_0+\lambda_1)/2$ and let $\eta > 0$. By Lemma 8.4(a) there exists a classical, annular Swiss cheese $A^{(1)} = ((a_n^{(1)}, r_n^{(1)}))$ with $a_0^{(1)} = a_1^{(1)} = a$, $\rho_{\text{ann}}(A^{(1)}) < \eta/16$, $r_0^{(1)} = \lambda_0$ and $\kappa - \eta/4 \le r_1^{(1)} \le \kappa - \eta/8$ and such that $R(X_{A^{(1)}})$ is regular. By Lemma 8.4(b) there exists a classical, annular Swiss cheese $A^{(2)} = ((a_n^{(2)}, r_n^{(2)}))$ with $a_0^{(2)} = a_1^{(2)} = a$, $\rho_{\text{ann}}(A^{(2)}) < \eta/16$, $r_1^{(2)} = \lambda_1$ and $\kappa \le r_0^{(2)} \le \kappa + \eta/4$ and such that $R(X_{A^{(2)}})$ is regular.

Let $(a_n^{(3)})_{n\ge 2}$ be a sequence containing all elements from the sequences $(a_n^{(1)})_{n\ge 2}$, and $(a_n^{(2)})_{n>2}$ exactly once and let $(r_n^{(3)})_{n\ge 2}$ be the corresponding sequence containing all elements from the sequences $(r_n^{(1)})_{n\ge 2}$, and $(r_n^{(2)})_{n\ge 2}$ exactly once. Let $a_0^{(3)} := a$, $a_1^{(3)} := a$ and $r_0^{(3)} := \lambda_0$, $r_1^{(3)} := \lambda_1$ and let $A^{(3)} = ((a_n^{(3)}, r_n^{(3)}))$ be the corresponding annular Swiss cheese. Then

$$\rho_{\mathrm{ann}}(A^{(3)}) = \sum_{n=2}^{\infty} r_n^{(3)} < \frac{\eta}{16} + \frac{\eta}{16} = \frac{\eta}{8}$$

.

Let $X := X_{A^{(3)}}$, then we can easily check that $X = \operatorname{int}_X X_{A^{(1)}} \cup \operatorname{int}_X X_{A^{(2)}}$ so, by Theorem 7.5, $R(X)$ is regular.

Choose $\eta>0$ small enough so that we have $\eta<(\lambda_0-\lambda_1)/4$, $\eta/2 < \varepsilon$ and $\eta<\lambda_1$. Let $M:=\eta/4$ and let $K:=\{z \in \mathbb{C}:\kappa-\eta/4\leq|z|\leq\kappa+\eta/4\}$. Let $A=((a_n,r_n)) \in N$ be obtained by applying Lemma 6.3. Then, for each open $U \subseteq \mathbb{C}$, $\rho_U(A) \leq \rho_U(A^{(3)})$. It is now easy to see that A,K and M satisfy the conditions of Corollary 5.8. Note that $U:=U(K,M)$ has $U\cap \overline{B}(a_1,r_1)=\emptyset$. Note that $X_A=X=X_{A^{(3)}}$.

Let $B=((b_n,s_n))$ be the classical abstract Swiss cheese obtained by applying Corollary 5.8 to A,K and M. Then $B \in L^*_A(C)$, where C is the controlling pair (K,U). Thus, by Lemma 5.5, there exists $\ell \in S_B$ such that $B(b_\ell,s_\ell)=B(a_1,r_1)$. Since the sequence (s_n) $n\geq1$ is non-increasing and $\rho_{\mathrm{ann}}(A)< r_1$, it follows that $\ell=1$. It follows that B is annular and has $s_0=\lambda_0$, $s_1=\lambda_1$ and $\rho_{\mathrm{ann}}(B)< \varepsilon$. Since $R(X_A)$ is regular, by Proposition 7.3, $R(X_B)$ is regular. This completes the proof.

We are now ready to construct a classical Swiss cheese set X such that $R(X)$ is regular and admits a non-degenerate bounded point derivation of infinite order.

THEOREM 8.6

Let $\varepsilon >0$. Then there exists a classical abstract Swiss cheese $B=((b_n,s_n))$ with $0 \in X_B$ and $\rho(B)<\varepsilon$ and such that $R(X_B)$ is regular and admits a non-degenerate bounded point derivation of infinite order at 0.

PROOF

We may assume that $\varepsilon < 2^{-5}$. For each $n \in \mathbb{N}$, let $\gamma_n=(2n)^{-n}\varepsilon$. Note that $\sum_{n=1}^{\infty} \gamma_n < \varepsilon$ Let $A^{(1)} = ((a_n^{(1)}, r_n^{(1)}))$ be a classical, annular Swiss cheese, given by Lemma 8.5, with $a_0^{(1)} = 0,\ r_0^{(1)} = 1,\ r_1^{(1)} = 2^{-1}$ and $\rho_{\mathrm{ann}}(A^{(1)})<\gamma 1 2^{-3}$, such that $R(X_1)$ is regular, where $X_1 := X_{A^{(1)}}$. For each $m \geq 2$, let $A^{(m)} = ((a_n^{(m)}, r_n^{(m)}))$ be a classical, annular Swiss cheese, given by Lemma 8.5, with the properties that $a_0^{(m)} = 0,\ r_0^{(m)} = (33/32)2^{1-m},\ r_1^{(m)} = 2^{-m}$, $\rho_{\mathrm{ann}}(A^{(m)})<\gamma m 2^{-m-2}$, and $R(X_m)$ is regular, where $X_m := X_{A^{(m)}}$. Note that, for each $m \in \mathbb{N}$, since $A^{(m)}$ is annular we have $a_1^{(m)} = 0$.

For each $m \in N$ we have

$$\rho_{\mathrm{ann}}(A^{(m)}) + \rho_{\mathrm{ann}}(A^{(m+1)}) \leq \frac{\gamma_m}{2^{m+2}} + \frac{\gamma_{m+1}}{2^{m+3}} < \frac{3}{2}\frac{\gamma_m}{2^{m+2}} \tag{7}$$

The set $\{(a_n^{(m)}, r_n^{(m)}) : m, n \in \mathbb{N}, n \geq 2\}$ is countably infinite so we may enumerate it as a sequence of pairs (a_n, r_n), so that each pair occurs exactly once. Let $a_0 := 0$ and $r_0 := 1$ and let $A = ((a_n, r_n))$ be the resulting abstract Swiss cheese. It is clear that $0 \in X_A$. It is easy to check that

$$\bigcup_{m=1}^{\infty} \mathrm{int}_{X_A} X_m = X_A \setminus \{0\}$$

so that $R(X_A)$ is regular by Corollary 7.6.

Using the notation for closed annuli from Section 6, for each $m \in \mathrm{N}$, let $W_m = K_{A(m)} \cup K_{A(m+1)}$. Then, by (7), we see that

$$\rho_{W_m}(A) < (3/2)\gamma_m 2^{-m-2} \tag{8}$$

We also have

$$\rho(A) < \sum_{m=1}^{\infty} 2^{-m-2}\gamma_m < \varepsilon \tag{9}$$

By an application of Lemma 4.3, we may assume that A is redundancy-free and $A \in N$ while preserving X_A, the regularity of $R(X_A)$ and the inequalities (8) and (9).

For each $m \in \mathrm{N}$, let $K_m := \{z \in \mathrm{C} : (15/16)2^{-m} \leq |z| \leq (17/16)2^{-m}\}$ and $M_m = 3\gamma_m/2^{m+2}$ so that, since $\gamma_m < 1$, we have

$$2^{-m-4} + M_m < 2^{-m-4}(1 + 3 \cdot 2^{-m}) \leq (5/32)2^{-m} \tag{10}$$

For each $m \in \mathrm{N}$, define $U_m := U(K_m, M_m)$ as in the statement of Theorem 5.7, and let C be the controlling collection $((K_n, U_n))_{n \in N}$. By (10)

$$U_m \subseteq \left\{ z \in \mathbb{C} : \frac{27}{32} 2^{-m} \leq |z| \leq \frac{37}{32} 2^{-m} \right\}$$

and hence $U_m \cap U_n = \emptyset$ for all $m, n \in \mathrm{N}$ with $m \neq n$. Clearly $E(A) \subseteq F(C)$. Since $U_m \subseteq W_m$, for each $m \in \mathrm{N}$, we see that

$$\rho_{U_m}(A) \leq \rho_{W_m}(A) < \frac{3}{2}\frac{\gamma_m}{2^{m+2}}.$$

This shows that the sequences $(K_n)_{n\geq 1}$ and $(M_n)_{n\geq 1}$ satisfy the conditions of Theorem 5.7.

Applying Theorem 5.7 to A, $(K_n)_{n\geq 1}$ and $(M_n)_{n\geq 1}$, there exists an abstract Swiss cheese $B=((b_n,s_n))$ in $L^*_A(C)$ with $b_0=0$, $s_0=1$, $X_B \subseteq X_A$ and $\rho(B)< \varepsilon$ such that B is classical and $X_B\setminus V(C)=X_A\setminus V(C)$; in particular $0 \in X_B$. Also, we have $\rho_{Um}(B) \leq \rho_{Um}(A)$ for each $m \in$ N. By Proposition 7.3, $R(X_B)$ is regular.

For each $m \in N_0$, let $E_m:=\{z \in C:(3/2)2^{-m-1} \leq |z| \leq (3/2)2^{-m}\}$, then $U_m \subseteq E_m$ and $U_j \cap E_m=\emptyset$ for all $j\neq m$. For each $m \in$N, let $I_A(m)$ denote the set of all $k \in H_A(E_m)\setminus H_A(V(C))$ and let $I_B(m)$ denote the set of all $k \in H_B(E_m)\setminus H_B(V(C))$. Note that, $\rho_{Em}(A) \leq \rho_{Wm}(A) < (3/2)\gamma_m 2^{-m-2}$. Since $\varepsilon<2^{-5}$, it follows that, if $k \in H_{Um}(A)$, we have

$$\bar{\mathrm{B}}(a_k,r_k)\subseteq\left\{z\in\mathbb{C}:\frac{3}{4}2^{-m}<|z|<\frac{5}{4}2^{-m}\right\}\subseteq \operatorname{int} E_m$$

.

Hence $H_A(E_0)=I_A(0)$ and $H_B(E_0)=I_B(0)$ and, for all $m \geq 1$, we have $I_A(m)=H_A(E_m)\setminus H_A(U_m)$ and $I_B(m)=H_B(E_m)\setminus H_B(U_m)$. Now since B $\in L^*$A (C) we see that $\sum_{n\in I_B(m)} s_n = \sum_{n\in I_A(m)} r_n$, for all $m \in$ N, by Lemma 5.5. Hence $\rho_{E_0}(B)=\rho_{E_0}(A)$ and for $m \geq 1$ we have

$$\rho_{E_m}(B)=\sum_{n\in I_B(m)} s_n+\rho_{U_m}(B)\leq \sum_{n\in I_A(m)} r_n+\rho_{U_m}(A)=\rho_{E_m}(A)$$

.

Since $\rho_{Em}(A) \leq \rho_{Wm}(A)<\gamma_m$, for each $m \in$ N, it now follows that

$$\sum_{m=1}^{\infty} m^m\rho_{E_m}(B)\leq\sum_{m=1}^{\infty} m^m\frac{\varepsilon}{2^m m^m}=\varepsilon<\infty$$

.

Since each disk meets at most two of the E_m, $R(X_B)$ admits a non-degenerate bounded point derivation of infinite order at 0 by Hallstrom's theorem [9] (see also [5]□).

We raise the following open question related to regularity and bounded point derivations.

QUESTION 8.7

Let X be a compact plane set such that R(X) has no non-zero bounded point derivations. Is R(X)necessarily regular?

ACKNOWLEDGMENT

We would like to thank the referee for helpful comments.

REFERENCES

1. Browder **introduction to function algebras** w.a. benjamin, inc., new york–amsterdam (1969)
2. H.g. dales **banach algebras and automatic continuity** london mathematical society monographs. New series, vol. 24, the clarendon press oxford university press, new york (2000)
3. H.g. dales, j.f. Feinstein **normed algebras of differentiable functions on compact plane sets** indian j. Pure appl. Math., 41 (1) (2010), pp. 153-187
4. H.g. dales, j.p. mcclure **higher point derivations on commutative banach algebras. I** j. Funct. Anal., 26 (2) (1977), pp. 166-189
5. A.g. o'farrell **a regular uniform algebra with a continuous point derivation of infinite order** bull. Lond. Math. Soc., 11 (1) (1979), pp. 41-44
6. J.f. feinstein, m.j. heath**swiss cheeses, rational approximation and universal plane curves** studia math., 196 (3) (2010), pp. 289-306
7. J.f. feinstein, d.w.b. somerset **non-regularity for banach function algebras** studia math., 141 (1) (2000), pp. 53-68
8. T.w. gamelan **uniform algebras** prentice-hall inc., englewood cliffs, n.j. (1969)
9. A.p. hallstrom**on bounded point derivations and analytic capacity** j. Funct. Anal., 4 (1969), pp. 153-165
10. J.w.d. mason **an inductive proof of the feinstein–heath swiss cheese "classicalisation" theorem** proc. Amer. Math. Soc., 138 (12) (2010), pp. 4423-4432
11. R. Mckissick**a nontrivial normal sup norm algebra** bull. Amer. Math. Soc., 69 (1963), pp. 391-395
12. Roth **approximationseigenschaften und strahlengrenzwerte meromorpher und ganzer funktionen** comment. Math. Helv., 11 (1) (1938), pp. 77-125
13. E.l. stout **the theory of uniform algebras** bogden & quigley, inc., tarrytown-on-hudson, n.y. (1971)
14. G.t. whyburn **topological characterization of the sierpiński curve** fund. Math., 45 (1958), pp. 320-324

CHAPTER 10

Mathematical analysis of drug resistance in vertical transmission of HIV/AIDS

Symon Chibaya[1], Moatlhodi Kgosimore[2], Estomih S. Massawe[3*]

[1]*Basic Sciences Programme, Natural Resources College of Malawi, Lilongwe, Malawi*

[2]*Department of Basic Sciences, Botswana College of Agriculture, Gaborone, Botswana*

[3]*Mathematics Department, University of Dar es Salaam, Dar es Salaam, Tanzania*

ABSTRACT

A nonlinear mathematical model of vertical transmission of HIV/AIDS is proposed to study the effects of drug resistance in the spread of the disease. The study assumes that treatment leads to the evolution of drug resistance in some pockets of the population. We use traditional methods to determine conditions for existence and stability of disease-free and endemic equilibrium points of the model. The study showed that the burden of the disease may be reduced if the reproduction number is reduced below unity and may persist if the reproduction number is raised above unity. Furthermore, evolution of drug resistance due to treatment may change the cause of the epidemic.

Keywords: Dynamics of HIV/AIDS; Vertical Transmission; Drug Resistance

1. INTRODUCTION

Diseases can be transmitted in various ways, some of which can be classified as horizontal or vertical. In the case of HIV/AIDS, horizontal transmission can result from direct physical contact between an infected individual and a susceptible individual. In this case, HIV infection is from the infected blood, semen and vaginal fluids to the exposure individuals. This often occurs when having unprotected sex with a person who has HIV. The other risks of HIV infection comprise of sharing needles, syringes or other equipment, illicit drugs for injection or during injection drug use (when needles are shared) with an infected HIV person. Any HIV infected individual can transmit the disease irrespective of the stage or treatment status. However, high risks of infections are in the early stages of infection and during full blown AIDS stages. These stages are associated with high viral loads.

Vertical transmission, on the other hand, results from direct transfer of a disease from an infected mother to an unborn or newly born offspring. Some of the diseases that can be transmitted vertically include chagas, dengue fever and hepatitis B. Vertical transmission of HIV/ AIDS generally occurs during pregnancy, delivery or breastfeeding and may be influenced by many factors such as maternal viral load and the type of delivery [1,2]. This type of transmission is sometimes called motherto-child transmission or ethically parent-to-child transmission. Research has shown that, not all pregnancies in HIV-infected women result in vertical transmission.

Pediatric HIV is a major contributor to high mortality rates among infants and children in sub-Saharan Africa and it remains a leading cause of deaths in children under the age of 5 years. It is estimated that one third of HIV positive infants die before their first birthday, while 50% die by their second birthday. Consequences of vertical transmission include reduction of life expectancy of HIV positive individuals [3]. Worldwide, one percent of pregnant women are HIV-positive and 95% of these HIV positive women live in sub-Saharan Africa. In the absence of treatment, approximately 25% - 50% of HIVpositive mothers will transmit the virus to their newborns during pregnancy, childbirth, or breastfeeding [4]. In sub-Saharan Africa, over 1000 newborns are infected with HIV every day, despite available medical interventions. The high rates of vertical transmission in Africa are attributed to low rates accessibility of intervention programmes and lack of adherence to the necessary medications to prevent mother-to-child transmission.

Prevention of mother-to-child transmission of HIV (PMTCT) can dramatically reduce the risk of vertical transmission. Success stories have been reported in the United States and Europe and it remains a serious challenge to Africa [3]. Prevention of mother-to-child transmission involves the use of antiretroviral therapy (ART), caesarean section [5,6] and refraining from breastfeeding. Antiretroviral drugs are administered during pregnancy, at delivery and postnatally to the child and mothers. In resource-poor countries, short course regimens based on single dose nevirapine to the mother and infant are being used [7]. In summary, the burden of vertical transmission of HIV can be reduced by enrolling parentto-child prevention programs which involve testing the mothers and blocking of transmission through the use of antiretroviral drugs, elective caesarean section and use of replacement infant feeding [8].

Drug resistance refers to the ability of microorganisms to adapt, survive and multiply in the presence of drugs that would normally kill or weaken them. In the case of HIV, the virus adapts and multiplies in the presence of antiretroviral (ARV) drugs often making mistakes of copying its genetic blueprint RNA. Some of these mistakes, called mutations, can make HIV resistant to one or more ARV drugs. With regard to vertical transmission, we are concerned with the potential selection of drug– resistant virus following short course of ARV treatment to prevent parent-to-child transmission (PPCT). Reports show that up to 20% of women who receive a single dose of nevirapine in HIVNET regime developed resistance [9] and this leads to the development of strains of HIV resistant to multiple combinations of ARVs.

This study therefore extends mathematical models of vertical transmission with treatment by incorporating the effects of drug resistance. The main objective of the study is to investigate the effects of vertical transmission and evolution of drug resistance in the HIV disease dynamics. The paper is arranged as follows: Section 1 introduces the study, model formulation is presented in Section 2, Section 3 presents model analysis, Section 4 presents numerical simulations while Section 5 discusses the results and presents the conclusions of the study.

2. MODEL FORMULATION

A mathematical model of a disease transmitted both horizontally and vertically in which the treatment programme results in the evolution of drug resistance is proposed and analysed. The proposed model sub-divides the population of interest into seven compartments, depending on their disease status and type of pathogen (i.e. resistant or wild strain). The sub-populations are the Susceptibles $S(t)$, drug resistant infectives I_R, drug sensitive infectives I_w, treated drug sensitive infective T_w, treated drug resistant infective T_R, AIDS drug sensitive A_w and AIDS drug resistant class A_R. The total population at time t is given by

$$N(t) = S(t) + I_w(t) + I_R(t) + T_w(t) + T_R(t) + A_w(t) + A_R(t) \tag{1}$$

To reduce the size of the model we relaxed the distinction between the asymptomatic (no clinical signs) and symptomatic (clinical signs) stages of HIV (WHO staging of HIV). We assumed mass action type of force of infection as opposed to standard incidence to allow for mathematical tractability of the model. This assumption is an over simplification of reality particularly highly infectious diseases like HIV. We further assume that treatment only reduces the level of infectiousness. **Table 1** below presents the definitions and descriptions of the parameters and state variables used in the model. It is worth stating that all model parameters non-negative and all have values in the open set (0, ∞) except ρ which picks values in the set [1, ∞).

The problem described above can be represented schematically by the flow diagram below (**Figure 1**):

The model description above leads to a system of nonlinear ordinary differential equations:

$$\begin{aligned}
\frac{dS}{dt} &= \pi_s + b\varepsilon_w I_{wf} + b\varepsilon_R I_{Rf} - \lambda_w S - \lambda_R S - \mu S,\\
\frac{dI_w}{dt} &= \lambda_w S + b\tau_w \varepsilon'_w I_{wf} - (\mu + \delta_{1w} + \alpha_w) I_w,\\
\frac{dT_w}{dt} &= b\varepsilon'_w \tau'_w I_{wf} + \alpha_w I_w\\
&\quad - (\nu_w + \mu + \delta_{2w} + \gamma_R + \rho\gamma_R) T_w,\\
\frac{dA_w}{dt} &= \nu_w T_w - (\mu + \delta_{3w} + \gamma_R) A_w,\\
\frac{dI_R}{dt} &= \lambda_R S + \gamma_R T_w + b\varepsilon'_R \tau_R I_{Rf}\\
&\quad - (\mu + \delta_{1R} + \alpha_R) I_R,\\
\frac{dT_R}{dt} &= \alpha_R I_R + \rho\gamma_R T_w + b\varepsilon'_R \tau'_R I_{Rf}\\
&\quad - (\nu_R + \mu + \delta_{2R}) T_R,\\
\frac{dA_R}{dt} &= \nu_R T_R + \nu_R A_w - (\mu + \delta_{3R}) A_R,
\end{aligned} \tag{2}$$

where

$$\lambda_w = \beta_{1w} I_w + \beta_{2w} T_w + \beta_{3w} A_w,$$
$$I_{Rf} = I_R + T_R + A_R,$$
$$\lambda_R = \beta_{1R} I_R + \beta_{2R} T_R + \beta_{3R} A_R,$$
$$\varepsilon'_R = 1 - \varepsilon_R,$$
$$I_{wf} = I_w + T_w + A_w,$$
$$\text{and } \varepsilon'_w = 1 - \varepsilon_w.$$

3. MODEL ANALYSIS

System (2) is analysed to find conditions for existence and stability of disease free and endemic equilibrium points. This analysis of the model allows us to determine the impact of drug resistant virus in the treatment and vertical transmission on the transmission of HIV/AIDS infection in a population characterized in terms of the reproductive number R_0.

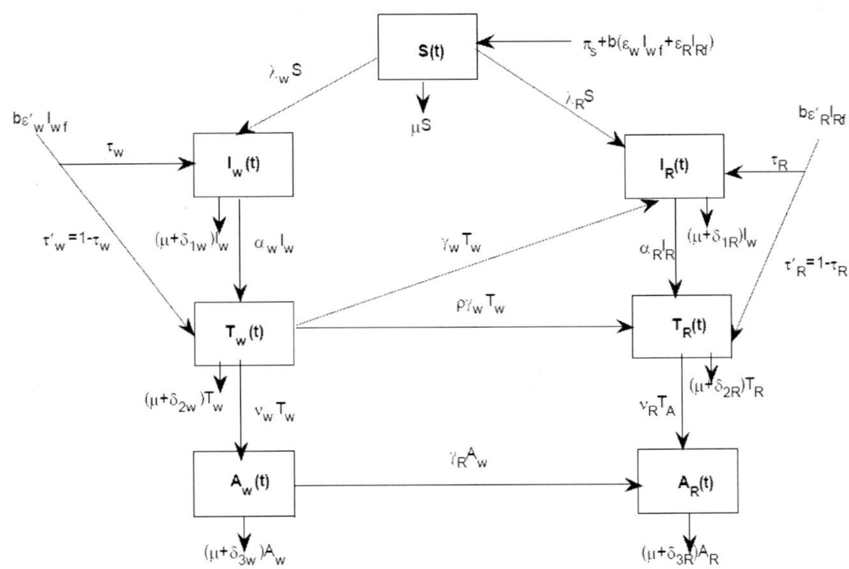

Figure 1. Vertical transmission model of HIV with drug resistance.

Table 1. Parameters and their descriptions.

Parameter	Description
ε_w	Proportion of offspring who are drug sensitive
ε_R	Proportion of offspring with drug resistance
λ_w	Force of infection of drug sensitive individuals
λ_R	Force of infection of drug resistant individuals
δ_{iw}, $i = 1, 2, 3$	AIDS related death of infected drug sensitive, treated drug sensitive and AIDS drug sensitive respectively
δ_{iR}, $i = 1, 2, 3$	AIDS related death of infected drug resistant, treated drug resistant and AIDS drug resistant respectively
ν_w	Progression rate from infected drug sensitive to AIDS drug sensitive
ν_R	Progression rate from infected drug resistant to AIDS drug resistant
α_w	Progression rate of infected drug sensitive to treated drug sensitive
α_R	Progression rate of infected drug resistant to treated drug resistant
γ_w	Evolution rate of drug resistant virus
ρ	Modification parameter

3.1. Positivity of Solutions

Adding equations of system (2) we obtain an equation governing changes in the total population

$$\dot{N} = \pi_s - \mu N - b\left(I_{wf} + I_{Rf}\right) - \alpha_w I_w - \alpha_R I_R$$
$$- \delta_{2w} T_w - \delta_{2R} T_R - \delta_{3w} A_w - \delta_{3R} A_R$$

In the absence of HIV infection the total population converges to π_s/μ It follows that the $\limsup_t N(t) \le \pi_s/\mu$ and it suffices to study system (2) in an invariant set

$$\Omega = \left\{ \left(S, I_w, I_R, T_w, T_R, A_w, A_R \right) \in \mathbb{R}_+^7 \left| \pi_s/\mu \right. \right\}$$

We now show that the initial data of (S, I_w, I_R, T_w, T_R, A_w, A_R) which are non-negative lead to solutions that remain non-negative for all $t \ge 0$.

Theorem 3.5

Let

$$\left\{ S(0) > 0, I_w(0) \ge 0, T_w(0) \ge 0, A_w(0) \ge 0, \right.$$
$$\left. I_R(0) \ge 0, T_R(0) \ge 0, A_R(0) \ge 0 \right\} \in F$$

where $F \subseteq \Omega$ Then the solutions $\{S(t), I_w(t), T_w(t), A_w(t), I_R(t), T_R(t), A_R(t)\}$ of the model system of Equation (2) are positive $\forall t \ge 0$.

Proof

From the first equation of (2), we have

$$\frac{dS}{dt} \ge -\left(\lambda_w + \lambda_R + \mu \right) S,$$
$$\int_0^t \frac{1}{S} dS \ge -\int_0^t \left(\lambda_w + \lambda_R + \mu \right) ds,$$

which gives

$$S(t) \ge S(0) e^{-\int_0^t (\lambda_w + \lambda_R + \mu) ds} = S(0) e^{-(\lambda_w + \lambda_R + \mu)s} \ge 0$$

As $t \to \infty$ we obtain $S(t) \ge 0$. Hence all feasible

solution of system (2) enter the region

$$\left\{ S(t), I_w(t), T_w(t), A_w(t), I_R(t), T_R(t), A_R(t) \right\} \in F.$$

Similar proofs can be established for the positivity of I_W, T_W, A_W, I_R, T_R, A_R.

Thus, the equations of system (2) are well posed.

3.2. Steady State Solutions

The equilibrium solutions of the model (2) are obtained by setting the right hand side of system (2) to zero, to solve the following system of non-linear equations:

$$\begin{aligned}
&\pi_s + b\varepsilon_w I_{wf}^* + b\varepsilon_R I_{Rf}^* - \lambda_w S^* - \lambda_R S^* - \mu S^* = 0,\\
&\lambda_w S^* + b\tau_w \varepsilon_w' I_{wf}^* - (\mu + \delta_{1w} + \alpha_w) I_w^* = 0,\\
&b\varepsilon_w' \tau_w' I_{wf}^* + \alpha_w I_w^* - (\nu_w + \mu + \delta_{2w} + \gamma_R + \rho\gamma_R) T_w^* = 0,\\
&\nu_w T_w^* - (\mu + \delta_{3w} + \gamma_R) A_w^* = 0,\\
&\lambda_R S^* + \gamma_R T_w^* + b\varepsilon_R' \tau_R I_{Rf}^* - (\mu + \delta_{1R} + \alpha_R) I_R^* = 0,\\
&\alpha_R I_R^* + \rho\gamma_R T_w^* + b\varepsilon_R' \tau_R' I_{Rf}^* - (\nu_R + \mu + \delta_{2R}) T_R^* = 0,\\
&\nu_R T_R^* + \nu_R A_w^* - (\mu + \delta_{3R}) A_R^* = 0.
\end{aligned} \quad . \qquad (3)$$

From the equation of A_w^* we have $A_w^* = P_0 T_w^*$ where

$$P_0 = \frac{\nu_w}{\mu + \gamma_R + \delta_{3w}}$$

Substituting A_w^* in the expressions for I_{wf}^* and λ_{wf} yields

$$I_{wf}^* = I_w^* + (1 + P_0) T_w^*$$

and

$$\lambda_{wf} = \beta_{1w} I_w^* + (\beta_{2w} + P_0 \beta_{3w}) T_w^*$$

which upon substitution in the equations for I_w^* and T_w^* gives $T_w^* = P_1 I_w^*$cwith $I_{wf}^* = P_2 I_w^*$ and $\lambda_{wf} = P_3 I_w^*$ where

$$P_2 = 1 + (1 + P_0)P_1$$

and

$$P_1 = \frac{(b\varepsilon'_w \tau'_w + \alpha_w)}{\nu_w + \mu + \delta_{2w} + \nu_R + \gamma_R + \rho\gamma_R - b\varepsilon'_w \tau'_w (1 + P_0)}.$$

Defining

$$\Re_{0w} = \frac{\beta_{1w} + (\beta_{2w} + P_0\beta_{3w})P_1}{\mu + \alpha_w + \delta_{1w} - b\tau_w \varepsilon'_w P_2}$$

to be threshold parameter due to the wild strain, and then substituting for I^*_{wf} and λ_{wf} in the equation for S^* we obtain $I^*_w = 0$ or

$$S^* = \frac{1}{\Re_{0w}}$$

The solution $I^*_w = 0$ yields $A^*_R = Q_0 T^*_R$ where

$$Q_0 = \frac{\nu_R}{\mu + \delta_{3R}}$$

Substituting A^*_R in the expressions for I^*_{Rf} and λ_R we get

$$\lambda_R = \beta_{1R} I^*_R + (\beta_{2R} + Q_0\beta_{3R})T^*_R$$

and

$$I^*_{Rf} = I^*_R + (1 + Q_0)T^*_R.$$

The expression for I^*_{wf} together with an equation for I^*_R give $T^*_R = Q_1 I^*_R$ where

$$Q_1 = \frac{\alpha_R + b\varepsilon'_R \tau'_R}{\mu + \delta_{2R} + \nu_R - b\varepsilon'_R \tau'_R (1 + Q_0)}$$

provided that

$$\mu + \delta_{2R} + \nu_R > b\varepsilon'_R \tau'_R (1 + Q_0).$$

With this expression for T_R^* the expressions for I_{Rf}^* and λ_R reduce to $\lambda_R = \eta_R I_R^*$ and $I_{Rf}^* = Q_2 I_R^*$ where

$$\eta_R = \beta_{1R} + (\beta_{2R} + Q_0 \beta_{3R}) Q_1$$

and

$$Q_2 = 1 + (1 + Q_0) Q_1$$

Substituting for I_{Rf}^* and λ_R in the equation for I_R^* we obtaining $I_R^* = 0$

or

$$S^* = \frac{1}{\Re_{0R}}.$$

3.2.1. Disease-Free Equilibrium

The solution $I_w^* = 0$and $I_R^* = 0$lead to the disease–free equilibrium point

$$E_0 = \left(\frac{\pi_s}{\mu}, 0, 0, 0, 0, 0, 0 \right).$$

3.2.2. Wild Strain-Free Equilibrium

The solutions $I_w^* = 0$

and

$$S^* = \frac{1}{\Re_{0R}}$$

lead to an drug resistant strain endemic equilibrium given by $E_R = \left(S^*, 0, 0, 0, I_R^*, T_R^*, A_R^*\right)$ with coordinates

$$S^* = \frac{1}{\Re_{0R}}, T_R^* = Q_1 I_R^*, A_R^* = Q_0 Q_1 I_R^*$$

and

$$I_R^* = \left(\frac{\mu}{\mu + \alpha_R + \delta_{1R} - b(\tau_R + \tau'_R \varepsilon_R) Q_2}\right)\left(\frac{1}{\Re_{0R}}\right)\left(\Re_0^R - 1\right);$$

$$\Re_0^R = \frac{\pi_s}{\mu} \Re_{0R}$$

.

3.2.3. Drug Resistant Strain-Free Equilibrium

The solutions $I_w^* \neq 0$ and

$$S^* = \frac{1}{\Re_{0R}}$$

lead to an drug resistant strain endemic equilibrium given by

$$E_w = \left(S^*, I_w^*, T_w^*, A_w^*, 0, 0, 0\right)$$

with coordinates

$$S^* = \frac{1}{\Re_{0w}}, T_w^* = P_1 I_w^*, A_w^* = P_0 P_1 I_w^*$$

,

and

$$I_w^* = \left(\frac{\mu}{\mu + \alpha_w + \delta_{1w} - b(\tau_w + \tau'_w \varepsilon_w) P_2}\right)\left(\frac{1}{\Re_{0w}}\right)\left(\Re_0^w - 1\right);$$

$$\Re_0^w = \frac{\pi_s}{\mu} \Re_{0w}$$

.

3.2.4. Coexistence of Wild and Resistant Strains

In this case, we consider $I_w^* \neq 0$ and

$$S^* = \frac{1}{\Re_{0w}}.$$

The equation for A_R^* leads to

$$A_R^* = Q_0 T_R^* + Q_0 P_0 P_1 I_w^*,$$

Where Q_0, P_0 and P_1 are as defined above. Substituting A_R^* in the expressions for I_{wf}^* and λ_w yields

$$\lambda_R = \beta_{1R} I_R^* + (\beta_{2R} + Q_0 \beta_{3R}) T_R^* + Q_0 P_0 P_1 \beta_{3R} I_R^*$$

and

$$I_{Rf}^* = I_R^* + (1 + Q_0) T_R^* + Q_0 P_0 P_1 I_w^*$$

Substitution of I_{Rf}^* and λ_R^* in the equation for I_R^* results in

$$T_R^* = Q_1 I_R^* + Q_3 I_w^*,$$

where

$$Q_1 = \frac{\alpha_R + b\tau_R' \varepsilon_R'}{\mu + \alpha_R + \delta_{2R} - b\tau_R' \varepsilon_R' (1 + Q_0)}$$

and

$$Q_3 = \frac{(\rho \gamma_R + Q_0 P_0 b \tau_R' \varepsilon_R') P_1}{\mu + \alpha_R + \delta_{2R} - b\tau_R' \varepsilon_R' (1 + Q_0)}$$

The expression for T_R^* gives to $\lambda_R = \eta_R I_R^* + \eta_{wR} I_w^*$

and

$$I_{Rf}^{*} = Q_2 I_R^{*} + Q_4 I_w^{*}$$

where

$$\eta_{wR} = Q_3 \beta_{2R} + Q_0 \left(P_0 P_1 + Q_3 \right) \beta_{3R}$$

and

$$Q_4 = Q_0 P_0 P_1 + Q_3 \left(1 + Q_0 \right)$$

Let

$$\Re_{0R} = \frac{\beta_{1R} + \left(\beta_{2R} + \left(1 + Q_0 \right) \beta_{3R} \right)}{\mu + \alpha_R + \delta_{1R} - b\tau_R \varepsilon_R' Q_2}$$

be the threshold parameter due to drug resistance and

$$Q_5 = \frac{\mu + \alpha_R + \delta_{1R} - b\tau_R \varepsilon_R' Q_2}{\eta_{wR} + \left(\gamma_R P_1 + b\tau_R \varepsilon_R' Q_2 \right) \Re_{0w}}$$

Then we establish a relationship between I_w^{*} and I_R^{*} as

$$I_w^{*} = Q_5 \left(\Re_{0w} - \Re_{0R} \right) I_R^{*}$$

Clearly, $\Re_{0w} > \Re_{0R}$and result confirms the findings in literature which claims that the virulence of resistant strain is weak than the wild strain. The results above and the equation for S^* yield

$$I_R^{*} = \frac{\mu \left(\Re_0^R - 1 \right)}{Q_6 \left(\Re_{0w} - \Re_{0R} \right) + Q_7},$$

where

$$Q_6 = Q_5 \left(\eta_w - b\varepsilon_R P_2 \Re_{0w} + \eta_{wR} - b\varepsilon_R Q_2 \Re_{0w} \right)$$

and

$$Q_7 = \eta_R - b\varepsilon_R Q_2 \Re_{0w}$$

satisfying the conditions

$$\eta_w > b\varepsilon_R P_2 \Re_{0w}, \eta_{wR} > b\varepsilon_R Q_2 \Re_{0w} \text{ and } \eta_R > b\varepsilon_R Q_2 \Re_{0w}$$

3.2.5. The Model Reproduction Numbers (Thresholds)

Suppose

$$\mu_2 = \frac{\alpha_w}{\mu + \alpha_w + \delta_{2w}}, \mu_3 = \frac{\nu_w}{\mu + \gamma_w + \rho\gamma_w + \nu_w + \delta_{2w}} \mu_2,$$

$$\Re_{1h}^w = \frac{K\beta_{1w}}{\mu + \alpha_w + \delta_{1w}}, \Re_{1v}^w = \frac{b\tau_R \varepsilon_R'}{\mu + \alpha_w + \delta_{1w}},$$

$$\Re_{2h}^w = \frac{K\beta_{2w}}{\mu + \gamma_w + \rho\gamma_w + \nu_w + \delta_{2w}},$$

$$\Re_{2v}^w = \frac{b\tau_R \varepsilon_R'}{\mu + \gamma_w + \rho\gamma_w + \nu_w + \delta_{2w}},$$

$$\Re_{3h}^w = \frac{K\beta_{3w}}{\mu + \delta_{3w}}, \Re_{3v}^w = \frac{b\tau_R \varepsilon_R'}{\mu + \delta_{3w}},$$

$$\Re_{1vT}^w = \frac{b\tau_R' \varepsilon_R'}{\mu + \alpha_w + \delta_{2w}},$$

$$\Re_{2vT}^w = \frac{b\tau_R' \varepsilon_R'}{\mu + \gamma_w + \rho\gamma_w + \nu_w + \delta_{2w}}, \text{ and } \Re_{3vT}^w = \frac{b\tau_R' \varepsilon_R'}{\mu + \delta_{3w}}.$$

Then we can express the generation of secondary infections due to horizontal transmission as

$$\Re_{0h}^w = \Re_{1h}^w + \mu_2 \Re_{2h}^w + \mu_3 \Re_{0h}^w$$

and vertical transmission as

$$\Re_{0v}^w = \Re_{1v}^w + \mu_2 \Re_{2v}^w + \mu_3 \Re_{0v}^w$$

with the net effects of the transmission due to the wild strain given by $\Re_0^w = \Re_{0h}^w + \Re_{0v}^w$. The reproduction number of the model is given by

$$R_0 = \max\left(R_0^w, R_0^w\right)\text{where } R_0^w = \frac{a_{1i} + \sqrt{a_{1i}^2 - 4a_{0i}a_{2i}}}{2a_{0i}}$$

with

$$a_{0i} = 1, \Re^w_{jhv} = \sum_{j=1}^{3}\left(\Re^w_{vh} + \Re^w_{jv}\right);$$

$$a_{1i} = \Re^i_{1hv} + \mu_2\Re^i_{2hv} + \mu_3\Re^i_{3hv} + \Re^i_{2v} + \mu_2\Re^i_{3v},$$

$$a_{2i} = \Re^i_{2vT}\Re^i_{1hv} - \Re^i_{1vT}\Re^i_{2hv} + \mu_3\left(\Re^i_{3vT}\Re^i_{1hv} - \Re^i_{1vT}\Re^i_{3hv}\right)$$

and $a_{1i}^2 > 4a_{2i}$ for $i = w, R$. The parameter R_0 is defined as the average number of secondary infections generated by introducing an infected individual (with either wild or resistant strain) in a wholly susceptible population. The threshold parameters $\Re^i_{jvT}$ for

$i = w, R$ and j=1,2,3, are the demographic replacement of secondary infections due to vertical transmission by individuals identified and enrolled on antiretroviral therapy (ART) from childhood, while $\Re^i_{jv}$ for $i = w, R$ and j=1,2,3 are the demographic replacement of secondary infections due to vertical transmission by offsprings who progress from birth to adulthood without treatment.

3.3. Local Stability of Disease-Free Equilibrium Point

Evaluating at the disease-free equilibrium point

$$E_0 = \left(\frac{\pi_s}{\mu}, 0, 0, 0, 0, 0, 0\right)$$

the system has the Jacobian matrix J_{E_0}, given by

$$\boldsymbol{J}_0 = \begin{pmatrix} \boldsymbol{A}_{11} & \boldsymbol{A}_{12} & \boldsymbol{A}_{13} \\ \boldsymbol{0} & \boldsymbol{A}_{22} & \boldsymbol{0} \\ \boldsymbol{0} & \boldsymbol{A}_{32} & \boldsymbol{A}_{33} \end{pmatrix}$$

where

$$\begin{aligned}
A_{11} &= -\mu, \\
A_{12} &= b\varepsilon_w - \beta_{1w}, b\varepsilon_w - \beta_{2w}, b\varepsilon_w - \beta_{3w}, \\
A_{13} &= b\varepsilon_R - \beta_{1R}, b\varepsilon_R - \beta_{2R}, b\varepsilon_R - \beta_{3R} \\
A_{22} &= \begin{pmatrix} \beta_{1w} - M_{1w} + b\tau_w\varepsilon'_w & \beta_{2w} + b\tau_w\varepsilon'_w & \beta_{3w} + b\tau_w\varepsilon'_w \\ b\varepsilon'_w\tau'_w + \alpha_w & b\varepsilon'_w\tau'_w - M_{2w} & b\varepsilon'_w\tau'_w \\ 0 & \nu_w & -M_{3w} \end{pmatrix}, \\
A_{33} &= \begin{pmatrix} \beta_{1R} - M_{1R} + b\tau_R\varepsilon'_R & \beta_{2R} + b\tau_R\varepsilon'_R & \beta_{3R} + b\tau_R\varepsilon'_R \\ b\varepsilon'_R\tau'_R + \alpha_R & b\varepsilon'_R\tau'_R - M_{2R} & b\varepsilon'_R\tau'_R \\ 0 & \nu_R & -M_{3R} \end{pmatrix}
\end{aligned}$$

and $A_{32} = \begin{pmatrix} 0 & \gamma_w & 0 \\ 0 & \rho\gamma_w & 0 \\ 0 & 0 & \nu_R \end{pmatrix}$.

We note from matrix $\boldsymbol{A}_{11}$ that $\lambda_1 = -\mu$ and the other eigenvalues can be found form solving the characteristic polynomials of the block matrices $\boldsymbol{A}_{22}$ and $\boldsymbol{A}_{33}$. The characteristic polynomial of the block matrix $\boldsymbol{A}_{22}$ is given by,

$$\lambda^3 - p_2\lambda^2 + p_1\lambda + p_0 = 0,$$

where

$$\begin{aligned}
p_2 &= M_{3w} - g_5 + M_{2w} - g_1 + M_{1w}, \\
p_1 &= -g_6\nu_w - g_5 M_{3w} + M_{2w}M_{3w} - g_1 M_{3w} + M_{1w}M_{3w} \\
&\quad - g_2 g_4 + g_1 g_5 - g_5 M_{1w} - g_1 M_{2w} + M_{1w}M_{2w}, \\
p_0 &= -\nu_w g_4 g_3 + \nu_w g_6 g_1 - \nu_w g_6 M_w - g_4 g_2 M_{3w} + g_1 g_5 M_{3w} \\
&\quad - g_5 M_w M_{3w} - g_1 M_{2w} M_{3w} + M_w M_{2w} M_{3w}
\end{aligned}$$

with

$$\begin{aligned}
&g_1 = \beta_{1w} + b\tau_w\varepsilon'_w, g_2 = b\varepsilon'_w\tau'_w, g_3 = b\varepsilon'_w\tau'_w, \\
&g_4 = b\varepsilon'_w\tau'_w + \alpha_w, g_5 = \beta_{2w} + b\tau_w\varepsilon'_w, g_6 = \beta_{3w} + b\tau_w\varepsilon'_w.
\end{aligned}$$

After some manipulations it can be shown that the expression for P_0 is written as

$$p_0 = Z_1\left(1 - R_{0.C}\left(w\right)\right) \tag{4}$$

where

$$Z_1 = v_w g_6 g_1 + g_1 g_5 M_{3w} + M_w M_{2w} M_{3w},$$

$$R_{0,C}\left(w\right) = \frac{v_w g_4 g_3 + v_w g_6 M_w + g_4 g_2 M_{3w} + g_5 M_{1w} M_{3w} + g_1 M_{2w} M_{3w}}{v_w g_6 g_1 + g_1 g_5 M_{3w} + M_{1w} M_{2w} M_{3w}} \tag{5}$$

The threshold $R_{0,C}\left(w\right)$ is called the basic reproduction number for wild virus type for the system (2).

The characteristic polynomial of block matrix $\boldsymbol{A}_{33}$ is given by,

$$\lambda^3 - h_2\lambda^2 + h_1\lambda + h_0 = 0,$$

where

$$\begin{aligned}
h_0 &= -v_R n_4 h_3 + v_R n_6 n_1 - v_w n_6 M_{1R} - n_4 n_2 M_{3R} + n_1 n_5 M_{3R} \\
&\quad - n_5 M_R M_{3R} - n_1 M_{2R} M_{3R} + M_{1R} M_{2R} M_{3R}, \\
h_1 &= -n_6 v_R - n_5 M_{3R} + M_{2R} M_{3R} - n_1 M_{3R} + M_{1R} M_{3R} \\
&\quad - n_2 n_4 + n_1 n_5 - n_5 M_{1R} - n_1 M_{2R} + M_{1R} M_{2R}, \\
h_2 &= M_{3R} - g_5 + M_{2R} - g_1 + M_{1R}, \\
n_1 &= \beta_{1R} + b\tau_w \varepsilon'_R, n_2 = b\varepsilon'_R \tau', n_3 = b\varepsilon'_R \tau'_R, \\
n_4 &= b\varepsilon'_R \tau'_R + \alpha_R, n_5 = \beta_{2R} + b\tau_R \varepsilon'_R, n_6 = \beta_{3R} + b\tau_R \varepsilon'_R.
\end{aligned}$$

The expression for h_0 can be written as

$$h_0 = Z_2\left(1 - \Re_{0.C}\left(R\right)\right) \tag{6}$$

where

$$Z_2 = v_R n_6 n_1 + n_1 n_5 M_{3R} + M_{1R} M_{2R} M_{3R},$$

$$\Re_{0,C}(R) = \frac{v_R n_4 n_3 + v_R n_6 M_{1R} + n_4 n_2 M_{3R} + n_5 M_{1R} M_{3R} + n_1 M_{2R} M_{3R}}{v_R n_6 n_1 + n_1 n_5 M_{3R} + M_{1R} M_{2R} M_{3R}} \tag{7}$$

The threshold $\Re_{0,C}(R)$ is called the basic reproduction number for drug resistant virus type for the system (2).

From Equation (5) and (7), we now have the effective reproduction number of the model (denoted by $\Re_{0,C}(wR)$)) given by

$$R_{0,C}(wR) = \max\{R_{0,C}(w), \Re_{0,C}(R)\}$$
$$M_{1w} = \mu + \delta_{3w} + \alpha_w, M_{1R} = \mu + \delta_{3R} + \alpha_R,$$
$$M_{2w} = v_w + \mu + \delta_{2w} + \gamma_R + \rho\gamma_R, M_{2R} = v_R + \mu + \delta_{2R},$$
$$M_{3w} = \mu + \delta_{3w} + v_R, M_{3R} = \mu + \delta_{3R}.$$

Theorem 1

The disease-free equilibrium point of the vertical transmission model (2) with drug resistance is locally asymptotically stable if $R_{0,C}(wR) < 1$ and unstable if $R_{0,C}(wR) > 1$.

4. NUMERICAL SIMULATIONS

In order to illustrate some of the analytical results of the study, numerous numerical simulations of the model (2) are carried out using a set of reasonable parameter values given in **Table 2** below.

However these parameters may (or may not) be biologically feasible.

Figure 2 below shows the variations of the seven classes of HIV/AIDS model with drug resistance.

Form **Figure 2** it can be seen is that, infective population is increasing. This leads into the decrease in susceptible class.

Table 2. Parameter values used in numerical simulation.

Parameters	Values	Source
π_S	100,000	[10]
b	0.06	[11]
μ	0.03	[12]
ε_w and ε_R	$\varepsilon_w = 0.7$	[11]
	$\varepsilon_R = 0.7$	[11]
τ_w and τ_R	$\tau_w = 0.2$	Estimated
	$\tau_R = 0.3$	Estimated
$\delta_{iw}, i = 1,2,3$	$\delta_{1w} = 0.111$	Estimated
	$\delta_{2w} = 0.222$	Estimated
	$\delta_{3w} = 0.333$	[10]
$\delta_{iR}, i = 1,2,3$	$\delta_{1R} = 0.4$	Estimated
	$\delta_{2R} = 0.5$	Estimated
	$\delta_{3R} = 0.6$	Estimated
ν_w and ν_R	$\nu_w = 0.05$	[12]
	$\nu_R = 0.1$	[13]
α_w and α_R	$\alpha_w = 0.6$	[14]
	$\alpha_R = 0.7$	Estimated
γ_w	$\gamma_w = 0.08$	Estimated
ρ	$\rho = 1.2$	Estimated
β_i	$\beta_{1w} = 0.2$	[15]
	$\beta_{2w} = 0.08$	[15]
	$\beta_{3w} = 0.12$	Estimated
	$\beta_{1R} = 0.3$	Estimated
	$\beta_{2R} = 0.09$	Estimated
	$\beta_{3R} = 0.13$	Estimated

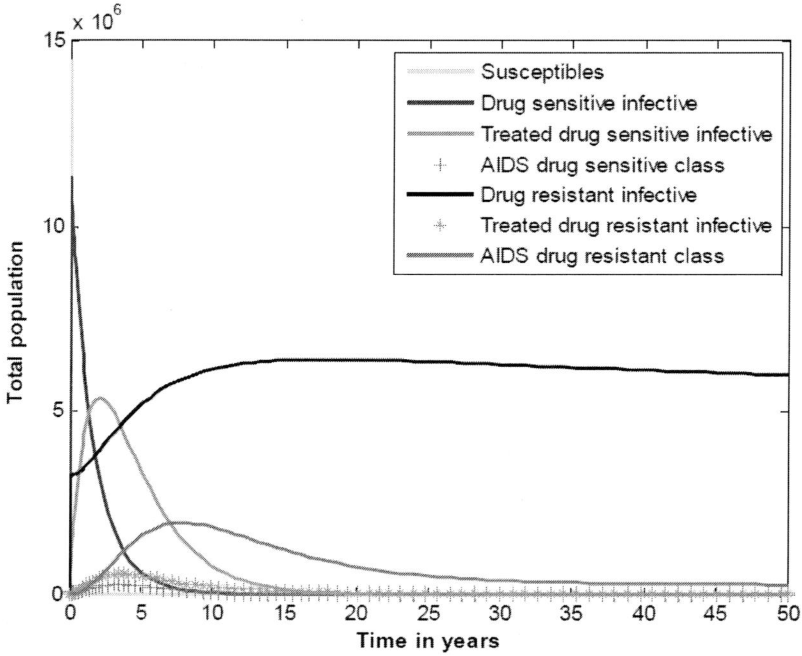

Figure 2. Illustration of the changes in seven state variables of full HIV/AIDS model with drug resistance.

Figure 3 below shows the effects of rate of HIV Free children born with wild sensitive virus.

Figure 3 shows that as ε_w increases, individuals in AIDS and treated classes with drug resistant virus decrease. However, the number of susceptible individuals increases.

Figure 4 below shows the effects of Rate of HIV Free Children Born with Drug Resistant Virus.

Figure 4 shows that as the proportion of children who are drug resistant born HIV negative increases, the numbers of populations in the drug resistant class and treated drug resistant class decrease.

Figure 5 below depicts the behaviour of the AIDS population with both wild and resistant strains, and treated population with resistant strain as influence by changes in the progression rates of drug resistant individuals to AIDS.

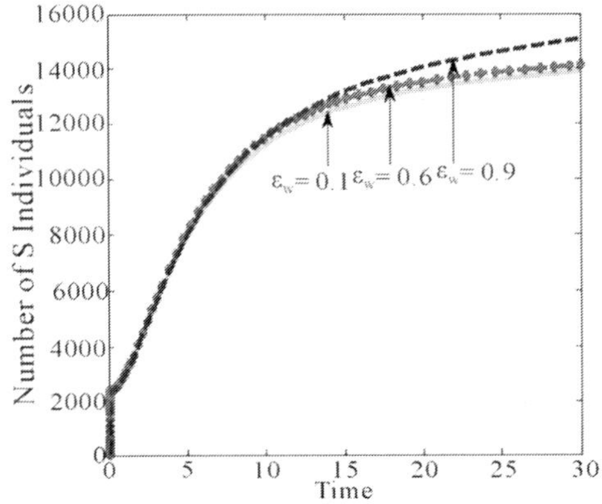

(a)

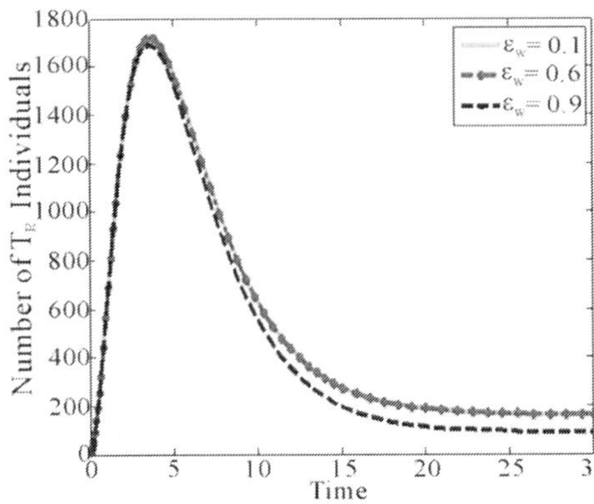

(b)

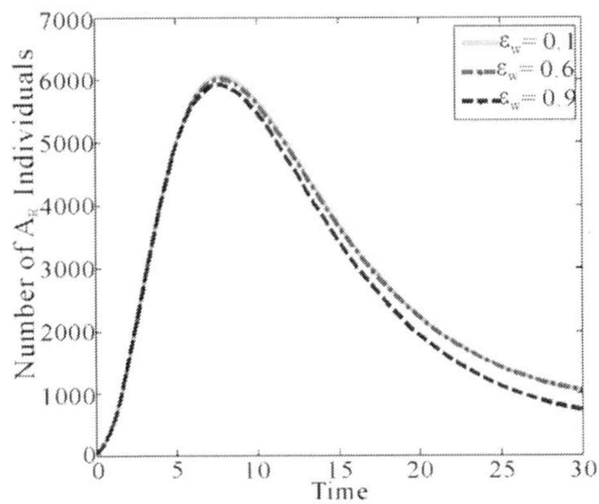

(c)

Figure 3. Effects of variation of proportion of children born with wild sensitive virus.

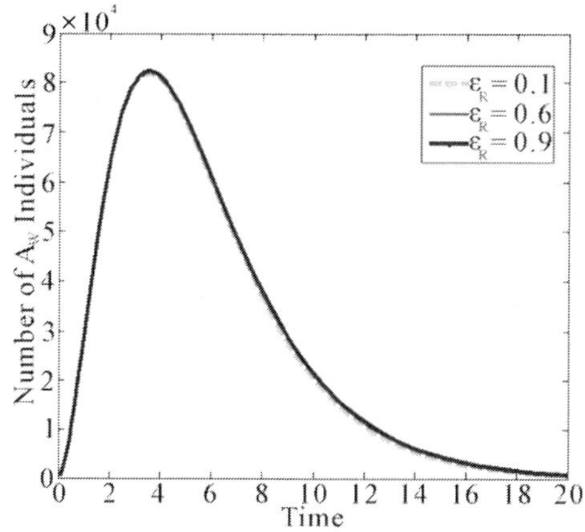

×10⁴
Number of A_R Individuals
Time
ε_R = 0.1
ε_R = 0.6
ε_R = 0.9

(a)

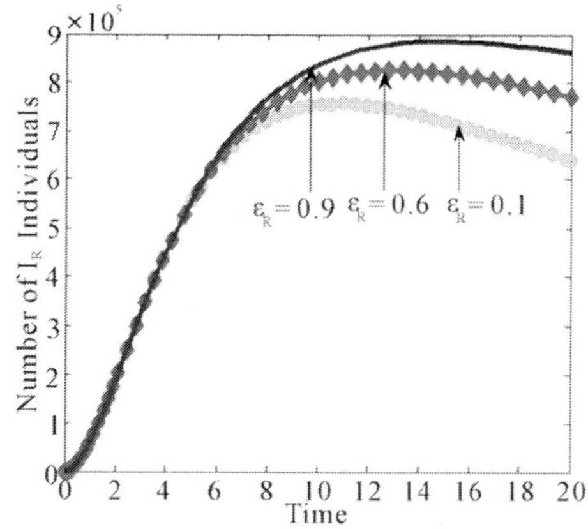

×10⁵
Number of I_R Individuals
Time
ε_R = 0.9
ε_R = 0.6
ε_R = 0.1

(b)

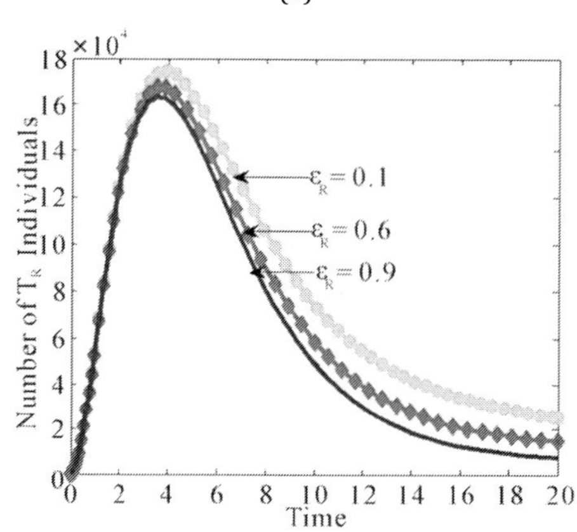

×10⁴
Number of T_R Individuals
Time
ε_R = 0.1
ε_R = 0.6
ε_R = 0.9

(c)

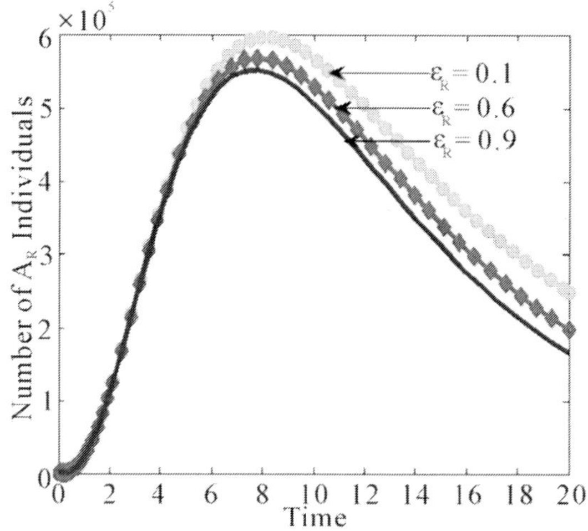

(d)

Figure 4. Dynamics of drug resistance influenced by vertical transmission with ε_R being varied.

From **Figure 5** it is seen that at high progression rates, the three classes reach their peaks much faster than at low values of progression rates, with AIDS class having the highest peak and infective population having the lowest peak.

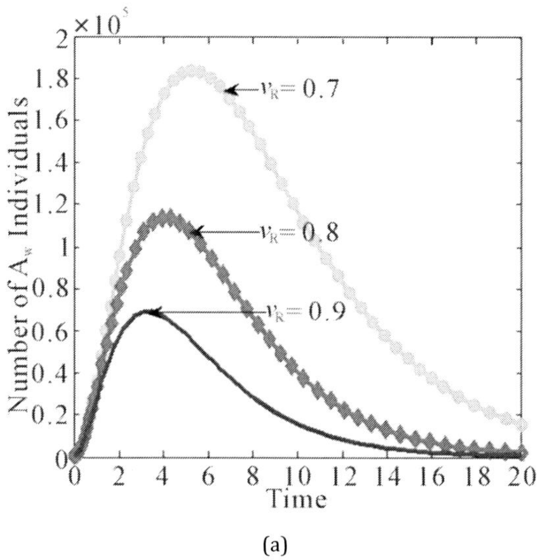

(a)

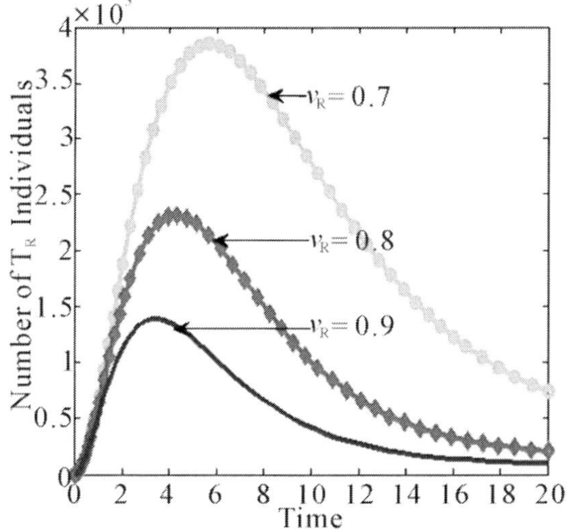

(b)

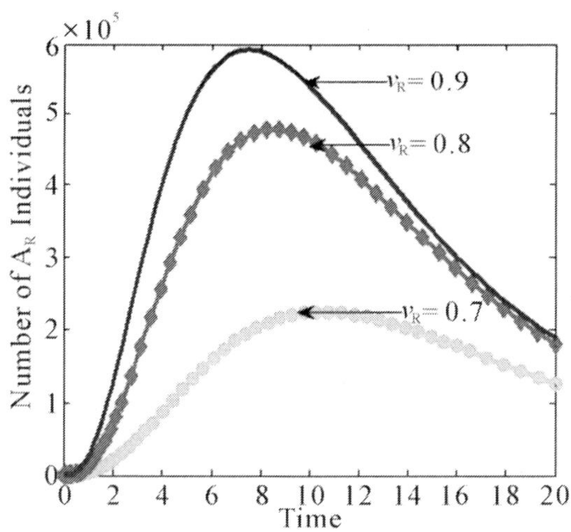

(c)

Figure 5. Effects of rate of progression to AIDS on vertical transmission.

Figure 6 shows the effects of variation of progression rates to I_R and T_R when γ_w is varied.

Figure 6 shows that increasing γ_w leads to decrease in number of individuals in the treated and infected with wild virus classes. However, individuals in I_R, T_R and A_R increase as γ_w increases.

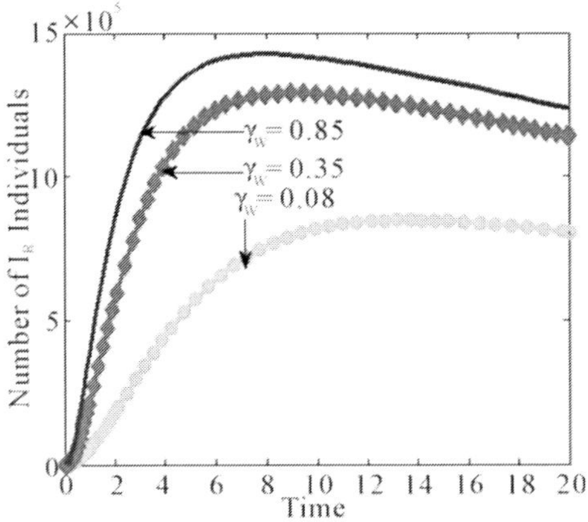
×10^5
Number of I_R Individuals
Time
γ_W = 0.85
γ_W = 0.35
γ_W = 0.08

(a)

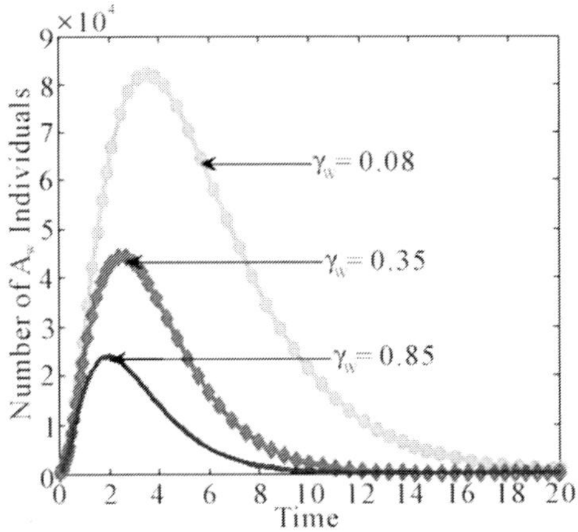
×10^4
Number of A_w Individuals
Time
γ_W = 0.08
γ_W = 0.35
γ_W = 0.85

(b)

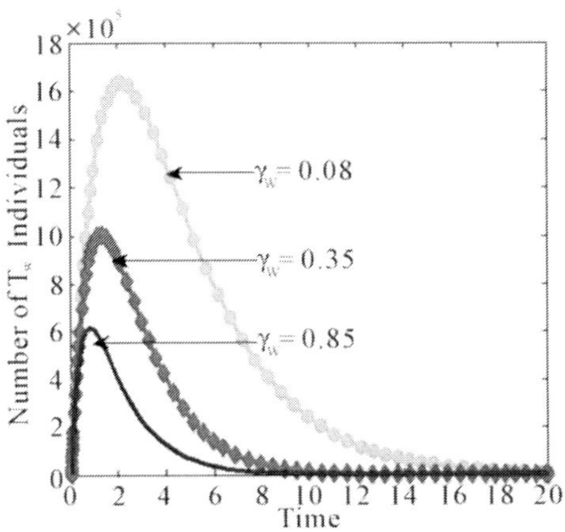
×10^5
Number of T_w Individuals
Time
γ_W = 0.08
γ_W = 0.35
γ_W = 0.85

(c)

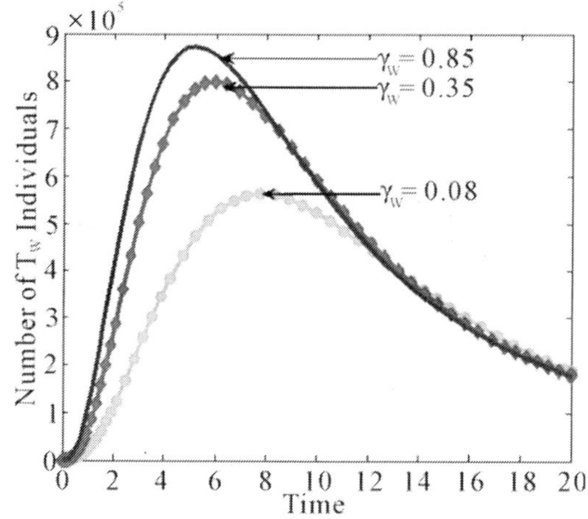

(d)

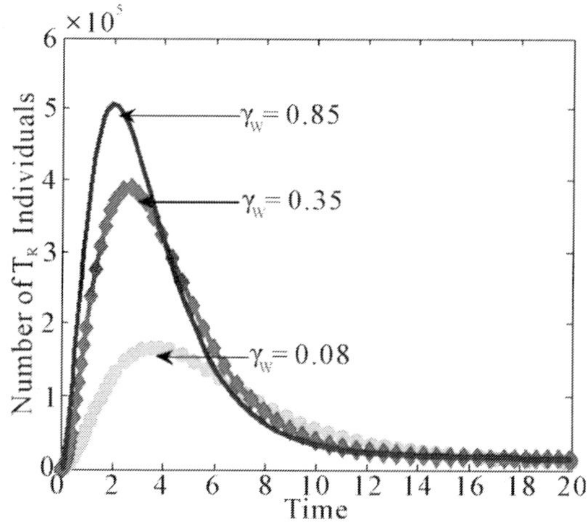

(e)

Figure 6. Effects of variation of progression rates to I_R and T_R when γ_w varied.

Figure 7 below show the effects of treatment rates of infected individuals with drug resistant virus as α_R varies.

As the rate of treatment, α_R increases, the number of individuals infected with drug resistant virus decreases. From **Figure 7**(c), as the rate of treatment increases, the number of AIDS patients with drug resistance increase. It is also noted that the peaks for different values of α_R occur at the same time as for I_R, T_R and A_R classes.

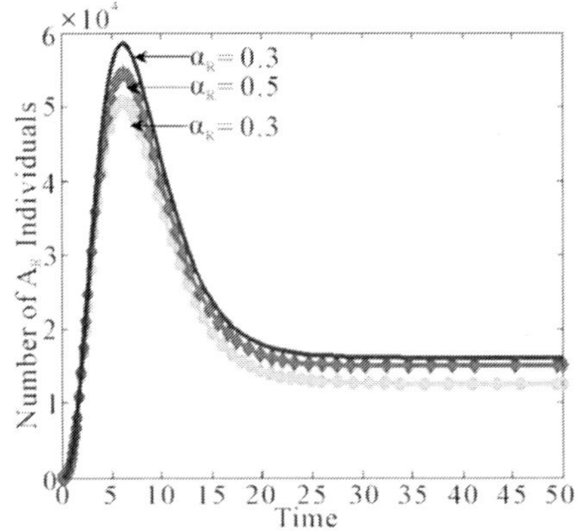

(a)

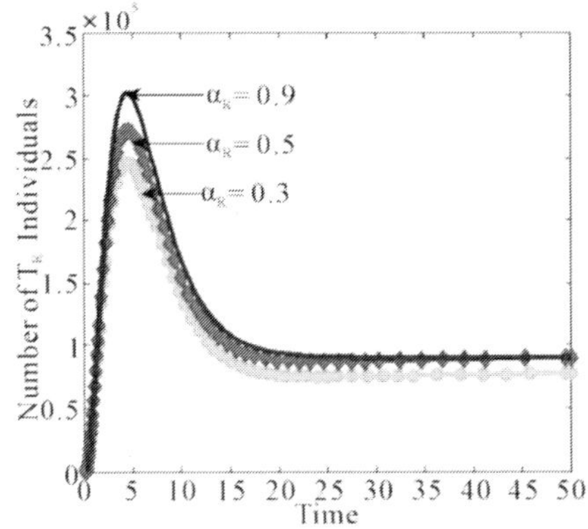

(b)

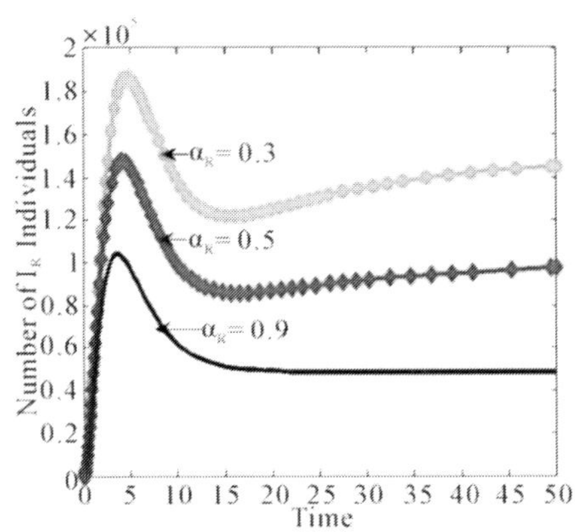

(c)

Figure 7. Effects of treatment rates of infected individuals with drug resistant virus as α_R varies.

5. DISCUSSION AND CONCLUSIONS

In this paper, a non-linear mathematical model to study the transmission of HIV/AIDS in a population of varying size with vertical transmission and drug resistance was proposed and analysed. Both qualitative and numerical analysis of the model was done. The disease-free and endemic equilibrium points were obtained and their stabilities were investigated. A numerical study of the model has been conducted to see the effect of certain key parameters on the spread of the disease. It was established that the disease-free equilibrium is locally asymptotically stable if the basic reproduction number $R_{0,c}(wR)<1$ and unstable if $R_{0,c}(wR)>1$ and the infection persists in the population.

The results have shown that treatment may increase drug resistance among treated individuals. The results further showed that increase in the rate of children born with HIV with drug resistance leads to increase of the infected individuals with drug resistance and will in long run lead to increase in the AIDS population with drug resistance. It was noted that drug resistant virus is emerging in some of the individuals who are on ARV treatment. This will in the long run cause treatment failure among people who are under ARV treatment. As a result of this, more people will be dying of AIDS related diseases. This clearly shows that drug resistance has an impact on the dynamics of HIV/AIDs transmission.

REFERENCES

1. Program for Appropriate Technology in Health (2001) Preventing HIV/AIDS in low-resource settings. Outlook, 19.
2. CDC (2011) Basic information about HIV and AIDS. http://www.cdc. gov/hiv/ topics/basic/
3. Hampanda, K. (2013) Vertical transmission of HIV in sub-Saharan Africa: Applying theoretical frameworks to understand social barriers to PMTCT. ISRN Infectious Diseases, 2013, Article ID: 420361.
4. Besser, M. (2010) HIV in pregnancy: Doing more with less: Mothers2Mothers.
5. Soderlund, N., Zwi, K., Kinghorn, A. and Grey, G (1999) Prevention of vertical transmission of HIV: Analysis of cost effectiveness of available in South Africa. British Medical Journal, 318, 1650-1656. doi:10.1136/bmj.318.7199.1650
6. Prasse, M. (2006) Mother to Child Transmission of HIV. http://the-aids-pandemic. blogspot.com/2006/12/mother-to-child-transmission-of-hiv.html
7. Gumbo, F.Z., Duri, K., Kandawasvika, G.Q., Kurewa, N.E., Mapangure, M.P., Munjoma, M.W., Rusakaniko, S., Chirenje, M.Z. and Pedersen, B.S (2010) Risk factors of HIV vertical transmission in a cohort of women under a PMCTCT program at per-urban clinics in the resourcepoor setting. Journal of Prinatology, 30, 717-723. doi:10.1038/jp.2010.31
8. Granish, R.M., Gilks, C.F., Dye, C., De Cock, K.M. and Williams, B.G (2009) Universal voluntary HIV testing with immediate antiretroviral therapy as a strategy for elimination of HIV transmission: A mathematical model. Lancent, 373, 48-59. doi:10.1016/S0140-6736(08)61697-9

9. WHO (2011) Basic information about HIV and AIDS. http://www.who.int/ mediacentre/ facsheets/fs360/en/inde.html
10. Malunguza, N., Mushayabasa, S., Chiyaka, C. and Mukandavire, Z (2010) Modelling the effects of condom use and antretroviral therapy in controlling HIV/AIDS among heterosexuals, homesexuals and bisexuals. Computational and Mathematical Methods in Medicine, 11, 201- 222. doi:10.1080/17486700903325167
11. Mukandavire, Z. and Garira, W (2007) Sex-structurerd HIV/AIDS model to analyse the effects of condom use with application to Zimbabwe. Journal of Mathematical Biology, 54, 669-699. doi:10.1007/s00285-006-0063-5
12. Kgosimore, M. and Lungu, E.W (2006) The effects of vertical transmission the spread of HIV/AIDS in the presence of treatment. Mathematical Biosciences and Engineering, 3, 297-312. doi:10.3934/mbe.2006.3.297
13. Naresh, R. and Sharma, D (2011) HIV/AIDS model with vertical transmission and time delay. World Journal of Modelling and Simulation, 7, 230-240.
14. Velasco-Hernadez, J.X., Gershengorn, H.B. and Blower, S.M (2002) Could widespread use of combination antiretroviral therapy eradicate HIV epidemics? Lancet Infectious Diseases, 2, 487-493. doi:10.1016/S1473-3099(02)00346-8
15. Hove-Musekwa S.D. and Nyabadza, F (2009) The dynamics of an HIV/AIDS model with screened disease carriers. Computational and Mathematical Methods in Medicine, 10, 287-305.

CHAPTER 11

Mathematical Analysis of Nipah Virus Infections Using Optimal Control Theory

Jakia Sultana[1], Chandra N. Podder[2]

[1]*Department of Computer Science and Engineering, Green University of Bangladesh, Dhaka, Bangladesh*

[2]*Department of Mathematics, University of Dhaka, Dhaka, Bangladesh*

ABSTRACT

The optimal use of intervention strategies to mitigate the spread of Nipah Virus (NiV) using optimal control technique is studied in this paper. First of all we formulate a dynamic model of NiV infections with variable size population and two control strategies where creating awareness and treatment are considered as controls. We intend to find the optimal combination of these two control strategies that will minimize the cost of the two control measures and as a result the number of infectious individuals will decrease. We establish the existence for the optimal controls and Pontryagin's maximum principle is used to characterize the optimal controls. The numerical simulation suggests that optimal control technique is much more effective to minimize the infected individuals and the corresponding cost of the two controls. It is also monitored that in the case of high contact rate, controls have to work for longer period of time to get the desired result. Numerical simulation reveals that the spread of Nipah virus can be controlled effectively if we apply control strategy at early stage.

Keywords: Nipah Virus (NiV), Optimal Control, Existence of the State, Existence of the Objective Functional, Pontryagin's Maximum Principle, Transversality Condition, Optimality Condition, Hamiltonian (H)

1. INTRODUCTION

Mathematical modeling has become an important tool for analyzing the spread as well as control of infectious diseases. It is also a useful tool for the measurement of the effect of different strategies for controlling the spread of infectious diseases within a population. In recent years epidemiological modeling of infectious disease transmission has had an increasing influence on the theory and practice of disease management and control [1]. There are a number of different methods for calculating the optimal control for a specific mathematical model. For example, Pontryagin's maximum principle [2] allows the calculation of the optimal control for a system of ordinary differential equation with a given constraint. Here the optimal control strategy is used to minimize the infected individuals and to maximize the total number of recovered individuals.

Nipah virus, a member of the genus Henipavirus, a new class of virus in the Paramyxoviridae family, has drawn attention as an emerging zoonotic virus in south-east and south-Asian region [3]. This emerging infectious disease has become one of the most alarming threats of the public health mainly due to its periodic outbreaks and the high mortality rate [4]. Epidemiology is the study of the distribution and determinants of health related states or events in specified populations and the application of epidemiology is to control of health problems. The crucial point is that epidemiology concerns itself with populations or groups of population in contrast to clinical medicine, which deals with individuals (patients). Therefore, epidemiology describes health and disease in terms of frequencies and distributions of determinants and conditions in a population or in a specific group of a population. Although Nipah virus has caused only a few outbreaks, it infects a wide range of animals and causes severe disease and death in people, making it a public health concern [5]. Treatment is mostly symptomatic and supportive as the effect of antiviral drugs is not satisfactory. So the very high case fatality addresses the need for adequate and strict control and preventive measures.

This paper deals with application of optimal control to a dynamic model of Nipah Virus (NiV) infections and its possible control and preventive strategy with the help of optimal control technique. Our aim is to minimize the total number of infectious individuals and the cost which is related for creating awareness and treatment.

2. FORMULATION OF MODEL

Nipah virus infection is a zoonotic virus and transmitted first from animal to human. Once it has been transmitted to human, then it continues to be transmitted through human to human (H2H) by the close contact of infected individuals due to its highly infectivity [6].

Let us suppose that $S(t)$, $I(t)$ and $R(t)$ denote the number of individuals in the susceptible, infectious and recovered classes at time t respectively. The total population at time t is represented by $N(t) = S(t) + I(t) + R(t)$.

We consider the following system of non-linear differential equation, is a type of standard SIR disease model, to describe the dynamics of Nipah Virus (NiV) infections in the community.

$$\begin{aligned} S'(t) &= \nu N(t) - \beta S(t)I(t) - \mu S(t) \\ I'(t) &= \beta S(t)I(t) - (\gamma + \mu + \alpha)I(t) \\ R'(t) &= \gamma I(t) - \mu R(t) \\ N'(t) &= \nu N(t) - \alpha I(t) - \mu N(t), \end{aligned} \qquad (1)$$

with initial conditions,

$$S(0) = S_0 \geq 0, I(0) = I_0 \geq 0, R(0) = R_0 \geq 0, N(0) = N_0, \qquad (2)$$

and where, the parameter β represents the effective contact rate, $\nu$is the natural birth rate, μ is the natural mortality rate, γ is the recovery rate and α represents the disease induced death rate.

Since there is no proper vaccination program or appropriate drugs for NiV infections, so in the model we introduce two control strategies, namely, creating awareness (u_1) among the community about the risky areas before outbreak of the disease and the treatment (u_2).

Here the control $u_1(t)$ measures the effort to be needed to increase awareness which results in the reduction of the transmission rate (β) and the control $u_2(t)$ measures the effort required for giving health cares for the infected people to reduce the infected individuals.

Now the NiV model with two control strategies is given below:

$$\begin{aligned}
S'(t) &= \nu N(t) - \beta S(t)I(t) - (\mu + u_1)S(t) \\
I'(t) &= \beta S(t)I(t) - (\gamma + \mu + \alpha)I(t) - u_2 I(t) \\
R'(t) &= \gamma I(t) - \mu R(t) + u_1 S(t) + u_2 I(t) \\
N'(t) &= \nu N(t) - \alpha I(t) - \mu N(t),
\end{aligned} \tag{3}$$

with initial conditions,

$$S(0) = S_0 \geq 0,\ I(0) = I_0 \geq 0,\ R(0) = R_0 \geq 0,\ N(0) = N_0. \tag{4}$$

Here our main objective is to minimize the total number of infected individuals and to reduce the cost which is needed for creating awareness and treatment on a specified time interval. For the fulfillment of our purpose, we work with the following objective function which is similar as [7].

$$\text{Minimize } J(u_1(t), u_2(t)) = \int_0^T A_1 I(t) + \frac{1}{2}\left(B_1 u_1^2 + B_2 u_2^2\right) dt$$

where, B_1 and B_2 are weight parameters that help to balance the corresponding costs. We define the control set as follows:

$$U = \left\{(u_1(t), u_2(t)) : 0 \leq u_1(t) \leq 1,\ 0 \leq u_2(t) \leq 1,\ t \in [0, T]\right\}.$$

In the objective function, $A_1 I$ represents the total number of infected individuals, $B_1 \frac{u_1^2}{2}$ represents the cost for creating awareness and $B_2 \frac{u_2^2}{2}$ represents the cost for treatment.

3. EXISTENCE OF THE OPTIMAL CONTROL FOR NIV MODEL

3.1. Existence of the State

Adding first three equations of the system (3) we get,

$$S' + I' + R' = \nu N(t) - \mu S(t) - \alpha I(t) - \mu I(t) - \mu R(t)$$
$$\Rightarrow S' + I' + R' = \nu N(t) - \mu(S + I + R)(t) - \alpha I(t)$$
$$\Rightarrow S' + I' + R' \le \nu N(t)$$
$$\Rightarrow S' + I' + R' \le \nu(S + I + R)(t)$$
$$\Rightarrow \frac{\mathrm{d}(S + I + R)}{S + I + R} \le \nu \mathrm{d}t.$$

On integrating we get,

$$S + I + R \le \mathrm{e}^{\nu T}\left[S_0 + I_0 + R_0\right] = M_1 \in \mathbb{R}_+, t \in [0, T].$$

So we have

$$S(t) \le M_1,\ I(t) \le M_1 \text{ and } R(t) \le M_1.$$

From the fourth equation of (3) we have

$$N'(t) \le (\nu - \mu)N(t),$$

and then

$$N(t) \le N_0 \mathrm{e}^{(\nu-\mu)T} = M_2 \in \mathbb{R}_+, t \in [0, T].$$

So, finally we get $N(t) \le M_2$.

Since $S(t)$, $I(t)$, $R(t)$ and $N(t)$ are bounded above, so there exists solution for the system (3).

3.2. Existence of the Objective Functional

By proving the following theorem we can establish the existence of the objective functional:

Theorem 1. Consider the control problem with system (3). Then there exists optimal controls (u_1^*, u_2^*) that minimize $J(u_1, u_2)$ over the control set U. i.e.,

$$J\left(u_1^*, u_2^*\right) = \min_{u_1, u_2 \in U} J\left(u_1, u_2\right). \tag{5}$$

Proof: To use an existence result in [8] , we must check the following properties [9].

1) The set of controls and corresponding state variables is non-empty.
2) The control set U is convex and closed.
3) The right-hand side of the state system is bounded by a linear function in the state and control variables.
4) The integrand of the objective functional is convex on U and is bounded below by $-k_2 + k_1 \left\|(u_1, u_2)\right\|^{\eta}$ with $k_1 > 0$, $k_2 > 0$and $\eta > 1$.

To prove the above theorem we need to use the following theorem 2 and 3.

Theorem 2. Let each of the functions $F_1, \ldots, F_n$ and the partial derivatives

$$\frac{\partial F_1}{\partial x_1}, \cdots, \frac{\partial F_1}{\partial x_n}, \cdots, \frac{\partial F_n}{\partial x_1}, \cdots, \frac{\partial F_n}{\partial x_n}$$

be continuous in a region $\mathcal{R}$ of $t, x_1, x_2, \ldots, x_n$ space defined by $\alpha < t < \beta$, $\alpha_1 < x_1 < \beta_1, \ldots, \alpha_n < x_1 < \beta_n$, and let the point $\left(t_0, x_1^0, x_2^0, \cdots, x_n^0\right)$ be in $\mathcal{R}$. Then there is an interval $[t\text{-}t_0] < h$ in which there exists a unique solution ($x_1 = \phi_1$ (t),..., $x_n = \phi_n$ (t))) of the system of differential equations

$$\begin{aligned} x_1' &= F_1\left(t, x_1, \cdots, x_n\right), \\ x_2' &= F_2\left(t, x_1, \cdots, x_n\right), \\ &\vdots \\ x_n' &= F_n\left(t, x_1, \cdots, x_n\right), \end{aligned}$$

that also satisfies the initial conditions

$$x_1\left(t_0\right) = x_1^0, x_2\left(t_0\right) = x_2^0, \cdots, x_n\left(t_0\right) = x_n^0.$$

Theorem 3. Let $x_i = F_i(t, x_1, \ldots, x_n)$ for $i \in [1, n]$ be a system of n differential equations with initial conditions $x_i\left(t_0\right) = x_i^0$ for $i \in [1, n]$. If each of the functions $F_1, \ldots, F_n$ and the partial derivatives

$$\frac{\partial F_1}{\partial x_1}, \cdots, \frac{\partial F_1}{\partial x_n}, \cdots, \frac{\partial F_n}{\partial x_1}, \cdots, \frac{\partial F_n}{\partial x_n}$$

are continuous in $\mathcal{R}^{n+1}$ space, then there exists a unique solution

$x_1 = \sigma_1(t), \ldots, x_n = \sigma_n(t)$ that satisfies the initial conditions.

Now with the help of the above two theorems we prove the four conditions of theorem (1).

Proof of theorem 1: To use an existence result in [8] , we must check the following properties [9].

1) The set of controls and corresponding state variables is non-empty.
2) The control set U is convex and closed.
3) The right-hand side of the state system is bounded by a linear function in the state and control variables.
4) The integrand of the objective functional is convex on U and is bounded below by $-k_2 + k_1 \left|(u_1, u_2)\right|^{\eta}$ with $k_1 > 0$, $k_2 > 0$, and $\eta > 1$.

Proof of 1): Let us consider,

$$\frac{dS}{dt} = F_1(t, S, I, R, N)$$

$$\frac{dI}{dt} = F_2(t, S, I, R, N)$$

$$\frac{dR}{dt} = F_3(t, S, I, R, N)$$

$$\frac{dN}{dt} = F_4(t, S, I, R, N),$$

where, F_1, F_2, F_3 and F_4 buildup the right hand side of the system (3). Let $u_1(t) = C_1$ and $u_2(t) = C_2$ for some constants C_1, C_2. F_1, F_2, F_3 and F_4 must be linear and they are also continuous everywhere. Moreover, the partial derivatives of F_1, F_2, F_3 and F_4 with respect to all state are constants and they are continuous everywhere.

So by following the theorem 3, we can say that there exists an unique solution $S = \sigma_1(t)$, $I = \sigma_2(t)$, $R = \sigma_3(t)$, $N = \sigma_4(t)$which satisfies the initial conditions. Therefore, the set of controls and corresponding state variables is non-empty. Hence the condition 1) is satisfied.

Proof of 2): By definition, U is closed. Take any controls u_1, $u_2 \in U$ and $\theta \in [0,1]$ Then

$$0 \leq \theta u_1 + (1-\theta) u_2.$$

Additionally, observe that $\theta u_1 \leq \theta$ and $(1-\theta)u_2 \leq (1-\theta)$. Then

$$\theta u_1 + (1-\theta)u_2 \leq \theta + (1-\theta) = 1.$$

Hence, $0 \leq \theta u_1 + (1-\theta)u_2 \leq 1$for all $u_1, u_2 \in U$ and $\theta \in [0,1]$. Therefore, U is convex, and condition 2) is satisfied.

Proof of 3):

We consider,

$$F_1 \leq \nu N - u_1 S$$
$$F_2 \leq K_1 I - u_2 I$$
$$F_3 \leq u_1 S + u_2 I + \gamma I$$
$$F_4 \leq \nu N.$$

The state system is given below:

$$\frac{dS}{dt} = F_1(t,S,I,R,N)$$
$$\frac{dI}{dt} = F_2(t,S,I,R,N)$$
$$\frac{dR}{dt} = F_3(t,S,I,R,N)$$
$$\frac{dN}{dt} = F_4(t,S,I,R,N).$$

Now we rewrite the system in matrix form:

$$\bar{F}(t,\bar{X},u) \leq \bar{m}\left(t, \begin{bmatrix} S \\ I \\ R \\ N \end{bmatrix}\right)\bar{X}(t) + \bar{n}\left(t, \begin{bmatrix} S \\ I \\ R \\ N \end{bmatrix}\right)\begin{pmatrix} u_1(t) \\ u_2(t) \end{pmatrix} \tag{6}$$

where,

$$\overline{m}\left(t,\begin{bmatrix}S\\I\\R\\N\end{bmatrix}\right)=\begin{bmatrix}0&0&0&v\\0&k_1&0&0\\0&\gamma&0&0\\0&0&0&v\end{bmatrix} \tag{7}$$

and

$$\overline{n}\left(t,\begin{bmatrix}S\\I\\R\\N\end{bmatrix}\right)=\begin{bmatrix}-S\\-I\\S+I\\0\end{bmatrix} \tag{8}$$

which gives a linear function of the controls u_1 and u_2 defined by time and state variables. Then we can find out the bound of the right hand side. It is noted that all parameters are constant and greater than or equal to zero. Therefore we can write,

$$\begin{aligned}\left|\overline{F}\left(t,\overline{X},u_1,u_2\right)\right| &\le \|\overline{m}\|\left|\overline{X}\right|+\left|\overline{S}+\overline{I}\right|\left|\left(u_1(t),u_2(t)\right)\right|\\ &\le p\left(\left|\overline{X}\right|+\left|\left(u_1(t),u_2(t)\right)\right|\right)\end{aligned}$$

Where $\overline{S}$, $\overline{I}$ are bounded and p includes the upper bound of the constant matrix. Hence we see that the right hand side is bounded by a sum of the state and the control. Therefore, condition 3) is satisfied.

Proof of 4):

For the proof of the condition 4) we use the result in [10] and [Fleming and Rishel (1975)]. The control and the state variables are non-negative values and are non-empty. In the minimization problem, the necessary convexity of the objective functional in u_1 is satisfied. The control variables u_1, $u_2 \in U$ is also convex and closed by definition. Furthermore, from [10] we see that the integrand in the objective functional which is

$$\left(AI(t)+A_2\frac{u_1^2}{2}+A_3\frac{u_2^2}{2}\right)$$

is convex on the control set U.

Now we have to prove that $J(u_1,u_2)\ge -k_2+k_1\left|\left(u_1,u_2\right)\right|^{\eta}$ with $k_1 > 0$, $k_2 > 0$ and $\eta >$ 1.

Here,

$$J(u_1,u_2)=A_1I(t)+B_1\frac{u_1^2}{2}+B_2\frac{u_2^2}{2}$$

$$J(u_1,u_2)\geq -A_1I(t)+B_1\frac{u_1^2}{2}+B_2\frac{u_2^2}{2}$$

$$J(u_1,u_2)\geq -A_1I(t)+\frac{1}{2}B(u_1^2+u_2^2)\quad [\text{let}, B_1+B_2=B]$$
$$=-k_2+k_1(u_1,u_2)^2$$

where, $k_2 > 0$ that depends on the upper bounds of $I(t)$. We can also see that $\eta = 2 > 1$, $k_1 > 0$. Hence, condition 4) is satisfied.

4. CHARACTERIZATION OF THE OPTIMAL CONTROL

In order to derive the necessary condition for the optimal control, we use pontryagin's maximum principle [2]. This principle converts the system and the objective functional into a problem minimizing pointwise a Hamiltonian H with respect to u_1 and u_2. In the objective function, the value A is the balancing parameter, B_1 and B_2 are the weight parameters balancing the cost. Here we can see from the system (3) that R appears only in the recovered class. So, when we build up the optimality system, we will ignore the recovered class.

By using pantraygin's Maximum principle we first derive the Hamiltonian which is given below

$$H(S,I,N)=A_1I(t)+B_1\frac{u_1^2}{2}+B_2\frac{u_2^2}{2}+\lambda_s(\nu N-\beta SI-\mu S-u_1S)$$
$$+\lambda_I(\beta SI-(\gamma+\mu+\alpha)I-u_2I)+\lambda_N((\nu-\mu)N-\alpha I) \tag{9}$$

where, λ_S, λ_I, λ_N are the associated adjoints for the state S, I, N respectively. By differentiating the Hamiltonian (H) with respect to each state variable, we find the differential equation for the associated adjoint. Hence, the adjoint system is,

$$\lambda_S' = \lambda_S(\mu + u_1 + \beta I) - \lambda_I \beta I$$
$$\lambda_I' = \lambda_S \beta S + \lambda_I(\gamma + \mu + \alpha) - \lambda_I \beta S + \lambda_I u_2 + \lambda_N \alpha - A_1$$
$$\lambda_N' = \lambda_S \nu + \lambda_N(\nu - \mu) \quad (10)$$

with the final conditions,

$$\lambda_S(T) = \lambda_I(T) = \lambda_N(T) = 0.$$

So by differentiating the Hamiltonian with respect to two controls u_1 and u_2 we obtain:

$$\left.\frac{\partial H}{\partial u_1}\right|_{u_1 = u_1^*} = 0$$

$$B_1 u_1^* - S\lambda_S = 0$$

$$u_1^* = \frac{S\lambda_S}{B_1}$$

and $\left.\frac{\partial H}{\partial u_2}\right|_{u_2 = u_2^*} = 0$

$$B_2 u_2^* - I\lambda_I = 0$$

$$u_2^* = \frac{I\lambda_I}{B_2}.$$

5. OPTIMALITY SYSTEM

State equations:

$$S'(t) = \nu N(t) - \beta S(t)I(t) - (\mu + u_1)S(t),$$
$$I'(t) = \beta S(t)I(t) - (\gamma + \mu + \alpha)I(t) - u_2 I(t),$$
$$R'(t) = \gamma I(t) - \mu R(t) + u_1 S(t) + u_2 I(t),$$
$$N'(t) = \nu N(t) - \alpha I(t) - \mu N(t), \quad (11)$$

with initial conditions,

$$S(0) = S_0 \geq 0, I(0) = I_0 \geq 0, R(0) = R_0 \geq 0, N(0) = N_0. \tag{12}$$

Adjoint equations:

$$\begin{aligned}
\lambda_S' &= \lambda_S(\mu + u_1 + \beta I) - \lambda_I \beta I. \\
\lambda_I' &= \lambda_S \beta S + \lambda_I(\gamma + \mu + \alpha) - \lambda_I \beta S + \lambda_I u_2 + \lambda_N \alpha - A_1. \\
\lambda_N' &= \lambda_S \nu + \lambda_N(\nu - \mu).
\end{aligned} \tag{13}$$

Transversality equations:

$$\lambda_S(T) = \lambda_I(T) = \lambda_N(T) = 0. \tag{14}$$

Characterization of the optimal control u_1^* and u_2^*:

$$u_1^* = \begin{cases} 0 & \text{if } \frac{S\lambda_S}{B_1} < 0, \\ \frac{S\lambda_S}{B_1} & \text{if } 0 \leq \frac{S\lambda_S}{B_1} \leq 1, \\ 1 & \text{if } \frac{S\lambda_S}{B_1} > 1. \end{cases} \tag{15}$$

and

$$u_2^* = \begin{cases} 0 & \text{if } \frac{I\lambda_I}{B_2} < 0, \\ \frac{I\lambda_I}{B_2} & \text{if } 0 \leq \frac{I\lambda_I}{B_2} \leq 1, \\ 1 & \text{if } \frac{I\lambda_I}{B_2} > 1. \end{cases} \tag{16}$$

In compact notion we can write,

$$u_1^* = \min\left[1, \max\left[0, \frac{S\lambda_S}{B_1}\right]\right] \tag{17}$$

and

$$u_2^* = \min\left[1, \max\left[0, \frac{I\lambda_I}{B_2}\right]\right]. \tag{18}$$

6. NUMERICAL RESULTS AND DISCUSSIONS

Numerical solutions to the optimal system are executed using MATLAB. The considered two controls (u_1, u_2) depend on the adjoints λ_S, λ_I and λ_N of the state variables S, I and N respectively. We simulate the model without control and with control and then we compare the results. We considered the numerical value of the controls u_1 and u_2 in between zero(0) and one(1) as they are not 100 percent effective. We also monitored the effectiveness of the weight parameter to see how the control is related to weight function. In this simulation we assumed the initial values of S, I and N as proportions instead of whole numbers.

The parameter values used in the simulations are presented in the following Table 1.

Figure 1 depicts the importance of the controls to the disease dynamics. From the graphs, we see that the control has a positive impact to reduce infection until the controls are effective enough. It is also clear from the figures that the disease can be controlled over finite period of time after imposing control strategies.

Figure 2 and Figure 3 monitored the impact of the parameter (a) of disease induced death rate. Here we see if a is at a low rate (a = 0.3) then the controls work effectively and as a result there is a significant reduction of the infections. When controls do not work it resulted the increase of infected individuals.

On the other hand for the higher rate of a (where awareness does not work, u_1 = 0 and the treatment u_2 works for a short period of time) there is a sharp decrease of infection due to death resulting the existence of fewer recovered people.

Table 1. Description and parameter values of the NiV model.

Variable	Description	Initial values
S_0	Initial susceptible individuals	0.90 [assumed]
I_0	Initial infected individuals	0.05 [assumed]
R_0	Initial recovered individuals	0.05 [assumed]
Parameters	**Description**	**Initial values**
ν	Birth rate	0.03 [assumed]
μ	Mortality rate	0.002 [assumed]
β	Contact rate	0.75 [7]
γ	Recovery rate	0.005 [assumed]
α	Disease induced death rate	0.01 [assumed]
A_1	Weight parameter	10 [7]
B_1	Weight parameter	1 [7]
B_2	Weight parameter	2 [7]
T	Number of years	6 [assumed]

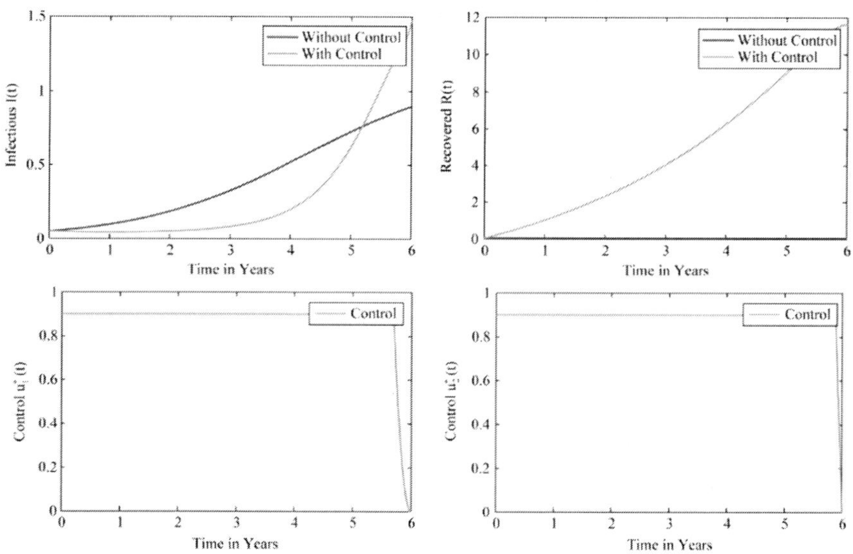

Figure 1. NIV model with control and without control, parameter values are taken from Table 1.

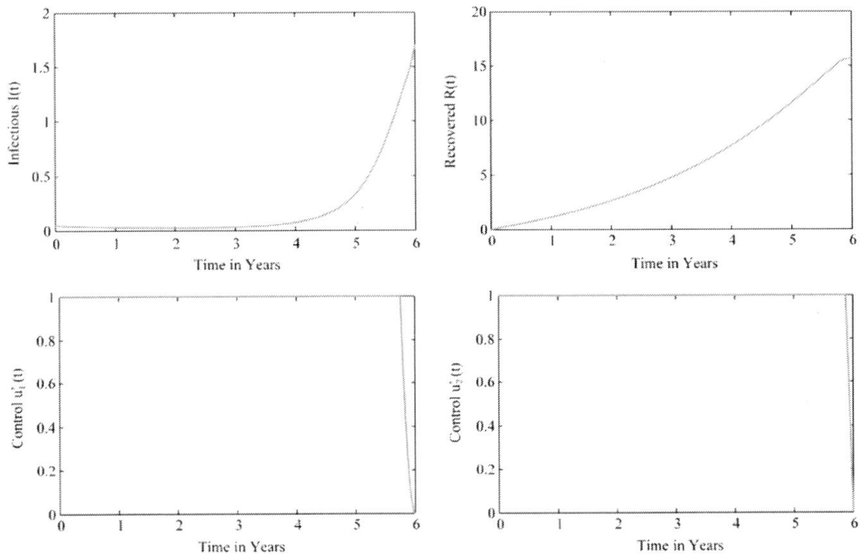

Figure 2. NiV model with low disease induced death rate, α = 0.3 and other parameter values are taken from Table 1.

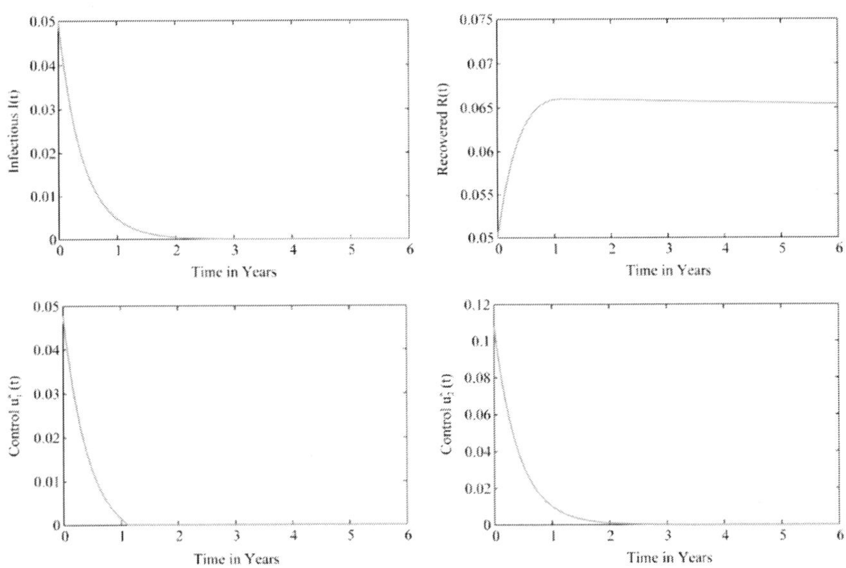

Figure 3. NiV model with high disease induced death rate, α = 3 and other parameter values are taken from Table 1.

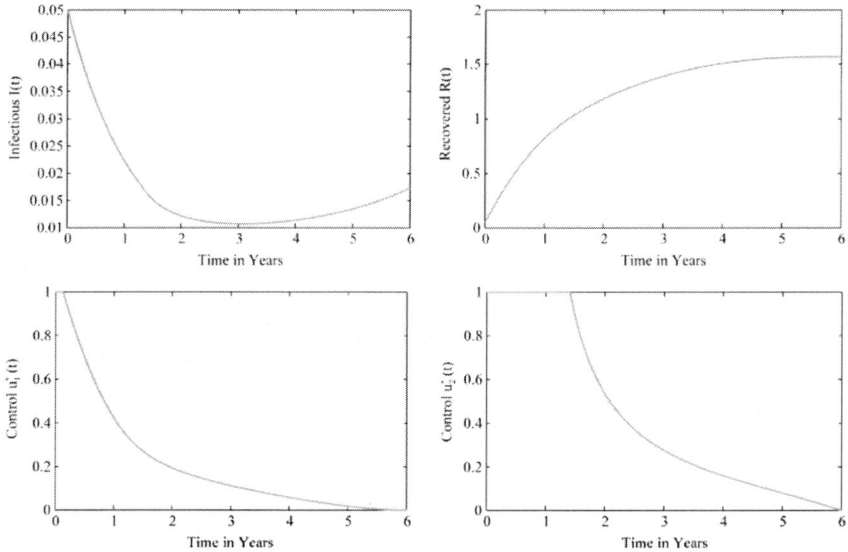

Figure 4. NiV model with low contact rate, β = 0.2 and other parameter values are taken from Table 1.

Figure 4 and Figure 5 show the comparative situation of the disease dynamics for low and high contact rates. In the case of low contact rate (b = 0.2), the infectious individuals decrease until the controls work effectively and as a result there is a notable increment of recovered individual.

On the other hand, for the very high contact rate (b = 2), which resulted a severe disease burden, the controls work for a longer period of time to reduce the disease burden.

Figure 6 and Figure 7 show the influence of the various weight parameters. Here we notice that for low weight parameters (B_1 = 0.2, B_2 = 0.3) the infectious individuals decrease sharply for first few years (as the controls work at maximum level).

It is also noticed that the infected individuals start to increase when the effectiveness of the controls start to decrease.

In the case of high weight parameter values (B_1 = 2, B_2 = 3) the high effectiveness of the controls are monitored and as a result there is a sharp reduction of infection during that effective level.

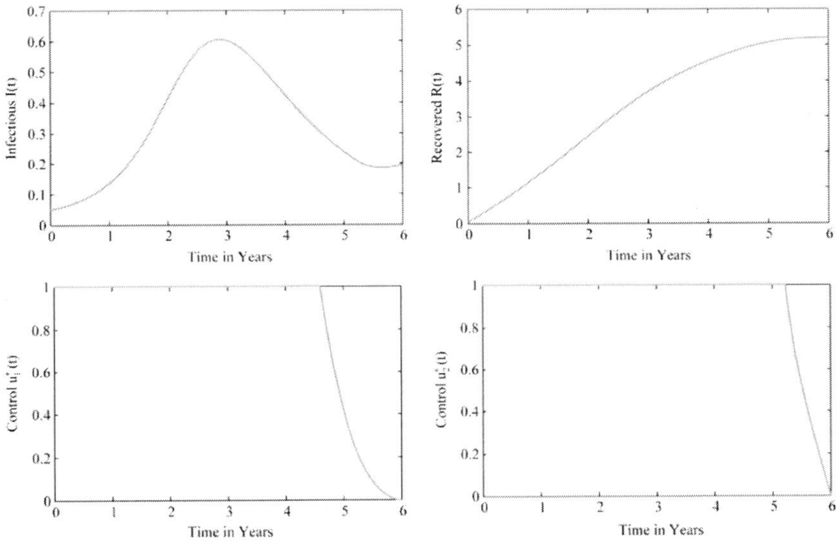

Figure 5. NiV model with high contact rate, β = 2 and other parameter values are taken from Table 1.

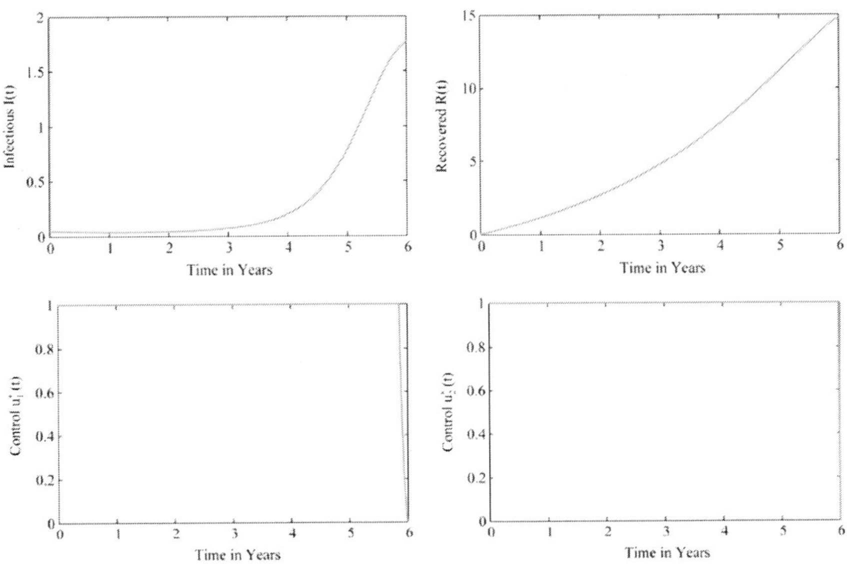

Figure 6. NiV model with low weight parameters, B_1 = 0.2, B_2 = 0.3 and other parameter values are taken from Table 1.

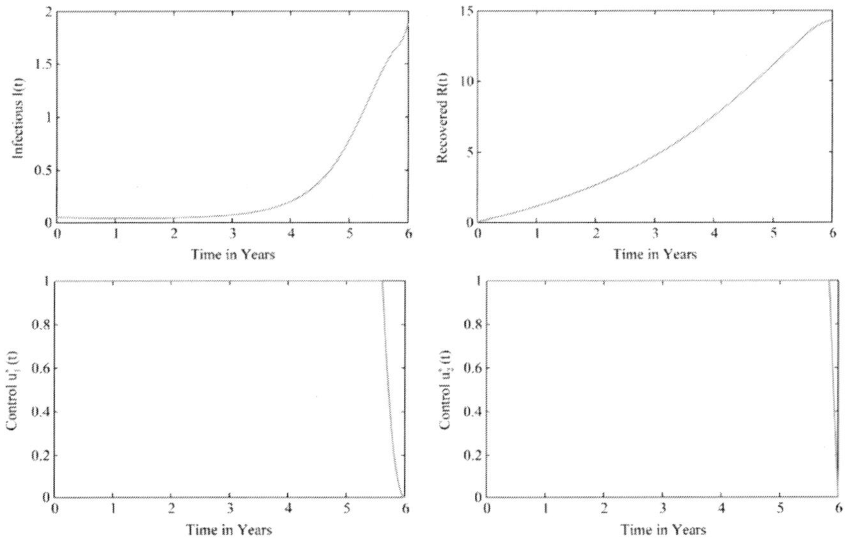

Figure 7. NiV model with high weight parameters, $B_1 = 2$, $B_2 = 3$ and other parameter values are taken from Table 1.

7. CONCLUSIONS

The important findings are given below:

- A comparison between with and without control strategy is monitored. The effect of control parameters is very much notable for reducing the infected individuals to control the disease dynamics.
- The controls need to be effective for longer period of time in case of high incidence.
- The optimal control is much more effective to minimize the infected individuals (as a result recovered individuals will be maximized) and also to minimize the cost of the two control measures.
- For low weight parameter values, the controls show their effectiveness at a maximum level.
- From the simulations it is monitored that the optimal combination of treatment and creating awareness is very prominent for disease elimination.

ACKNOWLEDGEMENTS

The author, JS acknowledge, with thanks, the support in part of the National Science and Technology (NST), Dhaka. The authors are grateful to the reviewers for their constructive comments.

REFERENCES

1. Chong, H.T., Hossain, M.J. and Tan, C.T. (2008) Differences in Epidemiologic and Clinical Features of Nipah Virus Encephalitis between the Malaysian and Bangladesh Outbreaks. Neurology Asia, 13, 23-26.
2. Pontryagin, L.S., Boltyanskii, V.G., Gamkrelize, R.V. and Mishchenko, E.F. (1962) The Mathematical Theory of Optimal Processes. New York, Wiley.
3. (2008) Nipah Virus Infections. WHO Report, Asia-Pacific Region, World Health Organization.
4. (2011) National Guideline for Management, Prevention and Control of Nipah Virus Infection including Encephalitis, Directorate General of Health Services. Ministry of Health and Family Welfare, Government of the People's Republic of Bangladesh.
5. (2004) Nipah Virus: Vaccination and Passive Protection Studies in a Hamster Model. Journal of Virology, 78, 834-840. http://dx.doi.org/10.1128/JVI.78.2.834-840.2004
6. Biswas, M.H.A. (2014) Optimal Control of Nipah Virus (NIV) Infections: A Bangladesh Scenario. Journal of Pure and Applied Mathematics: Advances and Applications, 12, 77-104.
7. Bakare, E.A., Nwagwo, A. and Danso-Addo, E. (2014) Optimal Control Analysis of an SIR Epidemic Model with Constant Recruitment. International Journal of Applied Mathematical Research, 3, 273-285. http://dx.doi.org/10.14419/ijamr.v3i3.2872
8. Fleming, W.H. and Rishel, R.W. (1975) Deterministic and Stochastic Optimal Control. Springer Verlag, New York. http://dx.doi.org/10.1007/978-1-4612-6380-7
9. Yusuf, T.T. and Benyah, F. (2012) Optimal Control of Vaccination and Treatment for an SIR Epidemiological Model. World Journal of Modelling and Simulation, 8, 194-204.
10. Hsieh, Y. and Sheu, S. (2001) The Effect of Density-Dependent Treatment and Behaviour Change on the Dynamics of HIV Transmission. Journal of Mathematical Biology, 43, 69-80. http://dx.doi.org/10.1007/s002850100087

CHAPTER 12

Isoperimetric inequalities for the logarithmic potential operator☆

MichaelRuzhansky[a,?], DurvudkhanSuraganb,[a]

[a]*Department of Mathematics, Imperial College London, 180 Queen's Gate, London SW7 2AZ, UnitedKingdom*

[b]*Institute of Mathematics and Mathematical Modeling, 125 Pushkin str., 050010 Almaty, Kazakhstan*

ABSTRACT

In this paper we prove that the disc is a maximiser of the Schatten p-norm of the logarithmic potential operator among all domains of a given measure in R^2, for all even integers $2 \leq p < \infty$. We also show that the equilateral triangle has the largest Schatten p-norm among all triangles of a given area. For the logarithmic potential operator on bounded open or triangular domains, we also obtain analogies of the Rayleigh–Faber–Krahn or Pólya inequalities, respectively. The logarithmic potential operator can be related to a nonlocal boundary value problem for the Laplacian, so we obtain isoperimetric inequalities for its eigenvalues as well.

Keywords: Logarithmic potential, Characteristic numbers, Schatten class, Isoperimetric inequality, Rayleigh–Faber–Krahn inequality, Pólya inequality

1. INTRODUCTION

Let $\Omega \subset R^2$ be an open bounded set. We consider the logarithmic potential operator on $L^2(\Omega)$ defined by

$$\mathcal{L}_\Omega f(x) := \int_\Omega \frac{1}{2\pi} \ln \frac{1}{|x-y|} f(y)dy, \quad f \in L^2(\Omega) \tag{1.1}$$

where ln is the natural logarithm and $|x - y|$ is the standard Euclidean distance between x and y. Clearly, L_Ω is compact and self-adjoint. Therefore, all of its eigenvalues and characteristic numbers are discrete and real. We recall that the characteristic numbers are the inverses of the eigenvalues. The characteristic numbers of L_Ω may be enumerated in ascending order of their modulus,

$$|\mu_1(\Omega)| \leq |\mu_2(\Omega)| \leq \ldots$$

where $\mu_i(\Omega)$ is repeated in this series according to its multiplicity. We denote the corresponding eigenfunctions by $u_1, u_2, \ldots$, so that for each characteristic number μ_i there is a unique corresponding (normalized) eigenfunction ui,

$$u_i = \mu_i(\Omega) L\Omega u_i, \; i=1,2,\ldots.$$

It is known, see for example Mark Kac [12] (see also [15]), that the equation

$$u(x) = \mathcal{L}_\Omega f(x) = \int_\Omega \frac{1}{2\pi} \ln \frac{1}{|x-y|} f(y)dy$$

is equivalent to the equation

$$-\Delta u(x) = f(x), \; x \in \Omega, \; (1.2)$$

with the nonlocal integral boundary condition

$$-\frac{1}{2}u(x) + \int_{\partial\Omega} \frac{\partial}{\partial n_y} \frac{1}{2\pi} \ln \frac{1}{|x-y|} u(y) dS_y - \int_{\partial\Omega} \frac{1}{2\pi} \ln \frac{1}{|x-y|} \frac{\partial u(y)}{\partial n_y} dS_y = 0, \; x \in \partial\Omega \tag{1.3}$$

where $\frac{\partial}{\partial n_y}$ denotes the outer normal derivative at a point y on the boundary $\partial\Omega$, which is assumed piecewise $C1$ here.

In general, the boundary value problem (1.2)–(1.3) has several interesting applications (see Kac [12,14], Saito [23] and [15]).

Spectral properties of the logarithmic potential have been considered in many papers (see [1,2,5,9,13,24,25]). In this paper we are interested in isoperimetric inequalities of the logarithmic potential L_Ω, that is also, in isoperimetric inequalities of the nonlocal Laplacian (1.2)–(1.3). For a recent general review of isoperimetric inequalities for the Dirichlet, Neumann and other Laplacians we refer to Benguria, Linde and Loewe in [4]. Isoperimetric inequalities for Schatten norms for double layer potentials have been recently considered by Miyanishi and Suzuki [19].

In Rayleigh's famous book "Theory of Sound" (first published in 1877), by using some explicit computation and physical interpretations, he stated that the disc minimizes (among all domains of the same area) the first eigenvalue of the Dirichlet Laplacian. The proof of this conjecture was obtained about 50 years later, simultaneously (and independently) by G. Faber and E. Krahn. Nowadays, the Rayleigh–Faber–Krahn inequality has been established for many other operators; see e.g. [11] for further references (see also [3] and [21]). Among other things, in this paper we also prove the Rayleigh–Faber–Krahn theorem for the integral operator L_Ω, i.e. it is proved that the disc is a minimizer of the first eigenvalue of the Laplacian (1.2)–(1.3) among all domains of a given measure in R_2.

By using the Feynman–Kac formula and spherical rearrangement Luttinger [18]proved that the disc D is a maximizer of the partition function of the Dirichlet Laplacian among all domains of the same area as D for all positive values of time, i.e.

$$\sum_{i=1}^{\infty} \exp(-t\mu_i^{\mathcal{D}}(\Omega)) \leq \sum_{i=1}^{\infty} \exp(-t\mu_i^{\mathcal{D}}(D)), \quad \forall t > 0, \ |\Omega| = |D|$$

where $\mu_i^{\mathcal{D}}, i = 1, 2, \ldots,$ are the characteristic numbers of the Dirichlet Laplacian. From here by using the Mellin transform one obtains

$$\sum_{i=1}^{\infty} \frac{1}{[\mu_i^{\mathcal{D}}(\Omega)]^p} \leq \sum_{i=1}^{\infty} \frac{1}{[\mu_i^{\mathcal{D}}(D)]^p}, \quad |\Omega| = |D| \quad , \tag{1.4}$$

when $p > 1$, $\Omega \subset R^2$. We prove an analogy of this Luttinger's inequality for the integral operator $L\Omega$. In our note [22] we obtained similar results for convolution type integral operators with positive nonincreasing kernels. In the present setting the main difficulty arises from the fact that the logarithmic kernel is not positive and that we cannot use the Brascamp–Lieb–Luttinger type rearrangement inequalities directly.

In Section 2 we present main results of this paper. Their proofs will be given in Section 4 and Section 3. In Section 5 we discuss shortly about isoperimetric inequalities for polygons and show that the Schatten p-norm is maximised on the equilateral triangle centered at the origin among all triangles of a given area.

The authors would like to thank Grigori Rozenblum and Rupert Frank for comments.

2. MAIN RESULTS AND EXAMPLES

Let H be a separable Hilbert space. By $S^\infty(H)$ we denote the space of compact operators P : H → H. Recall that the singular values $\{s_n\}$ of $P \in S\infty(H)$ are the eigenvalues of the positive operator $(P^*P)1/2$ (see e.g. [10]). The Schatten p-classes are defined as

$$S^p(H) := \{P \in S^\infty(H) : \{s_n\} \in \ell^p\}, \quad 1 \leq p < \infty.$$

In $S^p(H)$ the Schatten p-norm of the operator P is defined by

$$\|P\|_p := \left(\sum_{n=1}^{\infty} s_n^p \right)^{\frac{1}{p}}, \quad 1 \leq p < \infty \tag{2.1}$$

For $p = \infty$, we can set

$$\|P\|I := \|P\|$$

to be the operator norm of P on H. As outlined in the introduction, we assume that $\Omega \subset R^2$ is an open bounded set and we consider the logarithmic potential operator on $L2(\Omega)$ of the form

$$\mathcal{L}_\Omega f(x) = \int_\Omega \frac{1}{2\pi} \ln \frac{1}{|x-y|} f(y)dy, \quad f \in L^2(\Omega)$$

We also assume that the operator L_Ω is positive:

REMARK 2.1 In Landkof [16, Theorem 1.16, p. 80] the positivity of the operator L_Ω is proved in domains $\bar{\Omega} \subset U$, where U is the unit disc. In general, L_Ω is not a positive operator. For any bounded open domain Ω the logarithmic potential operator L_Ω can have at most one negative eigenvalue, see Troutman [24] (see also Kac [13]).

Note that for positive self-adjoint operators the singular values equal the eigenvalues. It is known that L_Ω is a Hilbert–Schmidt operator. By $|\Omega|$ we will denote the Lebesgue measure of Ω.

THEOREM 2.2 *Let D be a disc centered at the origin. Then*

$$||L_\Omega||p \le ||L_D||_p \tag{2.3}$$

for any even integer $2 \le p < \infty$ *and any bounded open domain* Ω *with* $|\Omega| = |D|$.

Note that for even integers p we do not need to assume the positivity of the logarithmic potential operator. For odd integers we have the following:

THEOREM 2.3 *Let D be a disc centered at the origin and let* Ω *be a bounded open domain with* $|\Omega| = |D|$. *Assume that the logarithmic potential operator is positive for* Ω *and D. Then*

$$||L_\Omega||p \le ||L_D||p \tag{2.4}$$

for any integer $2 \le p < \infty$.

Let us give several examples calculating explicitly values of the right hand side of (2.3) for different values of p.

EXAMPLE 2.4 Let $D \equiv U$ be the unit disc. Then by Theorem 2.2 we have

$$\|\mathcal{L}_\Omega\|_p \le \|\mathcal{L}_U\|_p = \left(\sum_{m=1}^{\infty} \frac{3}{j_{0,m}^{2p}} + \sum_{l=1}^{\infty} \sum_{m=1}^{\infty} \frac{2}{j_{l,m}^{2p}} \right)^{\frac{1}{p}} \tag{2.5}$$

for any even $2 \le p < \infty$ and any bounded open domain Ω with $|\Omega| = |D|$. Here j_{km} denotes the mth positive zero of the Bessel function J_k of the first kind of order k.

The right hand sight of the formula (2.5) can be confirmed by a direct calculation of the logarithmic potential eigenvalues in the unit disc, see Theorem 3.1 in [1].

We also obtain the following Rayleigh–Faber–Krahn inequality when $p = \infty$:

THEOREM 2.5 *The disc D is a minimizer of the characteristic number of the logarithmic potential* L_Ω *with the smallest modulus among all domains of a given measure, that is,*

$$||L_\Omega|| \leq ||L_D||$$

for an arbitrary bounded open domain $\Omega \subset R^2$ *with* $|\Omega| = |D|$. *Here*$||\cdot||$*is the operator norm on the space* L^2.

EXAMPLE 2.6 Let $D \equiv U$ be the unit disc. Then by Theorem 2.5 we have

$$||\mathcal{L}_\Omega|| \leq ||\mathcal{L}_U|| = \frac{1}{j_{01}^2} \tag{2.6}$$

for any bounded open domain Ω with $|\Omega| = |D|$.

From Corollary 3.2 in [1] we calculate explicitly the operator norm in the right hand sight of (2.6).

3. PROOF OF THEOREM 2.5

Let us first prove Theorem 2.5. To do it we first prove the following:

LEMMA 3.1 *The characteristic number*$\mu 1$*of the logarithmic potential* L_Ω *with the smallest modulus is simple, and the corresponding eigenfunction* u_1 *can be chosen nonnegative.*

PROOF The eigenfunctions of the logarithmic potential L_Ω may be chosen to be real as its kernel is real. First let us prove that u_1 cannot change sign in the domain Ω, that is,

$$u_1(x)u1(y) = |u_1(x)u1(y)|, x,y \in \Omega.$$

In fact, in the opposite case, by virtue of the continuity of the function $u_1(x)$, there would be neighborhoods $U(x_0, r) \subset \Omega$ such that

$$|u_1(x)u_1(y)| > u_1(x)u_1(y), x, y \in U(x_0, r) \subset \Omega.$$

On the other hand we have

$$\int_{\Omega} \frac{1}{2\pi} \ln \frac{1}{|x_0 - z|} \frac{1}{2\pi} \ln \frac{1}{|z - x_0|} dz > 0, \quad x_0 \in \Omega \tag{3.2}$$

From here by continuity it is simple to check that there exists $\rho > 0$ such that

$$\int_{\Omega} \frac{1}{2\pi} \ln \frac{1}{|x - z|} \frac{1}{2\pi} \ln \frac{1}{|z - y|} dz > 0, \quad x, y \in U(x_0, \rho) \subset U(x_0, r) \tag{3.2}$$

Now let us introduce a new function

$$\widetilde{u}_1(x) := \begin{cases} |u_1(x)|, \ x \in U(x_0, \rho), \\ u_1(x), \ x \in \Omega \backslash U(x_0, \rho) \end{cases} \tag{3.3}$$

Then we obtain

$$\begin{aligned} \frac{(\mathcal{L}_\Omega^2 \widetilde{u}_1, \widetilde{u}_1)}{\|\widetilde{u}_1\|^2} &= \frac{1}{\|\widetilde{u}_1\|^2} \int_\Omega \int_\Omega \int_\Omega \frac{1}{2\pi} \ln \frac{1}{|x - z|} \frac{1}{2\pi} \ln \frac{1}{|z - y|} dz \widetilde{u}_1(x) \widetilde{u}_1(y) dx dy \\ &> \frac{1}{\|u_1\|^2} \int_\Omega \int_\Omega \int_\Omega \frac{1}{2\pi} \ln \frac{1}{|x - z|} \frac{1}{2\pi} \ln \frac{1}{|z - y|} dz u_1(x) u_1(y) dx dy = \frac{1}{\mu_1^2} \end{aligned} \tag{3.4}$$

where μ_1^2 is the smallest characteristic number of $\mathcal{L}_\Omega^2$ and u_1 is the eigenfunction corresponding to μ_1^2, i.e.

$$u_1 = \mu_1^2 \mathcal{L}_\Omega^2 u_1$$

Therefore, by the variational principle we also have

$$\frac{1}{\mu_1^2} = \sup_{f \in L^2(\Omega)} \frac{(\mathcal{L}_\Omega^2 f, f)}{\|f\|^2} \tag{3.5}$$

This means that the strong inequality (3.4) contradicts the variational principle (3.5) because $\|\widetilde{u}_1\|_{L^2} = \|u_1\|_{L^2} < \infty$.

Since u_1 is nonnegative it follows that μ_1 is simple. Indeed, if there were an eigenfunction v_1 linearly independent of u_1 and corresponding to μ_1, then for all real c the linear combination $u_1 + cv_1$ also would be an eigenfunction corresponding to μ_1 and therefore, by what has been proved, it could not become negative in Ω. As c is arbitrary, this is impossible. □

PROOF OF THEOREM 2.5 Let Ω be a bounded open set in R^2. Its symmetric rearrangement $\Omega^* \equiv D$ is an open disc centered at 0 with the measure equal to the measure of Ω, i.e. $|D| = |\Omega|$. Let u be a nonnegative measurable function in Ω, such that all its positive level sets have finite measure. In the definition of the symmetric-decreasing rearrangement of u one can use the so-called layer-cake decomposition (see [17]), which expresses a nonnegative function u in terms of its level sets as

$$u(x) = \int_0^\infty \chi_{\{u(x)>t\}} dt \tag{3.6}$$

where χ is the characteristic function of the corresponding domain.

DEFINITION 3.2 Let u be a nonnegative measurable function in Ω. The function

$$u^*(x) := \int_0^\infty \chi_{\{u(x)>t\}^*} dt \tag{3.7}$$

is called the (radially) symmetric-decreasing rearrangement of a nonnegative measurable function u.

As in the proof of Lemma 3.1 $\mu_1^2(\Omega)$ is the smallest characteristic number of $\mathcal{L}_\Omega^2$ and u_1 is the eigenfunction corresponding to μ_1^2, i.e.

$$u_1 = \mu_1^2(\Omega)\mathcal{L}_\Omega^2 u_1.$$

By Lemma 3.1 the first characteristic number μ_1 of the operator L_Ω is simple; the corresponding eigenfunction u_1 can be chosen positive in Ω, and in view of Lemma 3.1 we can apply the above construction to the first eigenfunction u_1. Recall[1] the rearrangement inequality for the logarithmic kernel (cf. Lemma 2 in [7])

$$\int_\Omega\int_\Omega\int_\Omega u_1(y)\frac{1}{2\pi}\ln\frac{1}{|y-z|}\frac{1}{2\pi}\ln\frac{1}{|z-x|}u_1(x)dzdydx \leq$$
$$\int_D\int_D\int_D u_1^*(y)\frac{1}{2\pi}\ln\frac{1}{|y-z|}\frac{1}{2\pi}\ln\frac{1}{|z-x|}u_1^*(x)dzdydx. \tag{3.8}$$

In addition, for each nonnegative function $u \in L^2(\Omega)$ we have

$$\|u\|_{L2(\Omega)} = \|u^*\|_{L2(D)}. \tag{3.9}$$

Therefore, from (3.8), (3.9) and the variational principle for the positive operator $\mathcal{L}_D^2$, we get

$$\mu_1^2(\Omega) = \frac{\int_\Omega |u_1(x)|^2 dx}{\int_\Omega\int_\Omega\int_\Omega u_1(y)\frac{1}{2\pi}\ln\frac{1}{|y-z|}\frac{1}{2\pi}\ln\frac{1}{|z-x|}u_1(x)dzdydx} \geq$$
$$\frac{\int_D |u_1^*(x)|^2 dx}{\int_D\int_D\int_D u_1^*(y)\frac{1}{2\pi}\ln\frac{1}{|y-z|}\frac{1}{2\pi}\ln\frac{1}{|z-x|}u_1^*(x)dzdydx} \geq$$
$$\inf_{v\in L^2(D)} \frac{\int_D |v(x)|^2 dx}{\int_D\int_D\int_D v(y)\frac{1}{2\pi}\ln\frac{1}{|y-z|}\frac{1}{2\pi}\ln\frac{1}{|z-x|}v(x)dzdydx} = \mu_1^2(D)$$

Finally, note that 0 is not a characteristic number of L_D (cf. Corollary 1 in [24]). Therefore,

$$0 < |\mu_1(D)|.$$

This completes the proof. □

4. PROOFS OF THEOREM 2.2 AND THEOREM 2.3

First we prove the Brascamp–Lieb–Luttinger type rearrangement inequality for the logarithmic kernel (cf. [6]).

LEMMA 4.1 *Let D be a disc centered at the origin. Then*

$$\int_\Omega \cdots \int_\Omega \frac{1}{2\pi} \ln \frac{1}{|y_1 - y_2|} \cdots \frac{1}{2\pi} \ln \frac{1}{|y_p - y_1|} dy_1 \ldots dy_p \leq$$

$$\int_D \cdots \int_D \frac{1}{2\pi} \ln \frac{1}{|y_1 - y_2|} \cdots \frac{1}{2\pi} \ln \frac{1}{|y_p - y_1|} dy_1 \ldots dy_p,$$

for any $p = 2,3,\ldots$, *and for any bounded open set* Ω *with* $|\Omega| = |D|$.

PROOF. Here we prove it for $p = 2$ and the proof is based on the proof of Lemma 2 in [7]. The proof for arbitrary p is essentially the same as the case $p = 2$. Let us fix $r_0 > 0$ and consider the function

$$f(r) := \begin{cases} \frac{1}{2\pi} \ln \frac{1}{r}, & r \leq r_0, \\ \frac{1}{2\pi} \ln \frac{1}{r_0} - \frac{1}{2\pi} \int_{r_0}^{r} s^{-1} \frac{1+r_0^2}{1+s^2} ds, & r > r_0 \end{cases} \tag{4.2}$$

Let us see that the function $f(r)$ is strictly decreasing and has a limit as $r \to \infty$. If $r \leq r_0$ then

$$f(r_1) = \frac{1}{2\pi} \ln \frac{1}{r_1} > \frac{1}{2\pi} \ln \frac{1}{r_2} = f(r_2)$$

for $r_1 < r_2$. If $r > r_0$ then

$$f(r) = \frac{1}{2\pi} \ln \frac{1}{r_0} - \frac{1}{2\pi} \int_{r_0}^{r} s^{-1} \frac{1 + r_0^2}{1 + s^2} ds =$$

$$\frac{1}{2\pi} \ln \frac{1}{r_0} - \frac{1}{2\pi}(1 + r_0^2)[\ln r - \frac{1}{2} \ln(1 + r^2) - \ln r_0 + \frac{1}{2} \ln(1 + r_0^2)] \tag{4.3}$$

Thus $f(r_1) > f(r_2)$ for $r_1 < r_2$, that is, $f(r)$ is strictly decreasing. From (4.3) it is easy to see that

$$\lim_{r \to \infty} f(r) = \frac{1}{2\pi} \ln \frac{1}{r_0} - \frac{1}{2\pi}(1 + r_0^2)[- \ln r_0 + \frac{1}{2} \ln(1 + r_0^2)] \tag{4.4}$$

We use the notation

$$f_\infty := \frac{1}{2\pi}\ln\frac{1}{r_0} - \frac{1}{2\pi}(1+r_0^2)[-\ln r_0 + \frac{1}{2}\ln(1+r_0^2)]$$

By construction $\frac{1}{2\pi}\ln\frac{1}{r} - f(r)$ is decreasing. Thus if we define

$$h_1(r) = f(r) - f_\infty$$

we have the decomposition

$$\frac{1}{2\pi}\ln\frac{1}{r} = h_1(r) + h_2(r)$$

where h_1 is positive strictly decreasing function and h_2 is decreasing. Hence by the Brascamp–Lieb–Luttinger rearrangement inequality we have

$$\int_\Omega\int_\Omega h_1(|y_1-y_2|)h_1(|y_2-y_1|)dy_1dy_2 \le \int_D\int_D h_1(|y_1-y_2|)h_1(|y_2-y_1|)dy_1dy_2 \tag{4.5}$$

and

$$\int_\Omega\int_\Omega h_2(|y_1-y_2|)h_2(|y_2-y_1|)dy_1dy_2 \le \int_D\int_D h_2(|y_1-y_2|)h_2(|y_2-y_1|)dy_1dy_2 \tag{4.6}$$

Thus it remains to show

$$\int_\Omega\int_\Omega h_1(|y_1-y_2|)h_2(|y_2-y_1|)dy_1dy_2 \le \int_D\int_D h_1(|y_1-y_2|)h_2(|y_2-y_1|)dy_1dy_2 \tag{4.7}$$

which does not follow directly from the Brascamp–Lieb–Luttinger rearrangement inequality since h_2 is not positive. Define for $R > 0$

$$q_R(r) := \begin{cases} h_2(r) - h_2(R), & r \le R \\ 0, & r > R, \end{cases} \tag{4.8}$$

and note that by monotone convergence

$$I_\Omega(h_1, h_2) = \lim_{R\to\infty} [I_\Omega(h_1, q_R) + h_2(R) \int_\Omega \int_\Omega h_1(|y_1 - y_2|) dy_1 dy_2] \tag{4.9}$$

with the notation

$$I_\Omega(f, g) = \int_\Omega \int_\Omega f(|y_1 - y_2|) g(|y_2 - y_1|) dy_1 dy_2 \tag{4.10}$$

Since h_1 and q_R are positive and nonincreasing

$$I_\Omega(h_1, q_R) \le I_D(h_1, q_R)$$

by the Brascamp–Lieb–Luttinger rearrangement inequality. Noting that

$$\int_\Omega \int_\Omega h_1(|y_1 - y_2|) dy_1 dy_2 \le \int_D \int_D h_1(|y_1 - y_2|) dy_1 dy_2$$

we obtain

$$I_\Omega(h_1, h_2) = \lim_{R\to\infty} [I_\Omega(h_1, q_R) + h_2(R) \int_\Omega \int_\Omega h_1(|y_1 - y_2|) dy_1 dy_2] \le$$

$$\lim_{R\to\infty} [I_D(h_1, q_R) + h_2(R) \int_D \int_D h_1(|y_1 - y_2|) dy_1 dy_2] = I_D(h_1, h_2)$$

completing the proof. □

PROOF OF THEOREM 2.2 Since the logarithmic potential operator is a Hilbert–Schmidt operator, by using bilinear expansion of its iterated kernels (see, for example, [26]) we obtain for $p \geq 2$, $p \in N$,

$$\sum_{j=1}^{\infty} \frac{1}{\mu_j^p(\Omega)} = \int_\Omega \cdots \int_\Omega \frac{1}{2\pi} \ln \frac{1}{|y_1 - y_2|} \cdots \frac{1}{2\pi} \ln \frac{1}{|y_p - y_1|} dy_1 \ldots dy_p \tag{4.12}$$

Recalling the inequality (4.1) stating that

$$\int_{\Omega}\dots\int_{\Omega}\frac{1}{2\pi}\ln\frac{1}{|y_1-y_2|}\cdots\frac{1}{2\pi}\ln\frac{1}{|y_p-y_1|}dy_1\dots dy_p\leq$$
$$\int_{D}\dots\int_{D}\frac{1}{2\pi}\ln\frac{1}{|y_1-y_2|}\cdots\frac{1}{2\pi}\ln\frac{1}{|y_p-y_1|}dy_1\dots dy_p, \tag{4.13}$$

we obtain

$$\sum_{j=1}^{\infty}\frac{1}{\mu_j^p(\Omega)}\leq\sum_{j=1}^{\infty}\frac{1}{\mu_j^p(D)},\quad p\geq 2,\ p\in\mathbb{N} \tag{4.14}$$

for any bounded open domain $\Omega \subset R^2$ with $|\Omega| = |D|$. Taking even p in (4.14)we complete the proof of Theorem 2.2. □

PROOF OF THEOREM 2.3 The inequality (4.14) also proves Theorem 2.3 when the logarithmic potential operator is positive (see also Remark 2.1).

REMARK 4.2 It follows from the properties of the kernel that the Schatten p-norm of the operator L_Ω is finite when $p > 1$ see e.g. the criteria for Schatten classes in terms of the regularity of the kernel in [8]. The above techniques do not allow us to prove Theorem 2.2 for all $p > 1$. In view of the Dirichlet Laplacian case, it seems reasonable to conjecture that the Schatten p-norm is still maximised on the disc also for all $p > 1$.

5. ON THE CASE OF POLYGONS

We can ask the same question of maximizing the Schatten p-norms in the class of polygons with a given number n of sides. We denote by P_n the class of plane polygons with n edges. We would like to identify the maximizer for Schatten p-norms of the logarithmic potential L_Ω in P_n. According to Section 2, it is natural to conjecture that it is the n-regular polygon. Currently, we can prove this only for $n = 3$:

THEOREM 5.1 *The equilateral triangle centered at the origin has the largest Schatten p-norm of the operator L_Ω for any even integer $2 \leq p < \infty$ among all triangles of a given area. More precisely, if Δ is the equilateral triangle centered at the origin, we have*

$$\|L_\Omega\|p \leq \|L\Delta\|p \tag{5.1}$$

for any even integer $2 \leq p < \infty$ *and any bounded open triangle* Ω *with* $|\Omega| = |\Delta|$.

Similarly, we have the following analogy of Theorem 2.3:

THEOREM 5.2 *Let* Δ *be an equilateral triangle centered at the origin and let* Ω *be a bounded open triangle with* $|\Omega| = |\Delta|$. *Assume that the logarithmic potential operator is positive for* Ω *and* Δ. *Then*

$$||L_\Omega||p \leq ||L\Delta||p \tag{5.2}$$

for any integer $2 \leq p < \infty$.

Let u be a nonnegative, measurable function on R^2, and let x^2 be a line through the origin of R^2. Choose an orthogonal coordinate system in R^2 such that the x^1-axis is perpendicular to x^2.

DEFINITION 5.3 (See [6].) A nonnegative, measurable function $u\star(x|x^2)$ on R^2 is called a *Steiner symmetrization* with respect to x^2 of the function $u(x)$, if $u\star(x^1,x^2)$ is a symmetric decreasing rearrangement with respect to x^1 of $u(x^1,x^2)$ for each fixed x^2.

The Steiner symmetrization (with respect to the x^1-axis) $\Omega\star$ of a measurable set Ω is defined in the following way: if we write (x^1,z) with $z \in R$, and let $\Omega_z = \{x_1 : (x^1, z) \in \Omega\}$, then

$$\Omega^\star = \{(x^1,z) \in R \times R : x^1 \in \Omega^* z\}$$

where Ω^* is a symmetric rearrangement of Ω (see the proof of Theorem 2.5). We obtain:

LEMMA 5.4 *For a positive function u and a measurable* $\Omega \subset R^2$ *we have*

$$\int_\Omega\int_\Omega\int_\Omega u(y)\frac{1}{2\pi}\ln\frac{1}{|y-z|}\frac{1}{2\pi}\ln\frac{1}{|z-x|}u(x)dzdydx \leq$$
$$\int_{\Omega^\star}\int_{\Omega^\star}\int_{\Omega^\star} u^\star(y)\frac{1}{2\pi}\ln\frac{1}{|y-z|}\frac{1}{2\pi}\ln\frac{1}{|z-x|}u^\star(x)dzdydx$$

Where $\Omega^\star$ *and* $u^\star$ *are Steiner symmetrizations of* Ω *and u, respectively.*

PROOF. The proof is based on the proof of Lemma 2 in [7]. Let us fix $r_0 > 0$ and consider the function

$$f(r) := \begin{cases} \frac{1}{2\pi} \ln \frac{1}{r}, & r \leq r_0, \\ \frac{1}{2\pi} \ln \frac{1}{r_0} - \frac{1}{2\pi} \int_{r_0}^{r} s^{-1} \frac{1+r_0^2}{1+s^2} ds, & r > r_0 \end{cases} \tag{5.4}$$

The function $f(r)$ is strictly decreasing and has a limit as $r \to \infty$ (see the proof of Lemma 4.1)

$$\lim_{r \to \infty} f(r) = f_\infty$$

Since $f(r)$ is strictly decreasing $\frac{1}{2\pi} \ln \frac{1}{r} - f(r)$ is decreasing. Thus if we define

$$h_1(r) = f(r) - f_\infty$$

we have the decomposition

$$\frac{1}{2\pi} \ln \frac{1}{r} = h_1(r) + h_2(r)$$

where h_1 is positive strictly decreasing function and h_2 is decreasing. Hence by the Brascamp–Lieb–Luttinger rearrangement inequality for the Steiner symmetrization (see Lemma 3.2 in [6]) we have

$$\int_\Omega \int_\Omega \int_\Omega u(y) h_1(|y-z|) h_1(|z-x|) u(x) dz dy dx \leq \int_{\Omega^\star} \int_{\Omega^\star} \int_{\Omega^\star} u^\star(y) h_1(|y-z|) h_1(|z-x|) u^\star(x) dz dy dx \tag{5.5}$$

Thus it remains to show that

$$\int_\Omega \int_\Omega \int_\Omega u(y) h_2(|y-z|) h_2(|z-x|) u(x) dz dy dx \leq \int_{\Omega^\star} \int_{\Omega^\star} \int_{\Omega^\star} u^\star(y) h_2(|y-z|) h_2(|z-x|) u^\star(x) dz dy dx \tag{5.6}$$

and

$$\int_{\Omega}\int_{\Omega}\int_{\Omega} u(y)h_1(|y-z|)h_2(|z-x|)u(x)dzdydx \leq \int_{\Omega^\star}\int_{\Omega^\star}\int_{\Omega^\star} u^\star(y)h_1(|y-z|)h_2(|z-x|)u^\star(x)dzdydx \tag{5.7}$$

which does not follow directly from the Brascamp–Lieb–Luttinger rearrangement inequality since h_2 is not positive. Define for $R > 0$

$$q_R(r) := \begin{cases} h_2(r) - h_2(R), & r \leq R \\ 0, & r > R, \end{cases} \tag{5.8}$$

and note that by monotone convergence we have

$$I_\Omega(u,h_2) = \lim_{R\to\infty} [I_\Omega(u,q_R) + 2h_2(R)J_\Omega(u,q_R) + h_2^2(R)(\int_\Omega u(x)dx)^2] \tag{5.9}$$

with the notations

$$I_\Omega(u,g) = \int_{\Omega}\int_{\Omega}\int_{\Omega} u(y)g(|y-z|)g(|z-x|)u(x)dzdydx \tag{5.10}$$

and

$$J_\Omega(u,g) = \int_{\Omega}\int_{\Omega}\int_{\Omega} u(y)g(|z-x|)u(x)dzdydx \tag{5.11}$$

Since q_R is positive and nonincreasing and noting that

$$\int_\Omega u(x)dx = \int_{\Omega^\star} u^\star(x)dx$$

we obtain

$$I_{\Omega}(u, q_R) \le I_{\Omega^\star}(u^\star, q_R),$$
$$J_{\Omega}(u, q_R) \le J_{\Omega^\star}(u^\star, q_R),$$

by the Brascamp–Lieb–Luttinger rearrangement inequality. Therefore,

$$I_{\Omega}(u, h_2) = \lim_{R\to\infty}[I_{\Omega}(u, q_R) + 2h_2(R)J_{\Omega}(u, q_R) + h_2^2(R)(\int\limits_{\Omega} u(x)dx)^2] \le$$
$$\lim_{R\to\infty}[I_{\Omega^\star}(u^\star, q_R) + 2h_2(R)J_{\Omega^\star}(u^\star, q_R) + h_2^2(R)(\int\limits_{\Omega^\star} u^\star(x)dx)^2] = I_{\Omega^\star}(u^\star, h_2) \quad (5.12)$$

This proves the inequality (5.6). Similarly, now let us show that the inequality (5.7) is valid. We have

$$\widetilde{I}_{\Omega}(u, h_2) = \lim_{R\to\infty}[\widetilde{I}_{\Omega}(u, q_R) + h_2(R)\widetilde{J}_{\Omega}(u, h_1)] \quad (5.13)$$

with the notations

$$\widetilde{I}_{\Omega}(u, g) = \int\limits_{\Omega}\int\limits_{\Omega}\int\limits_{\Omega} u(y)h_1(|y-z|)g(|z-x|)u(x)dzdydx \quad (5.14)$$

and

$$\widetilde{J}_{\Omega}(u, h_1) = \int\limits_{\Omega}\int\limits_{\Omega}\int\limits_{\Omega} u(y)h_1(|y-x|)u(x)dzdydx \quad (5.15)$$

Since both q_R and h_1 are positive and nonincreasing

$$\widetilde{I}_{\Omega}(u, q_R) \le \widetilde{I}_{\Omega^\star}(u^\star, q_R) \quad (5.16)$$

$$\widetilde{J}_{\Omega}(u, h_1) \le \widetilde{J}_{\Omega^\star}(u^\star, h_1) \quad (5.17)$$

by the Brascamp–Lieb–Luttinger rearrangement inequality. Therefore, we obtain

$$\widetilde{I}_\Omega(u,h_2) = \lim_{R\to\infty}[\widetilde{I}_\Omega(u,q_R) + h_2(R)\widetilde{J}_\Omega(u,h_1)] \leq$$

$$\lim_{R\to\infty}[\widetilde{I}_{\Omega^*}(u^\star,q_R) + h_2(R)\widetilde{J}_{\Omega^*}(u^\star,h_1)] = \widetilde{I}_{\Omega^*}(u^\star,h_2)$$

This proves the inequality (5.7). □

Lemma 5.4 implies the following analogy of the Pólya theorem [20] for the operator L_Ω.

THEOREM 5.5 *The equilateral triangle* Δ *centered at the origin is a minimizer of the first characteristic number of the logarithmic potential* L_Ω *among all triangles of a given area, i.e.*

$$0 < |\mu_1(\Delta)| \leq |\mu_1(\Omega)|$$

for any triangle $\Omega \subset \mathbb{R}^2$ *with* $|\Omega| = |\Delta|$.

REMARK 5.6 In other words Theorem 5.5 says that the operator norm of L_Ω is maximized in an equilateral triangle among all triangles of a given area.

PROOF OF THEOREM 5.5 By Theorem 2.5 and Lemma 3.1 the first characteristic number μ_1 of the operator L_Ω is positive and simple; the corresponding eigenfunction u_1 can be chosen positive in Ω. Using the fact that by applying a sequence of the Steiner symmetrizations with respect to the mediator of each side, a given triangle converges to an equilateral one (see Fig. 3.2 in [11]), from (5.3) we obtain

$$\int_\Omega\int_\Omega\int_\Omega u_1(y)\frac{1}{2\pi}\ln\frac{1}{|y-z|}\frac{1}{2\pi}\ln\frac{1}{|z-x|}u_1(x)dzdydx \leq$$

$$\int_\Delta\int_\Delta\int_\Delta u_1^\star(y)\frac{1}{2\pi}\ln\frac{1}{|y-z|}\frac{1}{2\pi}\ln\frac{1}{|z-x|}u_1^\star(x)dzdydx. \tag{5.18}$$

Therefore, from (5.18) and the variational principle for the positive operator $\mathcal{L}_\Delta^2$, we get

$$\mu_1^2(\Omega) = \frac{\int_\Omega |u_1(x)|^2 dx}{\int_\Omega \int_\Omega \int_\Omega u_1(y) \frac{1}{2\pi} \ln \frac{1}{|y-z|} \frac{1}{2\pi} \ln \frac{1}{|z-x|} u_1(x) dz dy dx} \geq$$

$$\frac{\int_\Delta |u_1^\star(x)|^2 dx}{\int_\Delta \int_\Delta \int_\Delta u_1^\star(y) \frac{1}{2\pi} \ln \frac{1}{|y-z|} \frac{1}{2\pi} \ln \frac{1}{|z-x|} u_1^\star(x) dz dy dx} \geq$$

$$\inf_{v \in L^2(\Delta)} \frac{\int_\Delta |v(x)|^2 dx}{\int_\Delta \int_\Delta \int_\Delta v(y) \frac{1}{2\pi} \ln \frac{1}{|y-z|} \frac{1}{2\pi} \ln \frac{1}{|z-x|} v(x) dz dy dx} = \mu_1^2(\Delta)$$

Here we have used the fact that the Steiner symmetrization preserves the L^2-norm. □

PROOFS OF THEOREM 5.1 AND THEOREM 5.2 The proofs of Theorem 5.1 and Theorem 5.2 rely on the same techniques as the proofs of Theorem 2.2 and Theorem 2.3 with the difference that now the Steiner symmetrization is used instead of the symmetric-decreasing rearrangement. Since the Steiner symmetrization has the same property (4.13)(cf. Lemma 3.2 in [6]) as the symmetric-decreasing rearrangement, it is clear that any Steiner symmetrization increases (or at least does not decrease) the Schatten *p*-norms for integer $p \geq 2$. Thus, for the proof we only need to recall the fact that a sequence of Steiner symmetrizations with respect to the mediator of each side, a given triangle converges to an equilateral one. The rest of the proof is exactly the same as the proofs of Theorem 2.2 and Theorem 2.3. □

REMARK 5.7 A sequence of three Steiner symmetrizations allows us to transform any quadrilateral into a rectangle (see Fig. 3.3 in [11]). Therefore, it suffices to look at the maximization problem among rectangles for P_4. Unfortunately, for P_5(pentagons and others), the Steiner symmetrization increases, in general, the number of sides. This prevents us from using the same technique for general polygons.

REFERENCES

1. J.M. Anderson, D. Khavinson, V. Lomonosov **SPECTRAL PROPERTIES OF SOME INTEGRAL OPERATORS ARISING IN POTENTIAL THEORY** Q. J. Math. Oxford Ser. (2), 43 (172) (1992), pp. 387-407
2. J. Arazy, D. Khavinson **SPECTRAL ESTIMATES OF CAUCHY'S TRANSFORM IN** $L^2(\Omega)$ Integral Equations Operator Theory, 15 (6) (1992), pp. 901-919
3. C. Bandle **ISOPERIMETRIC INEQUALITIES AND APPLICATIONS** Monographs and Studies in Mathematics, vol. 7, Pitman (Advanced Publishing Program), Boston, MA, London (1980)
4. R.D. Benguria, H. Linde, B. Loewe **ISOPERIMETRIC INEQUALITIES FOR EIGENVALUES OF THE LAPLACIAN AND THE SCHRÖDINGER OPERATOR** Bull. Math. Sci., 2 (1) (2012), pp. 1-56

5. M.Š. Birman, M.Z. Solomjak **ESTIMATES FOR THE SINGULAR NUMBERS OF INTEGRAL OPERATORS** Uspekhi Mat. Nauk, 32 (1(193)) (1977), pp. 17-84 271
6. H.J. Brascamp, E.H. Lieb, J.M. Luttinger **A GENERAL REARRANGEMENT INEQUALITY FOR MULTIPLE INTEGRALS** J. Funct. Anal., 17 (1974), pp. 227-237
7. E. Carlen, M. Loss **COMPETING SYMMETRIES, THE LOGARITHMIC HLS INEQUALITY AND ONOFRI'S INEQUALITY ON** Sn Geom. Funct. Anal., 2 (1) (1992), pp. 90-104
8. J. Delgado, M. Ruzhansky **SCHATTEN CLASSES ON COMPACT MANIFOLDS: KERNEL CONDITIONS** J. Funct. Anal., 267 (3) (2014), pp. 772-798
9. M.R. Dostanić **THE ASYMPTOTIC BEHAVIOR OF THE SINGULAR VALUES OF THE CONVOLUTION OPERATORS WITH KERNELS WHOSE FOURIER TRANSFORMS ARE RATIONAL FUNCTIONS** J. Math. Anal. Appl., 395 (2) (2012), pp. 496-500
10. I.C. Gohberg, M.G. Kreĭn **INTRODUCTION TO THE THEORY OF LINEAR NONSELFADJOINT OPERATORS**
11. translated from Russian by A. Feinstein Translations of Mathematical Monographs, vol. 18, American Mathematical Society, Providence, RI (1969)
a. Henrot **EXTREMUM PROBLEMS FOR EIGENVALUES OF ELLIPTIC OPERATORS** Frontiers in Mathematics, Birkhäuser Verlag, Basel (2006)
12. M. Kac **ON SOME CONNECTIONS BETWEEN PROBABILITY THEORY AND DIFFERENTIAL AND INTEGRAL EQUATIONS** Proceedings of the Second Berkeley Symposium on Mathematical Statistics and Probability, 1950, University of California Press, Berkeley, Los Angeles (1951), pp. 189-215
13. M. Kac **ON SOME PROBABILISTIC ASPECTS OF CLASSICAL ANALYSIS** Amer. Math. Monthly, 77 (1970), pp. 586-597
14. M. Kac **INTEGRATION IN FUNCTION SPACES AND SOME OF ITS APPLICATIONS: LEZIONI FERMIANE (FERMI LECTURES)** Accademia Nazionale dei Lincei, Pisa (1980)
15. T.Sh. Kal'menov, D. Suragan **ON SPECTRAL PROBLEMS FOR THE VOLUME POTENTIAL** Dokl. Math., 80 (2) (2009), pp. 646-649
16. N.S. Landkof **FOUNDATIONS OF MODERN POTENTIAL THEORY** translated from Russian by A.P. Doohovskoy Die Grundlehren der Mathematischen Wissenschaften, Band 180, Springer-Verlag, New York–Heidelberg (1972)
17. E.H. Lieb, M. Loss **ANALYSIS** (second edition), Graduate Studies in Mathematics, vol. 14, American Mathematical Society, Providence, RI (2001)
18. J.M. Luttinger **GENERALIZED ISOPERIMETRIC INEQUALITIES** Proc. Natl. Acad. Sci. USA, 70 (1973), pp. 1005-1006
19. Y. Miyanishi, T. Suzuki **EIGENVALUES AND EIGENFUNCTIONS OF DOUBLE LAYER POTENTIALS** arXiv:1501.03627 (2015)
20. G. Pólya **ON THE CHARACTERISTIC FREQUENCIES OF A SYMMETRIC MEMBRANE** Math. Z., 63 (1955), pp. 331-337
21. G. Pólya, G. Szegö**ISOPERIMETRIC INEQUALITIES IN MATHEMATICAL PHYSICS** Annals of Mathematics Studies, vol. 27, Princeton University Press, Princeton, NJ (1951)
22. M. Ruzhansky, D. Suragan **ON SCHATTEN NORMS OF INTEGRAL OPERATORS OF CONVOLUTION TYPE**
23. Russian Math. Surveys (2015) in press
24. N. Saito **DATA ANALYSIS AND REPRESENTATION ON A GENERAL DOMAIN USING EIGENFUNCTIONS OF LAPLACIAN** Appl. Comput. Harmon. Anal., 25 (1) (2008), pp. 68-97
25. J.L. Troutman **THE LOGARITHMIC POTENTIAL OPERATOR** Illinois J. Math., 11 (1967), pp. 365-374
26. J.L. Troutman **THE LOGARITHMIC EIGENVALUES OF PLANE SETS** Illinois J. Math., 13 (1969), pp. 95-107
27. V.S. Vladimirov **EQUATIONS OF MATHEMATICAL PHYSICS** translated from Russian by Eugene Yankovsky Mir, Moscow (1984)

CHAPTER 13

Mathematical Analysis of an Optimal Control Problem of Surface Water Pollution

Djibo Moustapha, Hamidou Haoua, Saley Bisso

Department of Mathematics and Informatics, Abdou Moumouni University, Niamey, Niger

ABSTRACT

We present in this paper a new technique based on Gelfand's triplet [1] and include differential theory to make a theoretical analysis of an optimal control problem with constraints governed by coupled partial differential equations. This technique allowed us to give some theoretical results of existence and uniqueness of the solution of constraints and characterize the optimal control.

Keywords: Sand Dune Formation, Objective Function, Operators, State Equation

1. INTRODUCTION

The objective of this paper is to make a theoretical analysis of an optimal control problem with constraints governed by coupled partial differential equations.

The interest of this work is two-fold. Indeed firstly the resolution of optimal control problem with partial differential equations constraints is a challenge for the current research in the field both theoretically and numerically. Secondly, the mathematical model studied is of utmost importance in practice since the physical phenomenon is modeled through this paper concerning the pollution of surface water. The studied problem is very complex since the objective function is not explicitly defined in term of control directly. Hence there is a need to solve equations that govern constraints that are also complex as we couple the model which describes the pollutants dissolution with Navier-Stokes equations.

To overcome these difficulties, we built a new technique based on the Gelfand's triplet and theory of differential solutions to transform the partial differential equations into ordinary differential equations [1] [2].

The outline of this paper is as follows: in this second section we present the mathematical model; the third section is devoted to the main results of existence uniqueness and also estimation of the system; in the fourth section we conduct the characterization of the optimal control and we end this work with a conclusion and perspectives.

2. MATHEMATICAL MODEL

Let Ω be a bounded domain of R^3, with boundary Γ sufficiently regular. We denote by $x = (x, y, z)$ an element of Ω.

We consider the following model:

$$\min J(v) = \frac{1}{2}\int_0^T\int_\Omega |Dc(v) - z_d|^2 \, dx dt + \frac{N}{2}\int_0^T\int_\Omega |v|^2 \, dx dt, \tag{1}$$

subject to

$$\begin{cases} \dfrac{\partial c}{\partial t} + \sum_{i=1}^3 \dfrac{\partial (u_i c)}{\partial x_i} - d\Delta c = g + v & \text{in } \Omega \times]0,T[\quad (2.a) \\ c = 0 & \text{on } \Gamma \times]0,T[\quad (2.b) \\ c(\cdot,0) = c_0, & \text{in } \Omega \quad (2.c) \\ \dfrac{\partial u}{\partial t} + (u\cdot\nabla)u - \nu\Delta u + \nabla p = f & \text{in } \Omega \times]0,T[\quad (2.d) \\ \operatorname{div} u = 0 & \text{in } \Omega \times]0,T[\quad (2.e) \\ u = 0, & \text{on } \Gamma \times]0,T[\quad (2.f) \\ u(\cdot,0) = u_0 & \text{in } \Omega \quad (2.g) \end{cases} \tag{2}$$

where:

- z_d is an observation function; D is a given linear continuous operator of observation; N is a given positif real;
- c denotes the concentration of pollutants, g is a source term, d is a diffusion coefficient, v represented the control which allows to act on the system;
- $u = (u_1, u_2, u_3)$and $p = p\ (x, t)$ denote respectively the velocity and the pressure of water, f external forces, ν the kinematic viscosity;
- Equations (2.a)-(2.c) modelling the transport and dissolution of pollutants;
- Systems (2.d)-(2.g) is Navier-Stokes equations.

We rewrite problems (1)-(2) in the following compact form:

$$\begin{cases} \text{Find} \quad u \in U_{ad} \quad \text{such that :} \\ J(u) = \min\limits_{v \in U_{ad}} J(v) \end{cases} \tag{3}$$

where U_{ad} denotes set of admissible controls. In this form it is easier to establish results of existence and uniqueness concerning the control when knowing some properties of J and U_{ad} [3] [4] [5].

3. MAIN RESULTS

To make the theoretical study of system (2), we define the following Gelfand triplet:

$V \rightarrowtail H \rightarrowtail V^*$, where the injections are continuous, dense and compact.

We choose in this paper $V = H_0^1(\Omega)$, $H = L^2(\Omega)$ and $V^* = \left(H_0^1(\Omega)\right)' = H^{-1}(\Omega)$..

We assume that there exist two positive constants A and A^* such that the following inequality hold:

$$\frac{1}{A^*}\|\cdot\|_{V^*} \leq \|\cdot\|_H \leq A\|\cdot\|_V . \tag{4}$$

We also recall the following result concerning the Navier-Stokes equations whose proof can be found in [6].

Theorem 1. Let u_0 belonging to $L^2(\Omega)$ a divergence field. Then there exists a unique solution u of the Navier-Stokes equations associated to the initial data u_0, such that

$$u \in L^\infty\left(\left[0,T\right];L^2\left(\Omega\right)\right)\cap L^2\left(\left[0,T\right];H_0^1\left(\Omega\right)\right)$$

and

$$\forall t \geq 0,\quad \frac{1}{2}\left\|u\left(t\right)\right\|_{L^2(\Omega)}^2 + \int_0^T \left\|\nabla u\left(\theta\right)\right\|_{L^2(\Omega)}^2 \, \mathrm{d}\theta \leq \frac{1}{2}\left\|u_0\right\|_{L^2(\Omega)}^2 \tag{5}$$

3.1. Main Result 1

This first result concerning existence and uniqueness of solution of the model that governs the dissolution of pollutants:

$$\begin{cases} \dfrac{\partial c}{\partial t} + \displaystyle\sum_{i=1}^{3} \dfrac{\partial\left(u_i c\right)}{\partial x_i} - d\Delta c = g + v & \text{on } \Omega\times\left]0,T\right[\\ c = 0 & \text{in } \Gamma\times\left]0,T\right[\\ c\left(\cdot,0\right) = 0, & \text{on } \Omega \end{cases} \tag{6}$$

For this we introduce Gelfand's triplet for the following operators:

$$\Delta := \begin{pmatrix} -\dfrac{\partial^2}{\partial x^2} & 0 & 0 \\ 0 & -\dfrac{\partial^2}{\partial y^2} & 0 \\ 0 & 0 & -\dfrac{\partial^2}{\partial z^2} \end{pmatrix} : V \to V^*, \tag{7}$$

and

$$U \cdot \nabla := \begin{pmatrix} u_1 \frac{\partial}{\partial x} & 0 & 0 \\ 0 & u_2 \frac{\partial}{\partial y} & 0 \\ 0 & 0 & u_3 \frac{\partial}{\partial z} \end{pmatrix} : V \to H. \tag{8}$$

We consider a function C: $[0, T] \to V$ defined by:

$$C(t) := \begin{pmatrix} kc(t,\cdot) \\ kc(t,\cdot) \\ kc(t,\cdot) \end{pmatrix} \text{ where } k = \frac{1}{3} \tag{9}$$

and

$$G = \begin{pmatrix} k(g+v) \\ k(g+v) \\ k(g+v) \end{pmatrix} = G_0 + V \tag{10}$$

With these notations Equation (6) becomes:

$$\begin{cases} \dot{C}(t) + U \cdot \nabla C(t) + d\Delta C(t) = G(t) & a.e\ t \in [0,T] \\ C(0) = C_0 \quad \text{on} \quad H, \end{cases} \tag{11}$$

where $\dot{c}$ denotes the derivative in the sense of the distributions of C and $C_0 = \begin{pmatrix} kc_0 \\ kc_0 \\ kc_0 \end{pmatrix}$.

We define the space of solutions by:

$$W(0,T) = \left\{ w \in L^2\left(0,T; H_0^1(\Omega)\right), \dot{w} \in L^2\left(0,T; H^{-1}(\Omega)\right) \right\}, \tag{12}$$

where $L^2\left(0,T;H_0^1(\Omega)\right)$ and $L^2\left(0,T;H^{-1}(\Omega)\right)$ are equipped with the respective norms:

$$\|f\|_{L^2(0,T;H_0^1(\Omega))} = \left(\int_0^T \|f(t)\|_{H_0^1(\Omega)}^2 \,\mathrm{d}t\right)^{\frac{1}{2}} \tag{13}$$

and

$$\|f\|_{L^2(0,T;H^{-1}(\Omega))} = \left\{\int_0^T \left(\sup_{\phi\in H_0^1(\Omega)} \frac{(f(t),\phi(t))}{\|\phi\|_{H_0^1(\Omega)}}\right)^2 \mathrm{d}t\right\}^{\frac{1}{2}} \quad \phi(t)\neq 0. \tag{14}$$

Theorem 2. If $U\in L^\infty\left(0,T;\left[H_0^1(\Omega)\right]^3\right)$, $C_0\in H$ and $G\in L^2(0,T;H)$ then there exists a unique state:

$$C\in W(0,T)\cap L^\infty(0,T;H), \tag{15}$$

solution of the problem (11).

Proof. To proof uniqueness we first establish the following intermediate result:

$\forall U\in L^2\left(0,T;\left[H_0^1(\Omega)\right]^3\right)$ and $C\in L^2(0,T;V)$ then for all $t\in[0,T]$:

$$\left(U(t)\cdot\nabla C(t),C(t)\right)_{V^*,V} = 0. \tag{16}$$

Indeed,

$$\begin{aligned}
&\left(U(t)\cdot\nabla C(t),C(t)\right)_{V^*,V}\\
&=\int_\Omega U(t)\cdot\nabla C(t)C(t)=\int_\Omega\sum_{i=1}^3 U_i(t)\frac{\partial C(t)}{\partial x_i}C(t)\\
&=\int_\Gamma\sum_{i=1}^3 U_iC(t)C(t)-\int_\Omega\sum_{i=1}^3\left(U_i(t)\frac{\partial C(t)}{\partial x_i}+C(t)\frac{\partial U_i(t)}{\partial x_i}\right)C(t)\\
&=-\int_\Omega U(t).\nabla C(t)C(t)=-\left(U(t).\nabla C(t),C(t)\right)_{V^*,V}
\end{aligned} \tag{17}$$

$$2\left(U(t).\nabla C(t)C(t)\right)_{V^*,V}=0\Rightarrow\left(U(t).\nabla C(t),C(t)\right)_{V^*,V}=0.$$

Assuming existence of two solutions C_1 and C_2 of (11), let denote by $C = C_1 - C_2$ then Csatisfies the following equation:

$$\begin{cases} \dot{C}(t) + d\Delta C_1(t) = U_2 \cdot \nabla C_2(t) - U_1 \cdot \nabla C_1(t) & a.e\ t \in [0,T] \\ C(0) = 0 \quad \text{on} \quad H. \end{cases} \tag{18}$$

by multiplying the first equation of (18) by C and integrating over $[0,t]$, we obtain:

$$\begin{aligned} &\int_0^t \left(\dot{C}(\theta), C(\theta)\right) d\theta + d\int_0^t \left(\Delta C(\theta), C(\theta)\right) d\theta \\ &= \int_0^t \left(U_2 \cdot \nabla C_2(t) - U_1 \cdot \nabla C_1(t), C(\theta)\right) d\theta \end{aligned} \tag{19}$$

$$\begin{aligned} \frac{1}{2}|C(t)|^2 - \frac{1}{2}|C(0)|^2 = &-d\int_0^t \left(\Delta C_1(\theta) - \Delta C_2(\theta), C(\theta)\right) d\theta \\ &- \int_0^t \left(U_2.\nabla C_2(t), C_2(\theta)\right) d\theta + \int_0^t \left(U_1.\nabla C_1(t), C_1(\theta)\right) d\theta \end{aligned}$$

using the monotony of $\Delta\ (t,\cdot)$, we deduce that, $(\Delta C_1(\theta) - \Delta C_2(\theta),\ C(\theta)) \geq 0$ and using relation (16) we deduce that:

$$\frac{1}{2}|C(t)|^2 - \frac{1}{2}|C(0)|^2 \leq 0 \tag{20}$$

$$|C(t)|^2 \leq |C(0)|^2 \tag{21}$$

Since $C(0) = 0$, we have $|C(t)| = 0 \Leftrightarrow C = 0$ implying that $C1 = C2$.

To prove existence of the solution we first prove that the problem approached of the problem model has a unique solution and that this approximate solution converges to the exact solution. Let consider the following approached problem (P_m) of Equation (11) given by:

$$\begin{cases} \dot{C}_m(t) + U \cdot \nabla C_m(t) + \mathrm{d}\Delta C_m(t) = G_m(t) & a.e\ t \in [0,T] \\ C_m(0) = C_{0m} \end{cases} \tag{22}$$

where $G_m(:) = \sum_{i=1}^{m} \left(G(\cdot), e_j\right) e_j$ and $C_{0m} = \sum_{i=1}^{m} \left(C_0, e_j\right) e_j$.

We assume that this approximated problem is posed on the subspace $V_m = Vect\{e_1, e_2,...,e_m\}$ of the space V which is separable, then there exists a sequence $(e_m)_{m\in\mathbb{N}^*}$ such that:

- $\forall m \in \mathbb{N}^*$, The first terms $e_1, e_2,...,e_n$ are linearly independent;
- The set of linear combinations of elements of this sequence is dense in V^* (and therefore in H and V^*).

Proposition 3. Under the assumptions of theorem (2), $\forall m \geq 1$, (22) admits a unique solution C_m which belongs to V_m.

Proof. The proof is based on the Carathéodory's theorem which we apply to the following function:

$F_m : [0,T] \times V_m \to \mathbb{R}^m$ defined by

$$F_m(t,X) = \sum_{i=1}^{m} \left[\left(-d\Delta C_m(t) - U \cdot \nabla X(t), e_j \right) + G_m(t) \right]. \tag{23}$$

- F_m is measurable as $\Delta(\cdot,v)$ and $U \cdot \nabla(\cdot,v)$ are measurable respectively in V^* and H, this function is also continuous in W [7] [8] ;
- the condition of minoration on any compact of V_m is assured by the fact that $\Delta(t,\cdot)$ is bounded *a.e* $t \in [0, T]$ in V^* and $U \cdot \nabla(t,\cdot)$ is bounded in H *a.e* $t \in [0, T]$.

Then, for every $X_0 \in \Omega$, there exists at least one solution to the system:

$$\begin{cases} \dot{X} = F_m(t,X) \\ X(0) = X_0 \end{cases} \tag{24}$$

which is therefore the solution of (22).

In order to demonstrate the convergence of the approximated solution to the exact solution, we establish the following priori estimation results.

Proposition 4. If C is a solution of (11) then:

$$\|C\|_{L^2(0,T;V)} \leq \frac{1}{\sqrt{2d}} |C_0| + \frac{1}{d} \|G\|_{L^2(0,T;V^*)} \tag{25}$$

$$\|C\|_{L^\infty(0,T;H)} \leq |C_0| + \sqrt{\frac{2}{d}} \|G\|_{L^2(0,T;V^*)} \tag{26}$$

$$\|\dot{C}\|_{L^2(0,T;V^*)} \le \left[\sqrt{\frac{d}{2}}|C_0| + \|G\|_{L^2(0,T;V^*)}\right]\left[A + \sqrt{\frac{2}{d}}\|U\|_{L^2\left(0,T;\left[H_0^1(\Omega)\right]^3\right)}\right] + \|G\|_{L^2(0,T;V^*)} \quad (27)$$

Proof. To prove relation (25) we form the scalar product of the first equation of Equation (11) with *C* and integrate over [0, *T*], we obtain:

$$\left(\dot{C}(t), C(t)\right)_{V^*,V} + \left(U \cdot \nabla C(t), C(t)\right)_{V^*,V} + d\left(\Delta C(t), C(t)\right)_{V^*,V} = \left(G(t), C(t)\right)_{V^*,V} \quad (28)$$

$$\left(\dot{C}(t), C(t)\right)_{V^*,V} + d\left(\Delta C(t), C(t)\right)_{V^*,V} = \left(G(t), C(t)\right)_{V^*,V}.$$

Integrate over [0, *T*]

$$\int_0^T \left(\dot{C}(t), C(t)\right)_{V^*,V} \mathrm{d}t + d\int_0^T \left(\Delta C(t), C(t)\right)_{V^*,V} \mathrm{d}t = \int_0^T \left(G(t), C(t)\right)_{V^*,V} \mathrm{d}t.$$

then

$$\int_0^T \left(\dot{C}(t), C(t)\right)_{V^*,V} \mathrm{d}t = \left(C(T), C(T)\right)_{L^2(\Omega)} - \left(C(0), C(0)\right)_{L^2(\Omega)} - \int_0^T \left(\dot{C}(t), C(t)\right)_{V^*,V} \mathrm{d}t$$

$$\int_0^T \left(\dot{C}(t), C(t)\right)_{V^*,V} \mathrm{d}t = \frac{1}{2}\left|C(T)\right|^2 - \frac{1}{2}\left|C(0)\right|^2.$$

We have: $\|\Delta C\|_{V^*} \le \alpha_1(t) + \alpha_2(t)\|C(t)\|_V \le \|C(t)\|_V$:

$$\int_0^T \left(\Delta C(t), C(t)\right)_{V^*,V} \mathrm{d}t \le \int_0^T \|\Delta C\|_{V^*}^2 \mathrm{d}t \le \int_0^T \|C(t)\|_V^2 \mathrm{d}t = \|C\|_{L^2(0,T;V)}^2$$

and we deduce the following successive inequalities

$$\int_0^T \left(G(t)+V(t),C(t)\right)_{V^*,V} \mathrm{d}t \le \int_0^T \left\|G(t)\right\|_{V^*} \left\|C(t)\right\|_V \mathrm{d}t$$
$$\le \left\|G\right\|_{L^2\left(0,T;V^*\right)} \left\|C\right\|_{L^2(0,T;V)}.$$

$$\frac{1}{2}\left|C(T)\right|^2 - \frac{1}{2}\left|C(0)\right|^2 + d\left\|C\right\|^2_{L^2(0,T;V)} \le \left\|G\right\|_{L^2\left(0,T;V^*\right)} \left\|C\right\|_{L^2(0,T;V)}.$$

$$d\left\|C\right\|^2_{L^2(0,T;V)} - \left\|G\right\|_{L^2\left(0,T;V^*\right)} \left\|C\right\|_{L^2(0,T;V)} \le \frac{1}{2}\left|C(0)\right|^2 - \frac{1}{2}\left|C(T)\right|^2$$
$$\le \frac{1}{2}\left|C(0)\right|^2$$

$$\left[\sqrt{d}\left\|C\right\|_{L^2(0,T;V)} - \frac{1}{2\sqrt{d}}\left\|G\right\|_{L^2\left(0,T;V^*\right)}\right]^2 \le \frac{1}{2}\left|C(0)\right|^2 + \frac{1}{4d}\left\|G\right\|_{L^2\left(0,T;V^*\right)}$$
$$\le \left[\frac{1}{\sqrt{2}}\left|C(0)\right| + \frac{1}{2\sqrt{d}}\left\|G\right\|_{L^2\left(0,T;V^*\right)}\right]^2$$

$$\sqrt{d}\left\|C\right\|_{L^2(0,T;V)} - \frac{1}{2\sqrt{d}}\left\|G+V\right\|_{L^2\left(0,T;V^*\right)} \le \left[\frac{1}{\sqrt{2}}\left|C(0)\right| + \frac{1}{2\sqrt{d}}\left\|G\right\|_{L^2\left(0,T;V^*\right)}\right]$$

$$\sqrt{d}\left\|C\right\|_{L^2(0,T;V)} \le \left[\frac{1}{\sqrt{2}}\left|C(0)\right| + \frac{1}{\sqrt{d}}\left\|G\right\|_{L^2\left(0,T;V^*\right)}\right]$$

$$\left\|C\right\|_{L^2(0,T;V)} \le \frac{1}{\sqrt{2d}}\left|C_0\right| + \frac{1}{d}\left\|G\right\|_{L^2\left(0,T;V^*\right)}.$$

To establish relation (26) we integrate the scalar product over $[0, t] \subset [0, T]$:

$$\int_0^t \left(\dot{C}(s),C(s)\right)_{V^*,V} \mathrm{d}s + d\int_0^t \left(\Delta C(s),C(s)\right)_{V^*,V} \mathrm{d}s$$
$$= \int_0^t \left(G(s),C(s)\right)_{V^*,V} \mathrm{d}s. \tag{29}$$

$$\frac{1}{2}\left|C(t)\right|^2 - \frac{1}{2}\left|C(0)\right|^2 + d\left\|C\right\|^2_{L^2(0,T;V)} \le \left\|G\right\|_{L^2\left(0,T;V^*\right)} \left\|C\right\|_{L^2(0,T;V)}.$$

$$\frac{1}{2}\left|C(t)\right|^2 \le \frac{1}{2}\left|C(0)\right|^2 - d\left\|C\right\|^2_{L^2(0,T;V)} + \left\|G\right\|_{L^2\left(0,T;V^*\right)} \left\|C\right\|_{L^2(0,T;V)}$$
$$\le \frac{1}{2}\left|C(0)\right|^2 + \left\|G\right\|_{L^2\left(0,T;V^*\right)} \left\|C\right\|_{L^2(0,T;V)},$$

Using (25) we have:

$$\frac{1}{2}|C(t)|^2 \leq \frac{1}{2}|C(0)|^2 + \|G+V\|_{L^2(0,T;V^*)}\left[\frac{1}{\sqrt{2d}}|C_0| + \frac{1}{d}\|G\|_{L^2(0,T;V^*)}\right]$$

$$\leq \frac{1}{2}|C(0)|^2 + \frac{1}{\sqrt{2d}}|C_0|\|G\|_{L^2(0,T;V^*)} + \frac{1}{d}\|G\|^2_{L^2(0,T;V^*)}$$

$$|C(t)|^2 \leq |C(0)|^2 + \sqrt{\frac{2}{d}}|C_0|\|G\|_{L^2(0,T;V^*)}\left[\sqrt{\frac{2}{d}}\|G\|_{L^2(0,T;V^*)}\right]^2$$

$$\leq \left[|C(0)| + \sqrt{\frac{2}{d}}\|G\|_{L^2(0,T;V^*)}\right]^2$$

$$|C(t)|^2 \leq |C(0)| + \sqrt{\frac{2}{d}}\|G\|_{L^2(0,T;V^*)},$$

then we have

$$\|C\|_{L^\infty(0,T;H)} \leq | C_0 | + \sqrt{\frac{2}{d}}\|G\|_{L^2(0,T;V^*)}$$

To proof relation (27) we use the fact that the Laplacian is a bounded linear operator and the following lemma:

Lemma 3.1. $\forall w \in \left[H_0^1(\Omega)\right]^3$ and for all $z \in H_0^1(\Omega)$ there exist two constants β and γ such as:

$$|w \cdot \nabla z| \leq \gamma\beta \|w\|_{\left[H_0^1(\Omega)\right]^3} \|z\|_{H_0^1(\Omega)}. \tag{30}$$

Proof. For all $w \in \left[H_0^1(\Omega)\right]^3$ and for all $z \in H_0^1(\Omega)$, we have:

$$|w \cdot \nabla z|^2 = \int_\Omega \left|\sum_{i=1}^3 w_i \frac{\partial z}{\partial x_i}\right|^2 dx \leq \int_\Omega \left|\sum_{i=1}^3 w_i\right|^2 \left|\sum_{i=1}^3 \frac{\partial z}{\partial x_i}\right|^2 \leq \|w\|^2_{\left[L^2(\Omega)\right]^3} \|z\|^2_{H^1(\Omega)}. \tag{31}$$

According to Poincaré inequality there exist $\gamma > 0$ such as $\|w\|_{\left[L^2(\Omega)\right]^3} \leq \gamma \|\nabla w\|_{\left[L^2(\Omega)\right]^3}$:

$$|w\cdot\nabla z|^2 \le \gamma^2 |\nabla w|^2_{[L^2(\Omega)]^3} \|z\|^2_{H^1(\Omega)}$$

$$|w\cdot\nabla z|^2 \le \gamma^2 \|w\|^2_{[H_0^1(\Omega)]^3} \|z\|^2_{H^1(\Omega)}$$

Norms $\|\cdot\|_{H_0^1}$ and $\|\cdot\|_{H^1}$ are equivalent in $H^1(\Omega)$, so that there exist a positive constant β such as:

$$|w\cdot\nabla z|^2 \le \gamma^2\beta^2 \|w\|^2_{[H_0^1(\Omega)]^3} \|z\|^2_{H_0^1(\Omega)},$$

then we have

$$|w\cdot\nabla z| \le \gamma\beta \|w\|_{[H_0^1(\Omega)]^3} \|z\|_{H^1(\Omega)}$$

Lemma 3.2. If $U \in L^2\left(0,T;\left[H_0^1(\Omega)\right]^3\right)$ and $C \in L^2(0, T;V)$ then $U\nabla C \in L^2(0, T;V)\cdot$. Furthermore

$$\|U\cdot\nabla C\|_{L^2(0,T;H)} \le \gamma\beta \|U\|_{L^2\left(0,T;\left[H_0^1(\Omega)\right]^3\right)} \|C\|_{L^2(0,T;V)} \tag{32}$$

Proof. For all $U \in L^2\left(0,T;\left[H_0^1(\Omega)\right]^3\right)$ and for all $C \in L^2(0, T;V)$, we have:

$$\|U\cdot\nabla C\|^2_{L^2(0,T;H)} = \int_0^T |U\cdot\nabla C|^2 \,\mathrm{d}t. \tag{33}$$

According to Lemma (3.1) we have:

$$\|U\cdot\nabla C\|^2_{L^2(0,T;H)} \le \gamma^2\beta^2 \int_0^T \|U(t)\|^2_{[H_0^1(\Omega)]^3} \,\mathrm{d}t \int_0^T \|C(t)\|^2_{H_0^1(\Omega)} \,\mathrm{d}t,$$

$$\|U\cdot\nabla C\|_{L^2(0,T;H)} \le \gamma\beta \|U\|^2_{[H_0^1(\Omega)]^3} \,\mathrm{d}t \|C\|_{H_0^1(\Omega)}, \tag{34}$$

Consider $\phi \in L^2(0, T;V)$ and we form the scalar product of the first equation of (11) with ϕ:

$$\left(\dot{C}(t),\phi(t)\right)_{V^*,V}+\left(U\cdot\nabla C(t),\phi(t)\right)_{V^*,V}+d\left(\Delta C(t),\phi(t)\right)_{V^*,V}$$
$$=\left(G(t),\phi(t)\right)_{V^*,V}$$
$$\left(\dot{C}(t),\phi(t)\right)_{V^*,V}=-\left(U\cdot\nabla C(t),\phi(t)\right)_{V^*,V}-d\left(\Delta C(t),\phi(t)\right)_{V^*,V}$$
$$+\left(G(t),\phi(t)\right)_{V^*,V} \tag{35}$$

$$\left|\int_0^T\left(\dot{C}(t),\phi(t)\right)_{V^*,V}\right|\mathrm{d}t\le d\left|\int_0^T\left(\Delta C(t),\phi(t)\right)_{V^*,V}\mathrm{d}t\right|+\left|\int_0^T\left(U.\nabla C(t),\phi(t)\right)_{V^*,V}\mathrm{d}t\right|$$
$$+\left|\int_0^T\left(G(t),\phi(t)\right)_{V^*,V}\right|\mathrm{d}t$$

$$\left|\int_0^T\left(\Delta C(t),\phi(t)\right)_{V^*,V}\right|\mathrm{d}t\le\int_0^T\left|\Delta C(t)\right|\left\|\phi(t)\right\|\le\left\|\Delta C\right\|_{V^*}\le A\left\|C\right\|_{L^2(0,T;V)}$$

$$\left|\int_0^T\left(U\cdot\nabla C(t),\phi(t)\right)_{V^*,V}\mathrm{d}t\right|\le\gamma\beta\left\|U\right\|_{L^2\left(0,T;\left[H_0^1(\Omega)\right]^3\right)}\left\|C\right\|_{L^2(0,T;V)}.$$

$$\left|\int_0^T\left(U\cdot\nabla C(t),\phi(t)\right)_{V^*,V}\mathrm{d}t\right|\le\left\|U\right\|_{L^2\left(0,T;\left[H_0^1(\Omega)\right]^3\right)}\left\|C\right\|_{L^\infty(0,T;H)}.$$

Finally we get

$$\left|\int_0^T\left(\dot{C}(t),\varphi(t)\right)_{V^*,V}\right|\mathrm{d}t$$
$$\le\left[dA\left\|C\right\|_{L^2(0,T;V)}+\left\|U\right\|_{L^2\left(0,T;\left[H_0^1(\Omega)\right]^3\right)}\left\|C\right\|_{L^\infty(0,T;H)}+\left\|G\right\|_{L^2\left(0,T;V^*\right)}\right],$$

Using estimates (25) and (26), we obtain:

$$\left\|\dot{C}\right\|_{L^2\left(0,T;V^*\right)}\le\frac{Ad}{\sqrt{2d}}\left|C(0)\right|+\frac{Ad}{d}\left\|G\right\|_{L^2\left(0,T;V^*\right)}+\left\|U\right\|_{L^2\left(0,T;\left[H_0^1(\Omega)\right]^3\right)}$$
$$+\sqrt{\frac{2}{d}}\left\|U\right\|_{L^2\left(0,T;\left[H_0^1(\Omega)\right]^3\right)}\left\|G\right\|_{L^2\left(0,T;V^*\right)}+\left\|G\right\|_{L^2\left(0,T;V^*\right)},$$

To achieve the proof of Theorem 2 it must be shown that the approximate solution (C_m) converges to the exact solution in $W(0,T)$. By construction, the result of initial states $(C_{0m})_{m\in\mathbb{N}^*}$ converges to C_0 in H.

Using the relation (25), we deduce that the terms of the sequence (C_m) is bounded in L^2 (0, $T;V$) which is a reflexive Banach space. We can extract a subsequence that is weakly convergent in $\sigma(L^2(0, T;V), L^2(0, T;V^*))$.

Let denote by $C_{\phi(\nu)}$ this subsequence where $\phi : \mathrm{N} \to \mathrm{N}$ is a function strictly increasing such that:

$$C_{\phi(n)} \to C_1 \text{ weakly in } L^2(0,T;V) \tag{36}$$

From relation (26) we deduce that the sequence (C_m) are bounded in L^∞ (0, $T;H$) which is the dual of L^1 (0, $T;H$). Then there exists a sub-sequence which converges weakly towards C_2 such as:

$$C_{\phi(n)} \to C_2 \text{ weakly in } L^\infty(0,T;V); \tag{37}$$

Using relation (27) we obtain the weak convergence $C_{\phi(\nu)}$ to $\alpha \in L^2(0, T;V^*)$ i.e:

$$\dot{C}_{\phi(n)} \to \alpha \text{ weakly in } L^2(0,T;V^*) \tag{38}$$

According to [9] [10] , we deduce that $C = C_1 = C_2$ and $\dot{C} = \alpha$.

On the other hand let $j \in \mathrm{N}$ fixed, $\forall m \in \mathbb{N}$ denoting by (Cm) the unique solution of (Pm) then when $m \to +\infty$ we obtain:

$$\left(U \cdot \nabla C_m(t) - d\Delta C_m(t) - G_m(t), e_j\right) \to \left(U \cdot \nabla C(t) - d\Delta C(t) - G(t), e_j\right)$$

and

$$\left(\dot{C}_m(t), e_j\right) \to \left(\dot{C}(t), e_j\right). \tag{39}$$

Since the family $\{e_j, j \in \mathrm{N}^*\}$ is dense in V thus:

$$\dot{C}(t) + U \cdot \nabla C(t) + d\Delta C(t) = G(t) \; p.p \; t \in [0,T].$$

Also to [10] , we can extract a subsequence noted again C_n such as $p.p$ $t \in [0, T]$, $(C_n(t))_{\mathrm{N}^*}$ converges to $C(t)$. Considering the continuous representative C, we obtain strong convergence $(C_n(0))_{\mathrm{N}^*}$ towards $C(0)$.

3.2. Regularity of the Solution

Let $\overline{C}$ be the unique solution of (11) and $\overline{C}+q$ a perturbed one of (11). We suppose that C_0 and $\lambda \in H$, $F, G \in L^2(0, T;V^*)$ and U_a, $U_b \in L^\infty\left(0,T;\left[H_0^1(\Omega)\right]^3\right)$, then q satisfies the following equation:

$$\begin{cases} \dot{q}(t)+U_a\cdot\nabla q(t)+U_b\cdot\nabla\left(\overline{C}(t)+q(t)\right)+d\Delta q(t)=F(t) & a.e\ t\in[0,T] \\ q(0)=\lambda \end{cases} \tag{40}$$

Effect the scalar product of (40) by $q(t)$ and integrate over $[0, t] \subset [0, T]$, we obtain:

$$\begin{aligned}&\frac{1}{2}|q(t)|^2-\frac{1}{2}|\lambda|^2+\int_0^t\left(U_a\cdot\nabla q(s),q(s)\right)\mathrm{d}s+\int_0^t\left(U_b.\nabla\left(\overline{C}(s)+q(s)\right),q(s)\right)\mathrm{d}s\\&+d\int_0^t\left(\Delta q(t),q(t)\right)=\int_0^t\left(F(s),q(s)\right)\mathrm{d}s.\end{aligned} \tag{41}$$

Δ being monotonous then:

$$\frac{1}{2}|q(t)|^2-\frac{1}{2}|\lambda|^2+\int_0^t\left(U_b\cdot\nabla\left(\overline{C}(t)+q(t),q(t)\right)\right)\mathrm{d}s=\int_0^t\left(F(t),q(t)\right)\mathrm{d}s.$$

By Lemma 3.2, we have:

$$\begin{aligned}|q(t)|^2\le|\lambda|^2&+2\beta\gamma\int_0^t\|U_b\|_{\left[H_0^1(\Omega)\right]^3}\left\|\overline{C}(s)+q(s)\right\|_{H_0^1(\Omega)}\|q(s)\|_{H_0^1(\Omega)}\mathrm{d}s\\&+2\int_0^t\|F(s)\|_{H^{-1}(\Omega)}\|q(s)\|_{H_0^1(\Omega)}\end{aligned}$$

Let:

$$\eta(s)=2\beta\gamma\|U_b\|_{\left[H_0^1(\Omega)\right]^3}\left\|\overline{C}(s)+q(s)\right\|_{H_0^1(\Omega)}+\|F(s)\|_{H^{-1}(\Omega)}$$

then

$$|q(t)|^2\le|\lambda|^2+\int_0^t\eta(s)\|q(s)\|_{H_0^1(\Omega)}.$$

According to Willet-Wong's inequality we deduce:

$$|q(t)| \leq |\lambda| + \frac{1}{2}\int_0^t \eta(s)\,\mathrm{d}s,$$

Using relation (25), one can have a priori estimation of q solution of (40). There exist a constant $M_{\lambda,W,F}$ as if $|\lambda|$, $\|F\|_{L^2(0,T,H^{-1}(\Omega))}$ are increased then:

$$|q(t)| \leq |\lambda| + \beta\gamma\int_0^t M_{\lambda,W,H} \|U_b\|_{[H_0^1(\Omega)]^3}\,\mathrm{d}s + \int_0^t \|F(s)\|_{H^{-1}(\Omega)}\,\mathrm{d}s$$

$$|q(t)| \leq |\lambda| + \beta\gamma M_{\lambda,W,H} \|U_b\|_{L^\infty(0,T,[H_0^1(\Omega)]^3)} + \|F(s)\|_{L^2(0,T;H^{-1}(\Omega))}.$$

Let $\rho = \max\{1, \beta\gamma\, M_{\lambda,W,F}\}$, then:

$$|q(t)| \leq \rho|\lambda| + \|U_b\|_{L^\infty(0,T;[H_0^1(\Omega)]^3)} + \|F(s)\|_{L^2(0,T;H^{-1}(\Omega))} \tag{42}$$

3.3. Main Result 2

The space of admissible controls considered in this paper is $U_{ad} = L^2(0,T;H)$.

Theorem 5 The optimal control problem (3) has a unique solution $W \in U_{ad}$ And we have the following inequality: $\forall V \in U_{ad}$:

$$\|W - V\|_{L^2(0,T;H)} \leq \frac{4}{N}[J(V) - J(W)]. \tag{43}$$

Proof. to proof this result, one can proof that the functional J is inferior semi- continuous, strongly convex and differentiable.

Let us proof that J is inferior semi-continuous. We consider a sequence (C_n, V_n) in $L^2(0,T;H_0^1(\Omega)) \times L^2(0,T;H)$ which converges strongly to (C, V), we can extract a subsequence also denoted by (C_n, V_n) such that:

$$(C_n(t), V_n(t)) \to (C(t), V(t))\ a.e \text{ in } H_0^1(\Omega) \times L^2(\Omega). \tag{44}$$

Let denote by $L(C(t), V(t)) = X(C(t)) + Y(V(t))$ with:

$$X : H_0^1(\Omega) \to [0,+\infty[\quad C(t) \to X(C(t)) = \int_\Omega |DC(v) - z_d|^2 \,dx$$
$$Y : L^2(\Omega) \to [0,+\infty[\quad V(t) \to Y(V(t)) = \int_\Omega |V|^2 \,dx$$

X and *Y* are inferior semi-continuous then:

$$L(C(t),V(t)) \le \liminf_{n\to\infty} L(C_n(t),V_n(t)) \tag{45}$$

Apply Fatou's lemma to the sequence $L(C_n(t),V_n(t))$ we obtain:

$$\int_0^T \liminf_{n\to\infty} L(C_n(t),V_n(t))\,dt \le \liminf_{n\to\infty} \int_0^T L(C_n(t),V_n(t))\,dt,$$

according to (45) we have:

$$\frac{1}{2}\int_0^T\int_\Omega |DC(v) - z_d|^2 \,dx dt + \frac{N}{2}\int_0^T\int_\Omega |V|^2 \,dx, dt \le \frac{1}{2}\int_0^T \liminf_{n\to\infty} \int_\Omega |DC_n(v) - z_d|^2 \,dx dt$$
$$+ \frac{N}{2}\int_0^T \liminf_{n\to\infty} \int_\Omega |V_n|^2 \,dx dt$$

$$J(V) \le \liminf_{n\to\infty} J(V_n).$$

So that *J* is inferior semi-continuous.

To proof the strong convexity, we choose *W* and $V \in L^2(0,T;H)$, then:

$$J\left(\frac{W+V}{2}\right) = \frac{N}{2}\left\|\frac{W+V}{2}\right\|^2_{L^2(0,T,H)}$$
$$+ \frac{1}{2}\int_0^T\int_\Omega \left|\frac{1}{2}(DC(V) - z_d) + \frac{1}{2}(DC(W) - z_d)\right|^2 \,dx dt. \tag{46}$$

Applying the first unequally Clarkson to the first term in right member, we obtain:

$$J\left(\frac{W+V}{2}\right) \leq \frac{N}{2}\left[\frac{1}{2}\|W\|^2_{L^2(0,T;H)} + \frac{1}{2}\|V\|^2_{L^2(0,T;H)} - \left\|\frac{W-V}{2}\right\|^2_{L^2(0,T;H)}\right]$$
$$+\frac{1}{2}\int_0^T\int_\Omega \left|\frac{1}{2}\left(DC(v)-z_d\right)+\frac{1}{2}\left(DC(w)-z_d\right)\right|^2 \mathrm{d}x\mathrm{d}t$$

By applying the equality of the parallelogram to the second term the right member, we obtain:

$$J\left(\frac{W+V}{2}\right) \leq \frac{N}{4}\|W\|^2_{L^2(0,T;H)} + \frac{N}{4}\|V\|^2_{L^2(0,T;H)} - \frac{N}{8}\|W-V\|^2_{L^2(0,T;H)}$$
$$+\frac{1}{4}\int_0^T\int_\Omega \left|DC(W)-z_d\right|^2 \mathrm{d}x\mathrm{d}t + \frac{1}{4}\int_0^T\int_\Omega \left|DC(V)-z_d\right|^2 \mathrm{d}x\mathrm{d}t$$
$$-\frac{N}{8}\left\|DC(W)-DC(V)\right\|^2_{L^2(0,T;H)}.$$

$$J\left(\frac{W+V}{2}\right) \leq \frac{N}{4}\|W\|^2_{L^2(0,T;H)} + \frac{N}{4}\|V\|^2_{L^2(0,T;H)} - \frac{N}{8}\|W-V\|^2_{L^2(0,T;H)}$$
$$+\frac{1}{4}\int_0^T\int_\Omega \left|DC(W)-z_d\right|^2 \mathrm{d}x\mathrm{d}t + \frac{1}{4}\int_0^T\int_\Omega \left|DC(V)-z_d\right|^2 \mathrm{d}x\mathrm{d}t$$

$$J\left(\frac{W+V}{2}\right) \leq \frac{J(W)+J(V)}{2} - \frac{N}{8}\|W-V\|^2_{L^2(0,T;H)}. \tag{47}$$

And the differentiability of J is proved by calculating the following limit:

$$\lim_{\theta\to 0}\frac{J(V+\theta W)-J(V)}{\theta}. \tag{48}$$

Let us denoted by $\tilde{c}$ the trajectory corresponding to $V(t) + \theta W(t)$ then $\tilde{c}$ satisfies the following equation:

$$\begin{cases} \tilde{C}(t)+\tilde{U}\cdot\nabla\tilde{C}(t)+d\Delta\tilde{C}(t)=G_0(t)+V(t)+\theta W(t) & a.e\ t\in[0,T] \\ \tilde{C}(0)=\tilde{C}_0 \quad \text{in} \quad H. \end{cases} \tag{49}$$

$$J(V+\theta W)-J(V)=\frac{1}{2}\int_0^T\int_\Omega\left|D\tilde{C}-z_d\right|^2\mathrm{d}x\mathrm{d}t+\frac{N}{2}\int_0^T\left|V+\theta W\right|^2\mathrm{d}x\mathrm{d}t$$
$$-\frac{1}{2}\int_0^T\int_\Omega\left|DC-z_d\right|^2\mathrm{d}x\mathrm{d}t-\frac{N}{2}\int_0^T\left|V\right|^2\mathrm{d}x\mathrm{d}t$$

$$\begin{aligned}
&J(V+\theta W)-J(V)\\
&=\frac{1}{2}\int_0^T\int_\Omega\left[\left|D\tilde{C}-z_d\right|^2-\left|DC-z_d\right|^2\right]\mathrm{d}x\mathrm{d}t+\frac{N}{2}\int_0^T\left[\left|V+\theta W\right|^2-\left|V\right|^2\right]\mathrm{d}x\mathrm{d}t\\
&=\frac{1}{2}\int_0^T\left[\left(D\tilde{C}-z_d,D\tilde{C}-DC+DC-z_d\right)-\left(DC-z_d,DC-z_d\right)\right]\mathrm{d}t\\
&+\frac{N}{2}\int_0^T\left[(V+\theta W,V+\theta W)-(V,V)\right]\mathrm{d}t\\
&=\frac{1}{2}\int_0^T\left[\left(D\tilde{C}-z_d,D(\tilde{C}-C)+DC-z_d\right)-\left(DC-z_d,DC-z_d\right)\right]\mathrm{d}t\\
&+\frac{N}{2}\int_0^T\left[2\theta(V,W)+\theta^2(W,W)\right]\mathrm{d}t\\
&=\frac{1}{2}\int_0^T\left[\left(D\tilde{C}-z_d,D\left(\tilde{C}-C\right)\right)+\left(D\left(\tilde{C}-C\right),DC-z_d\right)\right]\mathrm{d}t\\
&+\frac{N}{2}\int_0^T\left[2\theta(V,W)+\theta^2(W,W)\right]\mathrm{d}t
\end{aligned}$$

Let us designed by

$$\hat{C}=\lim_{\theta\to 0}\frac{\tilde{C}-C}{\theta},$$

then

$$\lim_{\theta\to 0}\frac{J(V+\theta W)-J(V)}{\theta}=\frac{1}{2}\int_0^T\left[\left(D\tilde{C}-z_d,D\hat{C}\right)+\left(D\hat{C},DC-z_d\right)\right]\mathrm{d}t$$
$$+N\int_0^T(V,W)\mathrm{d}t$$

$$J'(V)W=\frac{1}{2}\int_0^T\left[\left(D\tilde{C}-z_d,D\hat{C}\right)+\left(D\hat{C},DC-z_d\right)\right]\mathrm{d}t+N\int_0^T(V,W)\mathrm{d}t$$

Note that if $\theta\to 0$ then $\tilde{c}\to C$

$$J'(V)W=\int_0^T\int_\Omega\left(DC-z_d\right)D\hat{C}\mathrm{d}x\mathrm{d}t+N\int_0^T\int_\Omega VW\mathrm{d}x\mathrm{d}t \tag{50}$$

According to these properties of J we conclude that problem (3) admits an unique minimum W on U_{ad}.

From relation (47) we have the inequality:

$$\|W-V\|_{L^2(0,T;H)} \leq \frac{4}{N}\left[J(V)-J(W)\right].$$

As $U_{ad} = L^2(0,T;H)$ then

$$J'(V)W = \int_0^T \int_\Omega (C-z_d)\hat{C}\mathrm{d}x\mathrm{d}t + N\int_0^T \int_\Omega VW\mathrm{d}x\mathrm{d}t \tag{51}$$

Search $\hat{C}$, i.e. the tangent model: In fact $\frac{(49)-(11)}{\theta}$, we obtain:

$$\frac{\tilde{C}-\dot{C}}{\theta}+\frac{\tilde{U}\cdot\nabla\tilde{C}-U.\nabla C}{\theta}-d\Delta\frac{\tilde{C}-C}{\theta}=w, \tag{52}$$

by tender $\theta \to 0$, we obtain

$$\begin{cases} \hat{C}+U\cdot\nabla\hat{C}-d\Delta\hat{C}=W \\ \hat{C}(\cdot,0))=0, \end{cases} \tag{53}$$

since $\tilde{U} \to U$ when $\theta \to 0$.

3.4. Main Result 3

Theorem 6. If D is a surjective operator of the space $L^2(0,T;V)$ in $L^2(0,T;V)$ or D is the identity application of $L^2(0,T;V)$ then the control V is given by:

$$V = -N^{-1}P \tag{54}$$

where P is the solution of the adjoint equation.

First Case: We assume that D is an injective operator define from the space in $L^2(0,T;V) \to L^2(0,T;H)$ and $Z_d \in L^2(0,T;V)$, then the adjoint equation is given by:

$$\begin{cases} -\dot{P} - U \cdot \nabla P - d\Delta P = C - z_d \\ P(T) = 0 \end{cases} \tag{55}$$

Seconde Case: D is the identity operator define from $L^2(0,T;V) \rightarrow L^2(0,T;V)$ when $Z_d \in L^2(0,T;V)$ then adjoint equation is given by:

$$\begin{cases} -\dot{P} - U \cdot \nabla P - d\Delta P = C - z_d \\ P(T) = 0 \end{cases} \tag{56}$$

when $Z_d \in L^2(0,T;V)$ the adjoint equation becomes:

$$\begin{cases} -\dot{P} - U \cdot \nabla P - d\Delta P = (\Delta_x + I)(C - z_d) \\ P(T) = 0 \end{cases} \tag{57}$$

Proof. The derivative given by (51) is unusable since for each test W we must solving the tangent model (53). The introduction of the adjoint state allows us to obtain a explicit expression of J'[11] [12].

We first perform the scalar product of the first equation of (53) with P and then integrate by parts we obtain:

$$\int_0^T \left(\dot{\hat{C}}(t), P\right) \mathrm{d}t = \hat{C}(T)P(T) - C(0)P(0) - \int_0^T \left(\dot{P}(t), \hat{C}(t)\right) \mathrm{d}t \tag{58}$$

$$\begin{aligned} \int_0^T \left(U \cdot \nabla \hat{C}(t), P\right) \mathrm{d}t &= \int_0^T \int_\Omega \nabla \left(U\hat{C}\right) P \mathrm{d}t = \int_0^T \int_\Gamma U\hat{C}P \mathrm{d}t - \int_0^T \int_\Omega \nabla (UP) \hat{C} \mathrm{d}x \mathrm{d}t \\ &= -\int_0^T \left(U \cdot \nabla P, \hat{C}\right) \mathrm{d}x \mathrm{d}t \end{aligned}$$

$$\begin{aligned} \int_0^T \left(\Delta\hat{C}, P\right) \mathrm{d}t &= \int_0^T \int_\Omega \Delta\hat{C} P \mathrm{d}x \mathrm{d}t = \int_0^T \int_\Gamma \hat{C} P \mathrm{d}x \mathrm{d}t - \int_0^T \int_\Omega \nabla\hat{C} \nabla P \mathrm{d}x \mathrm{d}t \\ &= -\int_0^T \int_\Gamma \hat{C} P \mathrm{d}x \mathrm{d}t + \int_0^T \int_\Omega \Delta P \hat{C} \mathrm{d}x \mathrm{d}t. \end{aligned}$$

with additional conditions on P given by:

$$P(\boldsymbol{x}, t = T) = 0, \quad P(\boldsymbol{x}, t) = 0 \quad \boldsymbol{x} \in \Gamma$$

we obtain:

$$\int_0^T \int_\Omega \left(-\dot{P} - U \cdot \nabla P - d\Delta P\right)\hat{C}\,\mathrm{d}x\mathrm{d}t = \int_0^T \int_\Omega WP\,\mathrm{d}x\mathrm{d}t$$

By identifying this result with the first term of $J'(V)W$, one obtains the adjoint equation according to the nature of the operator D given by (55)-(57)

To characterize the control we consider the two cases given by relations (56), and (57). Then if the adjoint equation is given by the proof is the same for both cases D. The adjoint equation is given by:

$$\begin{cases} -\dot{P} - U \cdot \nabla P - d\Delta P = (C - z_d) \\ P(T) = 0 \end{cases}$$

In this case relation (51) becomes:

$$J'(V)W = \int_0^T \int_\Omega (C - z_d)\hat{C}\,\mathrm{d}t + N\int_0^T \int_\Omega VW\,\mathrm{d}x\mathrm{d}t \tag{59}$$

We form the scalar product between the first equation of the adjoint equation with $\hat{C}$ and we integrate over [0, T], we obtain:

$$\int_0^T \left(\dot{P}(t), \hat{C}\right)\mathrm{d}t = C(T)P(T) - C(0)P(0) - \int_0^T \left(\hat{C}(t), P(t)\right) \tag{60}$$

$$\begin{aligned} \int_0^T \left(U \cdot \nabla P(t), \hat{C}\right)\mathrm{d}t &= \int_0^T \int_\Omega \nabla (UP)\hat{C}\,\mathrm{d}t = \int_0^T \int_\Gamma UP\hat{C}\,\mathrm{d}t - \int_0^T \int_\Omega \nabla\left(U\hat{C}\right)P\,\mathrm{d}x\mathrm{d}t \\ &= -\int_0^T \left(U \cdot \nabla \hat{C}, P\right)\mathrm{d}x\mathrm{d}t \end{aligned}$$

$$\begin{aligned} \int_0^T \left(\Delta P, \hat{C}\right)\mathrm{d}t &= \int_0^T \int_\Omega \Delta P\hat{C}\,\mathrm{d}x\mathrm{d}t = \int_0^T \int_\Gamma P\hat{C}\,\mathrm{d}x\mathrm{d}t - \int_0^T \int_\Omega \nabla P \nabla \hat{C}\,\mathrm{d}x\mathrm{d}t \\ &= -\int_0^T \int_\Gamma P\hat{C}\,\mathrm{d}x\mathrm{d}t + \int_0^T \int_\Omega \Delta\hat{C}P\,\mathrm{d}x\mathrm{d}t. \end{aligned}$$

We have therefore:

$$\int_0^T \left(-\dot{P} - U \cdot \nabla P - d\Delta P, \hat{C}\right) \mathrm{d}t = \int_0^T \left(P, \hat{C}(t) + U \cdot \nabla \hat{C} + \Delta \hat{C}\right) \mathrm{d}t$$
$$= \int_0^T (P, W) \mathrm{d}t = \int_0^T \int_\Omega PW \mathrm{d}x\mathrm{d}t.$$

So that thus one can write J' as follows (51) give:

$$J'(V)W = \int_0^T \int_\Omega \left(-\dot{P} - U \cdot \nabla P - d\Delta P\right) \hat{C} \mathrm{d}x\mathrm{d}t + N \int_0^T \int_\Omega VW \mathrm{d}x\mathrm{d}t$$
$$J'(V)W = \int_0^T \int_\Omega PW \mathrm{d}x\mathrm{d}t + N \int_0^T \int_\Omega VW \mathrm{d}x\mathrm{d}t$$
$$J'(V)W = \int_0^T \int_\Omega (P + NV) W \mathrm{d}x\mathrm{d}t = 0 \tag{61}$$

His equation had to be satisfied whatever the disturbance on W, especially if $W \neq 0$. Necessarily, we have:

$$P + NV = 0. \tag{62}$$

4. CONCLUSIONS

The aim of this paper is a mathematical analysis of an optimal control problem of surface water pollution by using a triplet of evolution adapted to order of derivations. This technique allowed us to propose theorems not only of existence and uniqueness of the control but also of the solution of the equation that governs the constraints which is nonlinear. Then we solved the control problem directly and gave an optimal control characterization by using the adjoint equation.

It is essential to remember that this study only allowed us to characterize the optimal control without giving an analytical expression of it. In a future work we will proceed to a numerical approximation of this model in order to propose solution to the decision makers.

REFERENCES

1. Gelfand, I.M. and Ya, N. (1964) Vilenkin; Generalized Functions. Applications of Harmonic Analysis. Vol. 4, Academic Press, New York.
2. Pyatnitsky, E.S. and Rapoport, L.B. (1996) Criteria of Asymptotic Stability of Differential Inclusions and Periodic Motions of Time-Varying Nonlinear Control Systems. IEEE Transactions on Circuits and Systems, 43, 219-229. https://doi.org/10.1109/81.486446
3. Ciarlet, P.G. (1982) Introduction à l'analyse numérique matricielle et à l'optimisation. Masson, Paris.
4. Culioli, J.C. (1994) Introduction à l'optimisation. Editions Ellipses, Paris.
5. Faurre, P. (1988) Analyse numérique. Notes d'optimisation, Ecole Polytechnique. Ellipse, Edition Marketing, Paris.
6. Cannone, M. (2000) Nombres de Reynolds, stabilité et Navier-Stokes. Vol. 52, Banach Center Publications, Warszawa.
7. Brezis, H. (1968) Equations et inéquations non linéaires dans les espaces vectoriels en dualité. Annales de l'Institut Fourier (Grenoble), 18, 115-175.
8. Bonnans, J.F. (1998) Second Order Analysis for Control Constrained Optimal Control Problems of Semilinear Elliptic Systems. Applied Mathematics & Optimization, 38, 303-325. https://doi.org/10.1007/s002459900093
9. Aubin, J.P. (1963) Un théorème de compacité. Comptes Rendus de l'Académie des Sciences, Paris, 256.
10. Clérin, J.-M. (2009) Problèmes de controle optimal du type bilinéaire gouvernés par des équations aux dérivées partielles d'évolution. Thèse, LANLG, Université d'Avignon.
11. Lions, J.L. (1968) Controle optimal de systèmes gouvernés par des équations aux dérivées partielles. Dunod, Paris.
12. Lions, J.-L. (1969) Quelques méthodes de résolution des problèmes aux limites non linéaires. Gauthier-Villars, Paris.

CHAPTER 14

Non-standard shocks in the Buckley–Leverett equation

HenrikKalisch[a,], DarkoMitrovic[b], Jan M.Nordbotten[a]

[a]*University of Bergen, Department of Mathematics, Postbox 7800, 5020 Bergen, Norway*

[b]*University of Montenegro, Faculty of Mathematics, Cetinjski put bb, Montenegro*

ABSTRACT

It is shown how delta shock waves which consist of Dirac delta distributions and classical shocks can be used to construct non-monotone solutions of the Buckley–Leverett equation. These solutions are interpreted using a recent variational definition of delta shock waves in which the Rankine–Hugoniot deficit is explicitly accounted for [6]. The delta shock waves are also limits of approximate solutions constructed using a recent extension of the weak asymptotic method to complex-valued approximations [15]. Finally, it is shown how these non-standard shocks can be fitted together to construct similarity and traveling-wave solutions which are non-monotone, but still admissible in the sense that characteristics either enter or are parallel to the shock trajectories.

Keywords: Conservation laws, Delta shocks, Riemann problem, Weak asymptotics, Traveling waves

1. INTRODUCTION

The Buckley–Leverett (BL) equation describes two-phase flow in a porous medium in the limit where capillary forces can be neglected. In non-dimensional variables, the equation can be written in normalized form as

$$\partial_t u + \partial_x \left(\frac{u^2}{u^2 + a(u-1)^2} \right) = 0 \tag{1.1}$$

where the unknown u represents the saturation of the wetting phase, and the constant $a > 0$ in the case of homogeneous systems represents the viscosity ratio between the fluids. In terms of the mathematical theory of hyperbolic conservation laws, the physical situation modeled by the equation is described by entropy admissible solutions of (1.1). As shown in [1], the mathematical entropy for this equation is given by the capillary energy, and admissible solutions with discontinuities must have characteristic curves which enter into or are parallel to the shock trajectories. For instance, one may consider the Riemann problem describing the evolution of two constant states separated by a single discontinuity. Such a configuration describes well the setup of many experiments started with initially homogeneous saturations, and where the proportion of the injected fluids is held constant during the experiment.

Mathematically, the Riemann problem consists of (1.1) supplemented with initial data given by

$$u(x,0) = \begin{cases} u_L, & x < 0, \\ u_R, & x > 0. \end{cases} \tag{1.2}$$

Depending on the size of uR and uL, the similarity solution to (1.1), (1.2)consists either of one shock wave, or of the combination of a shock wave followed by a rarefaction wave (see Figs. 1 and 2). The shock waves satisfy the usual Rankine–Hugoniot condition, and uniqueness of a solution to the Riemann problem can be established in the case of a combination of a shock and a rarefaction wave if it is required that the shock propagates at the same speed as the slowest part of the rarefaction wave [1,19,23].

In the current work, the focus is on solutions which consist of non-standard shocks which do not satisfy the Rankine–Hugoniot condition. These shocks feature a non-zero Rankine–Hugoniot deficit, and may be described with the help of Dirac delta distributions.

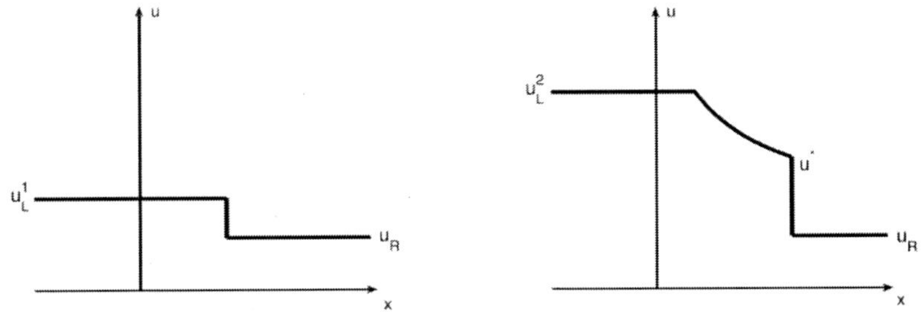

Figure. 1. The classical admissible solutions consisting of simple waves. In the left panel $u_L^1 < u^*$, and the Riemann problem is solved by a single shock. In the right panel $u_L^2 > u^*$, and the Riemann problem is solved by the combination of a rarefaction wave and a shock.

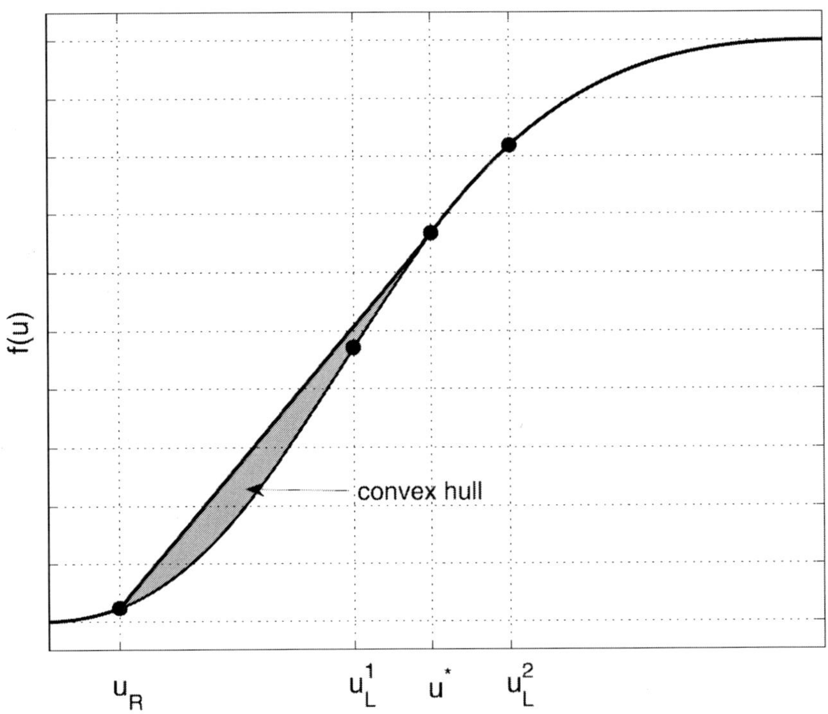

Figure. 2. The flux in the u–$f(u)$ plane. The right state is labeled u_R. The point at which the graph of f is tangent to the chord originating at $(u_R, f(u_R))$ is labeled $(u^*, f(u^*))$. Two possible left states are indicated as u_L^1 and u_L^2.

The study of such measure-valued solutions goes back to the work of Korchinski [18] and Keyfitz and Kranzer [17]. There are a number of reasonable ways to multiply Heaviside and Dirac distributions, such as [3–5], and using these, a number of definitions of measure-valued solutions of systems of conservation laws have been introduced, such as for example in [5,12,14,22].

More recently, a variational definition for delta shocks for systems of two conservation laws was put forward in [6]. The definition laid down in [6] is generally used in tandem with the weak asymptotics method, such as defined in [4], and is able to directly keep track of the Rankine–Hugoniot deficit, thus circumventing the need to include singular terms in the definition of the weak solution, and avoiding the problem of multiplication of distributions. The technique of using complex-valued corrections in the method of weak asymptotics was introduced in [15,16] extends the range of applicability of the variational approach to a larger class of fluxes. The method introduced in [15,16] also turns out to be crucial in the treatment of equations such as (1.1)which feature quotients of singular terms, and was also used in [2,24,25].

Solutions with non-monotone profiles were also found if the Buckley–Leverett equation is regularized with a third-order derivative term representing a physical effect known as dynamic capillarity [9,10]. In these works, solutions were found which do not satisfy the usual entropy conditions, but which are given as the limit of a regularized problem in which both equilibrium and dynamic capillarity were taken into account, and which satisfy the classical Rankine–Hugoniot conditions. In contrast, in the current work, we find solutions which are entropy-admissible in the classical sense, but which feature a non-zero Rankine–Hugoniot deficit.

The works [9,10] were motivated in part by laboratory experiments which indicate that under certain conditions, the saturation exhibits a non-monotone behavior. Indeed, as discussed in [11] (and the references contained therein) large fluxes often feature a behavior which is characterized by an initial overshoot of the concentration at the wetting front, and subsequent drainage of the wetting phase behind the front. Such a phenomenon was also observed in recent experiments on water infiltration into different types of sand by DiCarlo [7] in which the situation is essentially one-dimensional.

In the experiments presented in [7], a constant flux is applied to an initially dry medium, a situation which can be described mathematically by requiring that the saturation function u satisfies

$$u = \begin{cases} u_L, & x \to -\infty, \\ u = 0, & x \to +\infty. \end{cases}$$

These experiments clearly show that non-monotone behavior is observed for large enough fluxes. A number of plausible mathematical explanations of this phenomenon have been put forward, including modifications of the Richards equation [8,11] and the regularization of the Buckley–Leverett equation by dynamic capillarity [9,10]. However, one particular feature of the experimental results, namely the constant width of the overshoot region has not been explained by any of the proposed mathematical models.

In the present note, it is *not* our purpose to offer an explanation of the non-monotone behavior in physical terms, but rather to demonstrate that non-monotone solutions may be found directly in the Buckley–Leverett equation. The solutions found here are based on the variational theory laid down in [6], which defines solutions featuring a positive Rankine–Hugoniot deficit. The solutions are admissible in the sense that they are limits of regularized solutions in the weak asymptotic limit, and in addition are entropy-admissible in the classical sense. While the Buckley–Leverett equation may not be precisely the correct model for the study of sand infiltration such as described in [7], our approach shows that non-monotone solutions exist – at least in a mathematical sense – directly in the hyperbolic theory. One may therefore speculate if variational definition of weak solutions is connected to possible modifications to conservation laws which break down due to the inadequacies of the continuum hypothesis.

The disposition of the paper is as follows. In the next section, we formulate a variational framework for delta-shock solutions of the Buckley–Leverett equation which mirrors the variational concept defined for 2 × 2 systems in [6,15]. In Section 3, we justify the δ-shock solution mentioned above by the weak asymptotic method. Using the solution concepts laid down in Sections 2 and 3, we show how to construct non-monotone solutions of the Riemann problem in Section 4. Finally, in Section 5, we construct non-monotone traveling-wave solutions of the Buckley–Leverett equation.

2. VARIATIONAL FORMULATION OF DELTA-SHOCK SOLUTIONS

We define the flux

$$f(u) = \frac{u^2}{u^2 + a(u-1)^2} \tag{2.1}$$

and write the equation in the form

$$u_t + f(u)x = 0. \tag{2.2}$$

In order to allow solutions which are not admissible in the classical sense, we will make use of the concept of weak solutions which consist of shocks associated with Dirac masses placed on the path of the shock. The general definition follows the concept introduced in [6]. Suppose $\Gamma = \{\gamma_i \mid i \in I\}$ is a graph in the closed upper half plane, containing Lipschitz continuous arcs γ_i, $i \in$ I, where I is a finite index set. Let I_0 be the subset of I containing all indices of arcs that connect to the x-axis, and let $\Gamma_0 = \{x_k^0 \mid k \in I_0\}$ be the set of initial points of the arcs γk with $k \in I_0$. Define the singular part by $\alpha(x,t)\delta(\Gamma) = \sum_{i\in I}\alpha i(x,t)\delta(\gamma_i)$. Let u be a distribution of the form

$$u(x,t) = U(x,t) + \alpha(x,\mathrm{i})\delta(\Gamma),$$

where $U \in L\infty(R \times R_+)$. Let $\frac{\partial\varphi(x,t)}{\partial l}$ denote the tangential derivative of a function φ on the graph γ_i, and let $\int_{\gamma i}$ denote the line integral over the arc γ_i.

DEFINITION 2.1 The distribution $u(x,t) = U(x,t) + \alpha(x,t)\delta(\Gamma)$ is called a generalized δ-shock wave solution of Eq. (1.1) with the initial data $u(x,0) = U_0(x)+\sum_{I0}\alpha k(x^k{}_0,0)\delta(x - x^0{}_k)$ if the integral identity

$$\int_{\mathbb{R}_+}\int_{\mathbb{R}} (U\partial_t\varphi + f(U)\partial_x\varphi)\, dxdt + \int_{\mathbb{R}} U_0(x)\varphi(x,0)\, dx$$
$$+ \sum_{i\in I}\int_{\gamma_i} \alpha_i(x,t)\frac{\partial\varphi(x,t)}{\partial l}\, dx + \sum_{k\in I_0}\alpha_k(x_k^0,0)\varphi(x_k^0,0) = 0 \tag{2.3}$$

holds for all test functions $\varphi \in D(R{\times}R_+)$.

This definition applies to the case of general initial data. However, let us first look at the Riemann problem for (1.1) with initial data

$$U(x,0) = \begin{cases} u_L, & x < 0, \\ u_R, & x > 0. \end{cases} \tag{2.4}$$

Using Definition 2.1, it is not difficult to see that for any $c \in$ R, and any given u_L and u_R, a solution of the form $u(x,t) = U(x,t) + \alpha(t)\delta(x - ct)$ exists, where

$$U(x,t) = \begin{cases} u_L, & x < ct, \\ u_R, & x > ct, \end{cases}$$

and the amplitude of the singular part of the shock is given by

$$\alpha(t) = (c[u_R - u_L] - [f(u_R) - f(u_L)])t. \tag{2.6}$$

THEOREM 2.1 *Given any u_L, $u_R \in [0,1]$, and given any $c \in R$, define the distribution $u(x,t) = U(x,t) + \alpha(t)\delta(x - ct)$, where $U(x,t)$ is given by (2.5) and $\alpha(t)$ is given by (2.6). Then $u(x,t)$ is a solution of the Riemann problem (2.2),(2.4) in the sense of Definition 2.1.*

PROOF The proof of the theorem follows by substituting u into (2.3). After standard transformations, the identity

$$\int_{\mathbb{R}_+} (c[U] - [f(U)])\,\varphi(ct,t)\,dt - \int_{\mathbb{R}_+} \alpha'(t)\varphi(ct,t)\,dt = 0$$

appears, where $[U] = u_R - u_L$ and $[f(U)] = f(u_R) - f(u_L)$, and since $\alpha(0) = 0$, the conclusion follows immediately. □

In the next section, we will justify the solution given above by the method of weak asymptotics. In general, if the weak asymptotic method, such as defined in [4,6] is used in tandem with the above definition, the solutions are thought to be admissible, although this admissibility concept does not yield uniqueness. The weak asymptotic method was recently extended to the case where complex-valued corrections are allowed [15,16], and it will appear in the next section that the use of complex-valued corrections plays a crucial role in the construction of weak asymptotic solutions to the Buckley–Leverett equation.

3. WEAK ASYMPTOTICS

In this section, we shall construct an approximative solution to the Buckley–Leverett equation posed with piecewise constant initial data. We shall show how to accomplish this in the case of the Riemann problem since the case of multiple steps is treated similarly. We begin with a general definition of what we shall mean by an approximate solution. First we define a vanishing family of distributions.

DEFINITION 3.1

Let $f_\varepsilon(x) \in D'(R)$ be a family of distributions depending on $\varepsilon \in (0,1)$. We say that $f_\varepsilon = o_{D'}(1)$ if for any test function $\phi(x) \in D(R)$, the estimate

$$\langle f\varepsilon,\phi\rangle = o(1), \text{ as } \varepsilon \to 0$$

holds.

The estimate on the right-hand side is understood in the usual Landau sense. Thus we may say that a family of distributions approach zero in the sense defined above if for a given test function ϕ, the pairing $\langle f\varepsilon, \phi\rangle$ converges to zero as ε approaches zero.

DEFINITION 3.2 We say that the family of complex-valued distributions (u_ε) represents a weak asymptotic solution to (1.1) if there exist real-valued distribution $u \in C((0,\infty);D'(\mathrm{R}))$, such that for every fixed $t\in(0,\infty)$

$$u_\varepsilon \rightharpoonup u, \text{as } \varepsilon \to 0,$$

in the sense of distributions in $D'(\mathrm{R})$, and

$$\partial_t u\varepsilon + \partial_x f(u_\varepsilon),= {}^{o}D'(1). \tag{3.1}$$

Let us remark that it is usually assumed that the relations ${}^{o}D'(\mathrm{R})(1)$ hold uniformly with respect to $t\in(0,\infty)$. However, in the present case, Definition 3.2 is only used as an admissibility condition in tandem with Definition 2.1defining singular solutions. Since we also use further admissibility conditions for the solutions constructed in Sections 4 and 5, we settle for the slightly weaker concept that the distributions (u_ε) converge to u pointwise in t. With these definitions in hand, the following theorem can be proved.

THEOREM 3.1 *For any u_L, u R $\in [0,1]$, and every $c \in \mathrm{R}$, there exists a family of functions (u_ε) representing the weak asymptotic solution to (1.1) such that*

$$u_\varepsilon \rightharpoonup U(x,t) + \alpha(t)\delta(x - ct), \tag{3.2}$$

Where $U(x,t)$is given by(2.5) and $\alpha(t)$is given by(2.6).

PROOF In order to construct an approximate solution to (1.1) satisfying (3.2), we introduce a number of approximations. Denote by $\omega \in C\infty(\mathrm{R})$ a smooth non-decreasing function such that $0 \le \omega \le 1$ and

$$\omega(z) = \begin{cases} 0, & z \le -1, \\ 1, & z \ge 1. \end{cases}$$

We denote the approximate Heaviside function and the approximate delta distribution, respectively by

$$H_\varepsilon(x) = \omega(z/\varepsilon), \quad \delta_\varepsilon(x) = \frac{1}{\varepsilon}\omega'(x/\varepsilon), \quad x \in \mathbb{R}$$

In addition, we shall need the function χ_ε, defined by

$$\chi_\varepsilon(x) = \begin{cases} 1, & |x| \leq \varepsilon/2, \\ 0, & |x| \geq \varepsilon/2 + 2\varepsilon^2 \\ 1 - \frac{1}{\sqrt{2\varepsilon}}\sqrt{|x| - \varepsilon/2}, & \text{otherwise.} \end{cases}$$

This is a function which approximates the characteristic function of the interval $(-\varepsilon/2, \varepsilon/2) \subset R$ in the sense that it is equal to unity in the interval $(-\varepsilon/2,\varepsilon/2)$, and which has support on the interval $(-\varepsilon/2 - 2\varepsilon^2,\varepsilon/2+2\varepsilon^2)$. We start with the following ansatz:

$$\begin{aligned} u_\varepsilon(x,t) &= u_R H_\varepsilon(x - ct - 10\varepsilon) + u_L H_\varepsilon(-x + ct - 10\varepsilon) \\ &+ \frac{\alpha(t)}{2}\delta_\varepsilon(x - ct - 20\varepsilon) + \frac{\alpha(t)}{2}\delta_\varepsilon(x - ct + 20\varepsilon) + u_{p(t)/\varepsilon}\chi_\varepsilon(x - ct) \end{aligned} \quad (3.3)$$

where $u_p(t)/\varepsilon$ is chosen such that

$$f(u_{p(t)/\varepsilon}) = \frac{p(t)}{\varepsilon}$$

It is not difficult to see that

$$u_{p(t)/\varepsilon} = \frac{-a \pm \sqrt{\frac{a\varepsilon}{p(t)} - a}}{\frac{\varepsilon}{p(t)} - (a+1)} = \frac{a}{a + \sqrt{\frac{\varepsilon a}{p(t)} - a}} \quad (3.4)$$

if the plus sign is chosen. In this case, we clearly have $|u(x,t)| \leq 1$. Moreover, we may define $u_p(t)/\varepsilon(p = 0) = 0$. As will come to light, the function $p(t)$ will have to be chosen as $p(t) = c\alpha(t)$, and, for such a choice, $up(t)/\varepsilon$ will have a non-zero imaginary part for small enough ε. Notice also that

$$H\varepsilon(x - ct - 10\varepsilon)\, H\varepsilon(-x + ct - 10\varepsilon) = 0,$$

$$H\varepsilon(x - ct - 10\varepsilon)\ \delta\varepsilon(x - ct - 20\varepsilon) = \delta\varepsilon(x - ct - 20\varepsilon),$$

$$H\varepsilon(-x + ct - 10\varepsilon)\ \delta\varepsilon(x - ct + 20\varepsilon) = \delta\varepsilon(x - ct + 20\varepsilon),$$

$$H\varepsilon(-x + ct - 10\varepsilon)\ \chi\varepsilon(x - ct) = 0,$$

$$H\varepsilon(x - ct - 10\varepsilon)\ \chi\varepsilon(x - ct) = 0.$$

Moreover, we have

$$f\left(u_{p(t)/\varepsilon}\chi_\varepsilon(x - ct)\right) = \frac{p(t)}{\varepsilon},\quad x \in (ct - \varepsilon/2, ct + \varepsilon/2),\quad t > 0.$$

Now, it remains to insert u_ε into (3.1). Accordingly, for an arbitrary test function $\varphi \in C^\infty_c(\mathbb{R})$, we have

$$\begin{aligned}
&\int_{\mathbb{R}} \{\partial_t u_\varepsilon \varphi - f(u_\varepsilon)\partial_x\varphi\}dx \\
&= \int_{-\infty}^{ct-30\varepsilon} \{0 - f(u_L)\varphi'(x)\}dx + \int_{ct+30\varepsilon}^{\infty} \{0 - f(u_R)\varphi'(x)\}dx \\
&+ \int_{ct-30\varepsilon}^{ct-15\varepsilon} \left\{\left[\tfrac{\alpha'(t)}{2}\delta_\varepsilon(x - ct + 20\varepsilon) - \tfrac{\alpha(t)}{2}c\delta'_\varepsilon(x - ct + 20\varepsilon)\right]\varphi(x) - f(u_\varepsilon)\varphi'(x)\right\}dx \\
&+ \int_{ct+15\varepsilon}^{ct+30\varepsilon} \left\{\left[\tfrac{\alpha'(t)}{2}\delta_\varepsilon(x - ct - 20\varepsilon) - \tfrac{\alpha(t)}{2}c\delta'_\varepsilon(x - ct - 20\varepsilon)\right]\varphi(x) - f(u_\varepsilon)\varphi'(x)\right\}dx \\
&+ \int_{ct-15\varepsilon}^{ct-5\varepsilon} \left\{cu_L\delta_\varepsilon(-x + ct - 10\varepsilon)\varphi(x) - f(u_L H_\varepsilon(-x + ct - 10\varepsilon))\varphi'(x)\right\}dx \\
&+ \int_{ct+5\varepsilon}^{ct+15\varepsilon} \left\{-cu_R\delta_\varepsilon(x - ct - 10\varepsilon)\varphi(x) - f(u_R H_\varepsilon(x - ct - 10\varepsilon))\varphi'(x)\right\}dx \\
&+ \int_{ct-5\varepsilon}^{ct+5\varepsilon} \partial_t u_{p(t)/\varepsilon}\chi_\varepsilon(x - ct)\varphi(x)dx - cu_{p(t)/\varepsilon}\int_{ct-5\varepsilon}^{ct+5\varepsilon} \partial_x\chi_\varepsilon(x - ct)\varphi(x)dx \\
&- \int_{ct-\varepsilon/2}^{ct+\varepsilon/2} \frac{p(t)}{\varepsilon}\varphi'(x)dx - \int_{(ct-\varepsilon/2-2\varepsilon^2, ct+\varepsilon/2+2\varepsilon^2)\setminus(ct-\varepsilon/2, ct+\varepsilon/2)} f(u_{p(t)/\varepsilon}\chi_\varepsilon(x - ct))\varphi'(x)dx.
\end{aligned}$$

The first two integrals on the right-hand side of the above equality converge to $-f(u_L)\varphi(ct)$ and $f(u_R)\varphi(ct)$, respectively, as $\varepsilon \to 0$. The sum of the third and fourth integral converges to $\alpha'(t)\varphi(ct) + c\alpha(t)\varphi'(ct)$ as $\varepsilon \to 0$. The fifth and sixth integrals converge to $cu_L\varphi(ct)$ and $-cu_R\varphi(ct)$, respectively as $\varepsilon \to 0$. As for the seventh integral, we note that both $u_{\frac{p(t)}{\varepsilon}}$ and its time derivative are finite for any fixed $t > 0$ for small enough ε. Thus, since the size of the domain of integration of the seventh integral is 10ε, the integral tends to zero as $\varepsilon \to 0$. It will be shown in Appendix A that the eighth integral and the very last integral both tend to zero as $\varepsilon \to 0$. Then, it can be concluded that

$$\int_{\mathbb{R}} \{\partial_t u_\varepsilon \varphi - f(u_\varepsilon)\partial_x \varphi\} dx = \big(-c[u_R - u_L] + [f(u_R) - f(u_L)]\big)\varphi(ct)$$

$$+ \alpha'(t)\varphi(ct) + c\alpha(t)\varphi'(ct) - p(t) \int_{ct-\varepsilon/2}^{ct+\varepsilon/2} \frac{\varphi'(x)}{\varepsilon}\, dx + o_{\mathcal{D}'}(1)$$

where $o_{\mathcal{D}'}(1) \to 0$ as $\varepsilon \to 0$. The proof of the theorem is concluded by taking $p(t) = c\alpha(t)$ and letting $\varepsilon \to 0$. □

It should be noted that multiplication of distributions can also be studied in the context of Colombeau algebras, such as explained in [3,13,21]. In fact, Definition 3.2 can be understood as a variant of appropriate definitions in [4,20]. The main difference is that in the present case, a solution is found pointwise with respect to $t \in \mathrm{R}_+$, and it is required that the distributional limit of the weak asymptotic solution be a distribution.

Another comment is in order. Note that we have used the functional form of the flux given by (2.1) for values of u on the real line. However, from a physical point of view, only the values of the flux for $0 \le u \le 1$ are important, and the definition of the flux for values other than these is essentially arbitrary. For instance, one could define $f(u)$ to be monotone and unbounded outside of the interval $u \in [0,1]$. While such a definition would not change the physical applicability of Eq. (2.2), it would change the definition of delta-shock solutions. In particular, if $f(u)$ were unbounded, the use of imaginary corrections in (3.3)could possibly be avoided.

4. RIEMANN PROBLEM

The delta-shock solutions described in the previous section may be used to construct non-monotone solutions of the Riemann problem for (1.1). Let $u_I \in (0,1)$ denote the value of u at which the inflection point of the graph of $f(u)$ is located. Given initial data (1.2), with $0 \le u_R < u_I$, a solution can be constructed by taking a delta shock δS1 from u_L to a value u_M. Thanks to the Rankine–Hugoniot defect, this shock is slowed to have the same speed as the characteristic speed $f'(u_M)$. The solution continues with a rarefaction wave from u_M to u^*, where u^* denotes the value of u where the extension of the region under the graph of $f(u)$ to the convex hull to the point $\{u_R, f(u_R)\}$begins, and a classical shock from u* to u_R with shock speed determined by the usual Rankine–Hugoniot condition. This solution is depicted in Fig. 3, and the features are summarized in the following theorem.

THEOREM 4.1 *Suppose we are given Riemann data*(1.2), *such that* $0 < u_R < u_L$, *and* $0 < u_L \leq 1$. *If* $u_M >$ u* *is such that* $f'(u_M) < f'(u_L)$, *then there is a solution of the Riemann problem given by*

$$u(x,t) = U(x,t) + \alpha(t)\delta(x - c_1 t), \tag{4.1}$$

where $U(x,t)$*is given by*

$$U(x,t) = \begin{cases} u_L, & x < c_1 t, \\ u_M, & x = c_1 t, \\ [f']^{-1}(x/t), & c_1 t < x < c_2 t \\ u_R, & c_2 t < x. \end{cases}$$

The shock velocity of the delta shock δS_1 *is given by* $c_1 = f'(u_M)$, *and the strength of* δS_1 *is given by* $\alpha(t) = (c_1[u_M - u_L] - [f(u_M) - f(u_L)])t$. *The shock velocity of the shock* S_2 *is given by* $c_2 = \frac{[f(u_R) - f(u^*)]}{[u_R - u^*]} = f'(u^*)$. *The solution is admissible in the sense that both shocks are compressive.*

The proof of the theorem follows easily from piecing together the delta shock δS_1, the rarefaction wave, and the classical shock S_2. It is straightforward to check that this is a solution both in the sense of Definition 2.1 and in the sense of Definition 3.2 if the strength of the delta shock δS_1 is given by (2.6) as indicated. Moreover, as shown in the right panel of Fig. 3, the solution is admissible in the sense that characteristics are either entering the shock, or are parallel to the shock. Thus the solution is compressive, and it may also be shown that the solution is entropy admissible in the sense defined in [1].

It should be noted that the above construction works for both $u_L <$ u* and $u_L >$ u*. Note also that the admissibility condition is not strong enough to yield uniqueness, as the value of u_M is not uniquely determined from the initial data.

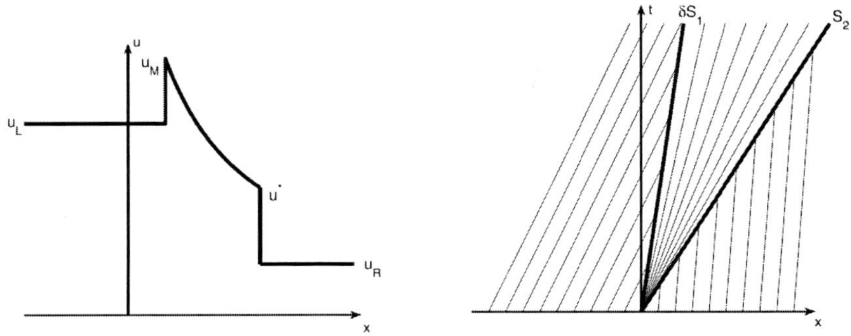

Figure. 3. Schematic picture of a non-monotone admissible solution to the Riemann problem. The shock δS_1 has been slowed to match the characteristic speed of u_M. Between u_M and u*, there is a rarefaction wave, and from u^* to u_R, there is another admissible (non-delta) shock *S*2.

5. TRAVELING WAVES

The solution in the previous section may be modified by extending the region in which the solution *u* takes the value uM to nonzero width. This region will then be sandwiched between the delta shock δS_1 on the left and the rarefaction wave on the right. However, in this case the solution is not the solution of a Riemann problem. On the other hand, inclusion of the rarefaction wave precludes the possibility of constant width of the overshoot region, which is observed experimentally. In this final section, we shall consider the possibility of traveling-wave solutions which are steady waves which propagate without altering the solution profile in time.

Suppose the values u_R and u_L are given such that $0 \leq u_R < u_I < u_L < 1$ and $f'(u_R) < f'(u_L)$. As before, let {u*, $f(u^*)$} be the point on the graph of $f(u)$which marks the right endpoint of the part of the convex hull which lies above the graph, as shown in Fig. 2. Suppose we have $u_R < u_L < u^*$. Then since uI<uL, we have Nf′(uL)>f′(uN), and the solution consists of a delta shock, a constant region $u = u^*$ and a regular shock connecting u^* and u_R. The solution is illustrated in Fig. 4. Defining the L∞-part of the solution by

$$U(x,t) = \begin{cases} u_L, & x < ct \\ u^*, & ct \leq x < ct + m \\ u_R, & ct + m \leq x, \end{cases} \tag{5.1}$$

for an arbitrary $m \in$ R, the following theorem can be formulated.

THEOREM 5.1 *Suppose we are given* u_R *and* u_L, *such that* $0 \leq u_R < u_I < u_L < u^*$, *and* $f'(u_R) < f'(u_L)$. *Then there exists a solution of* (1.1)*given by*

$$u(x, t) = U(x,t) + \alpha(t)\delta(x - ct), \tag{5.2}$$

where $U(x,t)$*is given by* (5.1), *the shock velocity of the delta shock* δS_1 *and the classical shock* S_2 *is given by* $c = f'(u^*)$, *and the strength of the* δS_1 *is given by*

$$\alpha(t) = (c[u^* - u_L] - [f(u^*) - f(u_L)])t.$$

The solution is admissible in the sense that all characteristics either enter into or are collinear to shock trajectories.

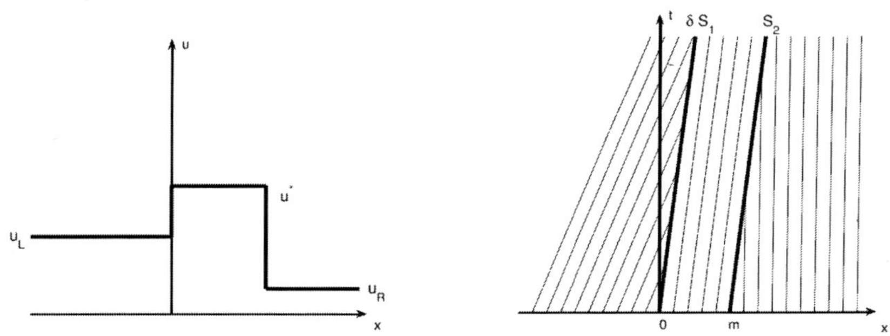

Figure. 4. Schematic picture of a traveling-wave solution. The shock δS_1 has been slowed to match the characteristic speed of u^*. The shock S_2 also matches the characteristic speed of u^*.

In this theorem, c represents the velocity of the profile, and m denotes the width of the overshoot region. The proof follows immediately from the proofs of Theorem 2.1 and Theorem 3.1 since the two shocks are separated by a region of nonzero width m.

Note that this solution is steady, but is not a similarity solution such as the weak solution of a Riemann problem. A certain measure of uniqueness for the above solution follows from the principle of minimizing the number of delta shocks (cf. [16]) because choosing a value other than u^* for the overshoot region would necessitate the inclusion of two delta shocks. However, this principle is not strong enough to provide overall uniqueness, as the width m is still undetermined.

Next, we treat the case where $u^* < u_L$. In this case, the solution contains two delta shocks. First, for an arbitrary m, define $U(x,t)$ by

$$U(x,t) = \begin{cases} u_L, & x < ct \\ u_M, & ct \le x < ct + m \\ u_R, & ct + m \le x. \end{cases} \tag{5.3}$$

Then the following theorem holds.

THEOREM 5.2

Suppose we are given u_R and u_L such that $0 \le u_R < u^* < u_L < 1$, and $f'(u_R) < f'(u_L)$. If u_M can be chosen such that $f'(u_R) < f'(u_M) < f'(u_L)$, then there exists a solution of (1.1) given by

$$u(x,t) = U(x,t) + \alpha_1(t)\delta(x - ct) + \alpha_2(t)\mathrm{i}(x - ct - m), \tag{5.4}$$

where $U(x,t)$ is given by (5.3). The shock velocity of the delta shocks δS_1 and δS_2 is given by $c = f'(u_M)$. The strength of δS_1 is given by $\alpha_1(t) = (c[u_M - u_L] - [f(u_M) - f(u_L)])t$, and the strength of δS_2 is given by $\alpha_2(t) = (c[u_R - u_M] - [f(u_R) - f(u_M)])t$.

The solution is admissible in the sense that all characteristics either enter into or are collinear to shock trajectories.

As explained above, the proof of this theorem also follows immediately from the proofs of Theorem 2.1 and Theorem 3.1 since the two shocks are separated by a region of nonzero width *m*. Note that there are two possibilities for choosing uM. One is in the interval (uR,Ul), where Ul<uL and f′(Ul)=f′(uL), and the other is in the interval (uL,Ur), where Ur>uL is such that f′(Ur)=f′(ur). The former choice yields a monotone solution while the latter yields a non-monotone solution (see Fig. 5).

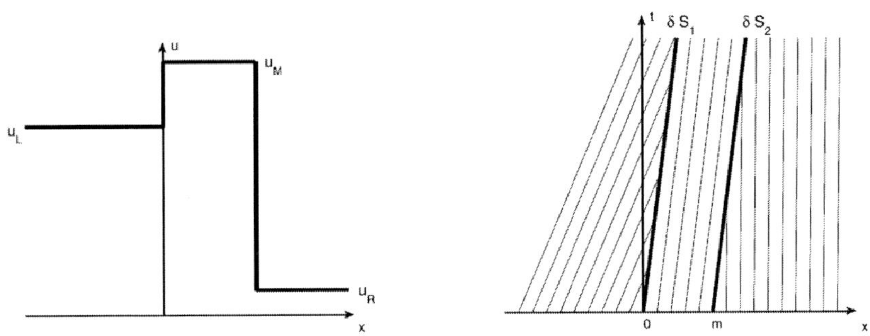

Figure. 5. Schematic picture of a non-monotone traveling-wave solution. The shock $\delta S1$ has been slowed to match the characteristic speed of u_M. The shock δS_2 also matches the characteristic speed of u_M.

As already mentioned in the introduction, experimental results show that if a saturation overshoot develops, then the speed of the two fronts will generally be the same, so that the width of the domain of maximum saturation should remain unchanged in time. While the solutions constructed in this section exhibit such a behavior, the link to the physical modeling of an infiltration problem is unclear since the non-zero Rankine–Hugoniot deficit appears to upset the principle of mass conservation. Nevertheless, it has been shown in this paper that it is at least mathematically possible to construct non-monotone admissible solutions of the Buckley–Leverett equation with various requisite properties.

ACKNOWLEDGMENTS

H.K. is supported by the Research Council of Norway through grant 213474/F20. D.M. is supported by the Ministry of Science of Montenegrothrough grant 01-471, and by the Croatian Science Foundation through grant 9780. J.M.N. is supported by the Research Council of Norway through grants 215641, 224936 and 233736, and by the Norwegian Academy of Science and Letters through VISTA – a basic research program funded by Statoil. The authors would like to thank Iuliu Sorin Pop for enlightening discussions, and an anonymous referee for a number of extremely helpful comments and suggestions.

APPENDIX A. CONVERGENCE OF INTEGRALS

In this appendix, it will be shown that the eighth and tenth integrals in the expression for $\int_R \{\partial_t u_\varepsilon \varphi - f(u_\varepsilon)\partial_x \varphi\}dx$ in the proof of Theorem 3.1 converge to zero. Let us first look at the eighth integral.

LEMMA A.1

$$c u_{p(t)/\varepsilon} \int_{ct-5\varepsilon}^{ct+5\varepsilon} \partial_x \chi_\varepsilon(x-ct)\varphi(x)dx \tag{A.1}$$

approaches zero as $\varepsilon \to 0$.

PROOF Note first of all that it follows from (3.4) that up(t)/ε is finite and uniformly bounded for all ε. Next, recall the definition of χ_ε,

$$\chi_\varepsilon(x) = \begin{cases} 1, & |x| \le \varepsilon/2, \\ 0, & |x| \ge \varepsilon/2 + 2\varepsilon^2, \\ 1 - \frac{1}{\sqrt{2\varepsilon}}\sqrt{|x| - \varepsilon/2}, & \text{otherwise}, \end{cases}$$

and note that we have

$$\int_{\varepsilon/2}^{\varepsilon/2+2\varepsilon^2} |\partial_x \chi_\varepsilon(x)|\, dx = -\int_{\varepsilon/2}^{\varepsilon/2+2\varepsilon^2} \partial_x \chi_\varepsilon(x)\, dx = \chi(\varepsilon/2) - \chi(\varepsilon/2 + 2\varepsilon^2) = 1 \tag{A.2}$$

Since the test function φ is smooth, we can use a finite Taylor expansion with the Lagrangian form of the remainder to write

$$\varphi(x) = \varphi(ct) + [x - ct]\varphi'(\xi),$$

where $\xi \in (ct - x, ct + x)$. Since the function $\partial_x\chi_\varepsilon(\cdot - ct)$ is odd around ct, we have

$$\begin{aligned} \int_{ct-5\varepsilon}^{ct+5\varepsilon} \partial_x\chi_\varepsilon(x - ct)\varphi(x)dx &= 2\int_{ct+\varepsilon/2}^{ct+\varepsilon/2+2\varepsilon^2} \partial_x\chi_\varepsilon(x - ct)\,[x - ct]\varphi'(\xi)\,dx \\ &= 2\int_{\varepsilon/2}^{\varepsilon/2+2\varepsilon^2} \partial_x\chi_\varepsilon(x)\,x\varphi'(\xi + ct)\,dx. \end{aligned}$$

Thus it follows that

$$\left|\int_{ct-5\varepsilon}^{ct+5\varepsilon} \partial_x\chi_\varepsilon(x - ct)\varphi(x)\,dx\right| \le 2\left[\varepsilon/2 + 2\varepsilon^2\right]\max|\varphi'| \int_{\varepsilon/2}^{\varepsilon/2+2\varepsilon^2} |\partial_x\chi_\varepsilon(x)|\,dx$$

Using (A.2), we see that (A.1) approaches zero as $\varepsilon \to 0$. □

Next, we show that the tenth integral in the expression for $\int_{\mathbb{R}} \{\partial_t u_\varepsilon \varphi - f(u_\varepsilon)\partial_x\varphi\}dx$ in the proof of Theorem 3.1 approaches zero.

LEMMA A.2

$$\int_{(ct-\varepsilon/2-2\varepsilon^2,ct+\varepsilon/2+2\varepsilon^2)\setminus(ct-\varepsilon/2,ct+\varepsilon/2)} f(u_{p(t)/\varepsilon}\chi_\varepsilon(x-ct))\varphi'(x)dx \tag{A.3}$$

approaches zero as $\varepsilon \to 0$.

PROOF Note first that

$$\int_{\varepsilon/2}^{\varepsilon/2+2\varepsilon^2} \frac{1}{1-\chi_\varepsilon(x)}\,dx = 2\sqrt{2}\varepsilon\,\sqrt{x-\varepsilon/2}\,\Big|_{\varepsilon/2}^{\varepsilon/2+2\varepsilon^2} \to 0 \tag{A.4}$$

as $\varepsilon \to 0$. The integral in question can be estimated as follows:

$$\left|\int_{(ct-\varepsilon/2-2\varepsilon^2,ct+\varepsilon/2+2\varepsilon^2)\setminus(ct-\varepsilon/2,ct+\varepsilon/2)} f(u_{p(t)/\varepsilon}\chi_\varepsilon(x-ct))\varphi'(x)dx\right|$$

$$\leq 2\max|\varphi'| \int_{ct+\varepsilon/2}^{ct+\varepsilon/2+2\varepsilon^2} \left|f\big(\chi_\varepsilon(x-ct)u_{p(t)/\varepsilon}\big)\right| dx$$

$$= 2\max|\varphi'| \int_{c\varepsilon/2}^{\varepsilon/2+2\varepsilon^2} \left|\frac{1}{1+a\big(1-\frac{1}{\chi_\varepsilon(x)u_{p(t)/\varepsilon}}\big)^2}\right| dx.$$

Now note that for any complex number, we have $|\mathrm{Im}(z)| \leq |z|$, so that the denominator of the integrand may be replaced by its imaginary part. Substituting the expression for $u_{p(t)/\varepsilon}$, we see that

$$\mathrm{Im}\left\{1+a\left(1-\frac{1}{\chi_\varepsilon(x)u_{p(t)/\varepsilon}}\right)^2\right\} = 2\sqrt{a}\sqrt{1-\varepsilon/p(t)}\,\frac{1}{\chi_\varepsilon(x)}\left[\frac{1}{\chi_\varepsilon(x)}-1\right].$$

Hence the integral (A.3) can be estimated by

$$\frac{2\max|\varphi'|}{2\sqrt{a}\sqrt{1-\varepsilon/p(t)}}\int_{\varepsilon/2}^{\varepsilon/2+2\varepsilon^2}\frac{\chi_\varepsilon(x)}{\left[\frac{1}{\chi_\varepsilon(x)}-1\right]}\,dx$$

Finally, we have

$$\begin{aligned}\lim_{\varepsilon\to 0}\frac{2\max|\varphi'|}{2\sqrt{a}\sqrt{1-\varepsilon/p(t)}}\int_{\varepsilon/2}^{\varepsilon/2+2\varepsilon^2}\frac{\chi_\varepsilon(x)}{\frac{1}{\chi_\varepsilon(x)}-1}\,dx &= \lim_{\varepsilon\to 0}\frac{2\max|\varphi'|}{2\sqrt{a}\sqrt{1-\varepsilon/p(t)}}\int_{\varepsilon/2}^{\varepsilon/2+2\varepsilon^2}\frac{\chi_\varepsilon^2(x)}{1-\chi_\varepsilon(x)}\,dx\\ &\le \lim_{\varepsilon\to 0}\frac{2\max|\varphi'|}{2\sqrt{a}}\int_{\varepsilon/2}^{\varepsilon/2+2\varepsilon^2}\frac{1}{1-\chi_\varepsilon(x)}\,dx,\end{aligned}$$

since $\chi_\varepsilon(x)$ is at most equal to unity. Invoking (A.4), we see that the integral does indeed approach zero. □

REFERENCES

1. Aavatsmark **KAPILLARENERGIE ALS ENTROPIEFUNKTION** Z. Angew. Math. Mech., 69 (1989), pp. 319-327
2. A.P. Choudhury **SINGULAR SOLUTIONS FOR** 2×2 **SYSTEMS IN NONCONSERVATIVE FORM WITH INCOMPLETE SET OF EIGENVECTORS** Electron. J. Differential Equations, 58 (2013), p. 11
3. J.-F. Colombeau **NEW GENERALIZED FUNCTIONS AND MULTIPLICATION OF DISTRIBUTIONS** North-Holland, Amsterdam (1984)
4. J.-F. Colombeau, M. Oberguggenberger **ON A HYPERBOLIC SYSTEM WITH A COMPATIBLE QUADRATIC TERM: GENERALIZED SOLUTIONS, DELTA WAVES, AND MULTIPLICATION OF DISTRIBUTIONS** Comm. Partial Differential Equations, 15 (1990), pp. 905-938
5. G. Dal Maso, P. LeFloch, F. Murat **DEFINITION AND WEAK STABILITY OF NON-CONSERVATIVE PRODUCTS**
6. J. Math. Pures Appl., 74 (1995), pp. 483-548
7. V.G. Danilov, V.M. Shelkovich **DYNAMICS OF PROPAGATION AND INTERACTION OF δ-SHOCK WAVES IN CONSERVATION LAW SYSTEM** J. Differential Equations, 211 (2005), pp. 333-381
8. D.A. DiCarlo **EXPERIMENTAL MEASUREMENTS OF SATURATION OVERSHOOT ON INFILTRATION** Water Resour. Res., 40 (2004), p. W04215 (9 pages)
9. D.A. DiCarlo **MODELING OBSERVED SATURATION OVERSHOOT WITH CONTINUUM ADDITIONS TO STANDARD UNSATURATED THEORY** Adv. Water Resour., 28 (2005), pp. 1021-1027
10. C.J. van Duijn, L.A. Peletier, I.S. Pop **A NEW CLASS OF ENTROPY SOLUTIONS OF THE BUCKLEY–LEVERETT EQUATION** SIAM J. Math. Anal., 39 (2007), pp. 507-536
11. C.J. van Duijn, Y. Fan, L.A. Peletier, I.S. Pop **TRAVELLING WAVE SOLUTIONS FOR DEGENERATE PSEUDO-PARABOLIC EQUATIONS MODELLING TWO-PHASE FLOW IN POROUS MEDIA** Nonlinear Anal. Real World Appl., 14 (2013), pp. 1361-1383

12. M. Eliassi, R.J. Glass **ON THE CONTINUUM-SCALE MODELING OF GRAVITY-DRIVEN FINGERS IN UNSATURATED POROUS MEDIA: THE INADEQUACY OF THE RICHARDS EQUATION WITH STANDARD MONOTONIC CONSTITUTIVE RELATIONS AND HYSTERETIC EQUATIONS OF STATE** Water Resour. Res., 37 (2001), pp. 2019-2035
13. R.F. Espinosa, G.A. Omel'yanov **WEAK ASYMPTOTICS FOR THE PROBLEM OF INTERACTION OF TWO SHOCK WAVES** Nonlinear Phenom. Complex Syst., 8 (2005), pp. 331-341
14. M. Grosser, M. Kunzinger, M. Oberguggenberger, R. Steinbauer **GEOMETRIC THEORY OF GENERALIZED FUNCTIONS WITH APPLICATIONS TO GENERAL RELATIVITY** Mathematics and Its Applications, vol. 537, Kluwer Academic Publishers, Dordrecht (2001)
15. F. Huang **WELL POSEDNESS FOR PRESSURELESS FLOW** Comm. Math. Phys., 222 (2001), pp. 117-146
16. H. Kalisch, D. Mitrović **SINGULAR SOLUTIONS OF A FULLY NONLINEAR** 2×2 **SYSTEM OF CONSERVATION LAWS** Proc. Edinb. Math. Soc., 55 (2012), pp. 711-729
17. H. Kalisch, D. Mitrović **SINGULAR SOLUTIONS FOR THE SHALLOW-WATER EQUATIONS** IMA J. Appl. Math., 77 (2012), pp. 340-350
18. B. Keyfitz, H.C. Kranzer **A VISCOSITY APPROXIMATION TO A SYSTEM OF CONSERVATION LAWS WITH NO CLASSICAL RIEMANN SOLUTION** Proc. Int. Conf. on Hyperbolic Problems, Lecture Notes in Mathematics, vol. 1402, Bordeaux, 1988 (1989), pp. 185-197
19. C. Korchinski **SOLUTION OF A RIEMANN PROBLEM FOR A** 2×2 **SYSTEM OF CONSERVATION LAWS POSSESSING NO CLASSICAL WEAK SOLUTION** PhD Thesis Adelphi University (1977)
20. R.J. LeVeque **NUMERICAL METHODS FOR CONSERVATION LAWS** (second edition), Lectures in Mathematics ETH Zürich, Birkhäuser Verlag, Basel (1992)
21. M. Nedeljkov **DELTA AND SINGULAR DELTA LOCUS FOR ONE-DIMENSIONAL SYSTEMS OF CONSERVATION LAWS** Math. Methods Appl. Sci., 27 (2004), pp. 931-955
22. M. Nedeljkov, S. Pilipović, D. Scarpalézos **THE LINEAR THEORY OF COLOMBEAU GENERALIZED FUNCTIONS** Pitman Research Notes in Mathematics, vol. 385, Longman, Harlow (1998)
23. E.Y. Panov, V.M. Shelkovichδ'**-SHOCK WAVES AS A NEW TYPE OF SOLUTIONS TO SYSTEMS OF CONSERVATION LAWS** J. Differential Equations, 228 (2006), pp. 49-86
24. J.W. Sheldon, B. Zondek, W.T. Cardwell Jr. **ONE-DIMENSIONAL, INCOMPRESSIBLE, NONCAPILLARY, TWO-PHASE FLUID FLOW IN A POROUS MEDIUM** Trans. AIME, 207 (1956), pp. 136-143
25. M. Sun **INTERACTIONS OF DELTA SHOCK WAVES FOR THE CHROMATOGRAPHY EQUATIONS** Appl. Math. Lett., 26 (2013), pp. 631-637
26. G. Wang **ONE-DIMENSIONAL NONLINEAR CHROMATOGRAPHY SYSTEM AND DELTA-SHOCK WAVES** Z. Angew. Math. Mech., 64 (2013), pp. 1451-1469

CHAPTER 15

Analytic Solution of Linear Fractional Differential Equation with Jumarie Derivative in Term of Mittag-Leffler Function

Uttam Ghosh[1,*], Srijan Sengupta[2], Susmita Sarkar[2], Shantanu Das[3,4]

[1]*Department of Mathematics, Nabadwip Vidyasagar College, Nabadwip, Nadia, West Bengal, India*

[2]*Department of Applied Mathematics, University of Calcutta, Kolkata, West Bengal, India*

[3]*Reactor Control System Design Section Bhabha Atomic Research Centre, Mumbai, India*

[4]*Department of Physics, Jadavpur University Kolkata, West Bengal, India*

ABSTRACT

There is no unified method to solve the fractional differential equation. The type of derivative here used in this paper is of Jumarie formulation, for the several differential equations studied. Here we develop an algorithm to solve the linear fractional differential equation composed via Jumarie fractional derivative in terms of Mittag-Leffler function; and show its conjugation with ordinary calculus. In these fractional differential equations the one parameter Mittag-Leffler function plays the role similar as exponential function used in ordinary differential equations.

Keywords: jumarrie fractional derivative, riemann-liouvelli fractional derivative, mittag-leffler function, fractional differential equations

1. INTRODUCTION

The analytical solutions of the fractional differential equation are emerging branch of applied science also in basic science. Different methods are developing to solve the fractional differential equations. Since the definition of fractional derivative is modifying to relate it with the classical derivative. Mathematicians are trying to develop the formulas of fractional calculus but geometry of fractional derivative has no concrete shape [1]. Depending on different type of derivatives different methods of solution are developing [2-7][2]. Riemann-Liouville definition the fractional derivative of a constant is non-zero which creates a difficulty to relate between the classical calculus. To overcome this difficulty Jumarie [2, 3, 4, 5] modified the definition of fractional derivative of Riemann-Liouvell type and with this new formulation, we obtain the derivative of a constant as zero. Thus using this definition linking between the fractional and classical calculus becomes easier. There is no unique method to solve the linear fractional differential equations. Using the Jumarie modified definition of fractional derivative we obtain the derivative of Mittag-Leffler function as Mittal-Leffler function, as in case of classical whole-order derivative the derivative of $\exp(x)$ is itself exponential function. Thus via use of Jumarie modified Riemann-Liouvelli derivative, there exists conjugation with classical calculus, which eases in many cases to solve fractional differential equation composed with Jumarie fractional derivative. Here we want to develop an algorithm to solving the linear fractional differential equation using the Mittag-Leffler function. We have obtained applied this method to homogeneous fractional differential equations and got corresponding fundamental solution.

Organization of the paper is as follows. In section 2.0 some definition of fractional derivative is reproduced with essential examples. In section 3.0 and 4.0 some properties of Mittag-Leffler function is described. Finally in section 5.0 the methods for solving the linear fractional differential equation composed by Jumarie fractional derivative is developed using the Mittag-Leffler function.

2. SOME DEFINITIONS OF FRACTIONAL

There are many definition of fractional derivative. Grunwald-Letnikov fractional derivative [6], Liouville fractional derivative [8], Riemann-Liouville fractional derivative [10], Caputo fractional derivative [8, 9, 10, 11], Kolwanker-Gangal local fractional derivative [12, 13, 14, 15, 16], Jumarie modified fractional derivative [2]. Here we use the Riemann-Liouville fractional derivative and its modified form by Jumarie [2].

2.1. Riemann-Liouville Definition of Fractional Derivative

Let the function $f(t)$ is one time integrable then the integro-differential expression

$$ {}_aD_t^{\alpha} f(t) = \frac{1}{\Gamma(-\alpha+m+1)} \left(\frac{d}{dt}\right)^{m+1} \int_a^t (t-\tau)^{m-\alpha} f(\tau) d\tau $$

is known as the Riemann-Liouville (R-L)definition of fractional derivative [6] with m as integer with condition $m \le \alpha < m + 1$

In Riemann-Liouville definition the function, $f(t)$ is integrated $(m - \alpha)$ fold and then differentiate $m + 1$ times. We can re-write the above as follows

The left R-L fractional derivative is defined by

$$ {}_aD_t^{\alpha} f(t) = \frac{1}{\Gamma(k-\alpha)} \left(\frac{d}{dt}\right)^{k} \int_a^t (t-\tau)^{k-\alpha-1} f(\tau) d\tau $$

And the right R-L derivative is

$$ {}_tD_b^{\alpha} f(t) = \frac{1}{\Gamma(k-\alpha)} \left(-\frac{d}{dt}\right)^{k} \int_t^b (\tau-t)^{k-\alpha-1} f(\tau) d\tau $$

Where in above k is integer such that $(k - 1) \le \alpha < k$ that is just greater than fractional number α.

Using the left R-L derivative we get the fractional derivative of the function $f(t) = k$ as non-zero, as demonstrated below.

$$
{}_aD_t^\alpha f(t) = \frac{1}{\Gamma(1-\alpha)}\frac{d}{dt}\int_a^t (t-\xi)^{-\alpha} K d\xi
$$

$$
= -\frac{K}{\Gamma(1-\alpha)}\frac{d}{dt}\left.\frac{(t-\xi)^{1-\alpha}}{1-\alpha}\right]_a^t
$$

$$
= K\frac{(t-a)^{1-\alpha}}{\Gamma(1-\alpha)}
$$

Similarly the right R-L derivative of $f(t) = k$ is

$$
{}_tD_b^\alpha f(t) = K\frac{(b-t)^{1-\alpha}}{\Gamma(1-\alpha)}
$$

This shows that the fractional derivative of a constant (K) is non-zero but in classical calculus derivative of a constant is zero which is contradiction between the classical derivative and the fractional derivative of a constant. To overcome this difference Jumarie [2] modified the left R-L fractional derivative.

We get the R-L left derivative of a power function as

2.2 Jumarie Modified Definition of the Fractional Derivative Is

$$
D_t^\alpha f(t) \overset{\text{def}}{=} \begin{cases} \dfrac{1}{\Gamma(-\alpha)}\displaystyle\int_0^t (t-\xi)^{-\alpha-1} f(\xi)d\xi, & \alpha < 0. \\ \dfrac{1}{\Gamma(1-\alpha)}\dfrac{d}{dt}\displaystyle\int_0^t (t-\xi)^{-\alpha}\begin{bmatrix} f(\xi) \\ -f(0) \end{bmatrix} d\xi, & 0 < \alpha < 1 \\ [f^{(\alpha-n)}(t)]^{(n)}, & n \le \alpha < n+1, n \ge 1. \end{cases}
$$

Using this definition we get $D^\alpha\{K\} = 0, 0 \le \alpha < 1$

The above formula in line-1, becomes fractional order integration if we replace α by $-\alpha$ which is

$$
{}_aD_t^{-\alpha} f(t) = \frac{1}{\Gamma(\alpha)}\int_a^t (t-\tau)^{\alpha-1} f(\tau)d\tau
$$

Using the above formula we get for $f(t) = (t - a)^{\gamma}$, the fractional integral for order α.

$$ {}_aD_t^{-\alpha}(t-a)^{\gamma} = \frac{1}{\Gamma(\alpha)}\int_a^t (t-\tau)^{\alpha-1}(\tau-a)^{\gamma}\,d\tau $$

Using the substitution $\tau = a + \xi(t - a)$ we have for; $\tau = a$, $\xi = 0$ and for $\tau = t$, $\xi = 1$; $d\tau(t - a) = (t - a)d\xi$, $(t - \tau) = t - a - \xi(t - a) = (t - a)(1 - \xi)$, we get the following

$$ \begin{aligned} {}_aD_t^{-\alpha}(t-a)^{\gamma} &= \frac{1}{\Gamma(\alpha)}\int_a^t (t-\tau)^{\alpha-1}(\tau-a)^{\gamma}\,d\tau \\ &= \frac{1}{\Gamma(\alpha)}\int_0^1 (t-a)^{\alpha-1}(1-\xi)^{\alpha-1}\xi^{\gamma}(t-a)^{\gamma}(t-a)d\xi \\ &= \frac{(t-a)^{\gamma+\alpha}}{\Gamma(\alpha)}\int_0^1 \xi^{\gamma}(1-\xi)^{\alpha-1}d\xi \\ &= \frac{(t-a)^{\gamma+\alpha}}{\Gamma(\alpha)}\mathrm{B}(\alpha,\gamma+1) \\ &= \frac{\Gamma(\gamma+1)}{\Gamma(\gamma+1+\alpha)}(t-a)^{\gamma+\alpha}, \quad (\alpha<0, \gamma>-1) \end{aligned} $$

We used Beta-function $\mathrm{B}(\alpha,\gamma+1) = \int_0^1 \xi^{\gamma}(1-\xi)^{\alpha-1}d\xi = \frac{\Gamma(\alpha)\Gamma(\gamma+1)}{\Gamma(\alpha+\gamma+1)}$ defined as

$$ \mathrm{B}(p,q) \overset{\text{def}}{=} \int_0^1 u^{p-1}(1-u)^{q-1}du = \frac{\Gamma(p)\Gamma(q)}{\Gamma(p+q)} $$

Applying the above obtained result the fractional integral of order $(1 - v)$, with $0 \le v < 1$ is

$$ {}_aD_t^{-(1-\upsilon)}(t-a)^{\gamma} = \frac{\Gamma(\gamma+1)}{\Gamma(\gamma+2-\upsilon)}(t-a)^{\gamma+1-\upsilon} $$

Taking one whole derivative of the above we get

$$D^1\left[{}_aD_t^{-(1-\upsilon)}(t-a)^\gamma\right]={}_aD_t^{\upsilon}(t-a)^\gamma$$
$$=\frac{d}{dt}\left[\frac{\Gamma(\gamma+1)}{\Gamma(\gamma+2-\upsilon)}(t-a)^{\gamma+1-\upsilon}\right]$$
$$=\frac{\Gamma(\gamma+1)}{\Gamma(\gamma-\upsilon+2)}(\gamma+1-\upsilon)(t-a)^{\gamma-\upsilon}$$
$$=\frac{\Gamma(\gamma+1)}{(\gamma-\upsilon+1)\Gamma(\gamma+1-\upsilon)}(\gamma+1-\upsilon)(t-a)^{\gamma-\upsilon}$$
$$=\frac{\Gamma(\gamma+1)}{\Gamma(\gamma+1-\upsilon)}(t-a)^{\gamma-\upsilon}$$

We have therefore calculated fractional derivative by R-L left formula for v such that $0 \le v < 1$ thus our nearest integer is one that is $k = 1$ and we write that below

$${}_aD_t^{\upsilon}f(t)=\frac{1}{\Gamma(1-\upsilon)}\left(\frac{d}{dt}\right)\int_a^t(t-\tau)^{-\upsilon}f(\tau)d\tau$$
$$=D^1\left[D_t^{-(1-\upsilon)}f(t)\right].$$

Thus for $a = 0$, the fractional RL derivative of $f(t) = t^\gamma$ is

$${}_0D_t^{\upsilon}t^\gamma=\frac{\Gamma(\gamma+1)}{\Gamma(\gamma+1-\upsilon)}t^{\gamma-\upsilon}$$

For a constant function $f(t) = 1$, putting in above expression $\gamma = 0$, we get

$${}_0D_t^{\upsilon}[1]=\frac{1}{\Gamma(1-\upsilon)}t^{-\upsilon}$$

Let us now see what Jumarrie derivative is from above R-L derivative obtained for $f(t) = t^\gamma$. The composition of the Jumarie derivative, with start point of integration as $t = a$ and $f(a) = a^\gamma$ is

$$f^{(\alpha)}[t^{\gamma}]_a^t = \frac{1}{\Gamma(1-\alpha)}\frac{d}{dt}\int_a^t (t-\xi)^{-\alpha}\left[\xi^{\gamma} - a^{\gamma}\right]d\xi$$

$$D^1\left[{}_aD_t^{-(1-\alpha)}t^{\gamma} - {}_aD_t^{-(1-\alpha)}a^{\gamma}\right]$$

$$= {}_aD_t^{\alpha}t^{\gamma} - {}_aD_t^{\alpha}a^{\gamma}.$$

The above expression show that this Jumarrie derivative is composed of two RL derivatives those are ${}_aD_t^{\alpha}t^{\gamma}$ minus RL derivative of a constant ${}_aD_t^{\alpha}a^{\gamma}$. Going by similar steps as done for ${}_0D_t^{\alpha}t^{\gamma}$, we get first the fractional integral in terms of incomplete Gamma function as

$${}_aD_t^{-(1-\alpha)}t^{\gamma} = \frac{t^{\gamma+1-\alpha}}{\Gamma(1-\alpha)}\int_0^{(t-a)/t} z^{-\alpha}(1-z)^{\gamma}\,\mathrm{d}z$$

$$= \frac{t^{\gamma+1-\alpha}}{\Gamma(1-\alpha)}\mathrm{B}_{\eta}(1-\alpha,\gamma+1) \quad \eta = \frac{t-a}{t}.$$

The fractional derivative of t^{γ} is by taking one whole derivative of above expression we get the following

$${}_aD_t^{\alpha}t^{\gamma} = \frac{d}{dt}\left[\frac{t^{\gamma+1-\alpha}}{\Gamma(1-\alpha)}\mathrm{B}_{\eta}(1-\alpha,\gamma+1)\right] \quad \eta = \frac{t-a}{t}.$$

The fractional derivative of a^{γ} is

$${}_aD_t^{\alpha}a^{\gamma} = \frac{a^{\gamma}}{\Gamma(1-\alpha)}(t-a)^{-\alpha}$$

Therefore

$$f^{(\alpha)}[t^{\gamma}]_a^t = \frac{1}{\Gamma(1-\alpha)}\frac{d}{dt}\int_a^t (t-\xi)^{-\alpha}\left[\xi^{\gamma} - a^{\gamma}\right]d\xi$$

$$= D^1\left[{}_aD_t^{-(1-\alpha)}t^{\gamma} - {}_aD_t^{-(1-\alpha)}a^{\gamma}\right] = {}_aD_t^{\alpha}t^{\gamma} - {}_aD_t^{\alpha}a^{\gamma}$$

$$= \frac{d}{dt}\left[\frac{t^{\gamma+1-\alpha}}{\Gamma(1-\alpha)}\mathrm{B}_{\eta}(1-\alpha,\gamma+1)\right] - \frac{a^{\gamma}}{\Gamma(1-\alpha)}(t-a)^{-\alpha}$$

For a = 0, we have

$$f^{(\alpha)}[t^{\gamma}]_0^t = \frac{1}{\Gamma(1-\alpha)}\frac{d}{dt}\int_0^t (t-\xi)^{-\alpha}\left[\xi^{\gamma}-0\right]d\xi$$

$$= D^1\left[{}_0D_t^{-(1-\alpha)}t^{\gamma}\right] = {}_0D_t^{\alpha}t^{\gamma} = \frac{\Gamma(\gamma+1)}{\Gamma(\gamma+1-\alpha)}(t)^{\gamma-\alpha}$$

We will be using Jumarie derivative for power function t^{γ} with start point of differentiation as a = 0, in subsequent sections. When start point of differentiation is non-zero we will be shifting the origin to that non-zero point and use the above formula.

3. MITTAG-LEFFLER FUNCTION

In 1903 Mittag-Leffler [17, 18, 19, 20] was introduce a function defined by $E_{\alpha}(z) = \sum_{k=0}^{\infty}\frac{z^k}{\Gamma(1+\alpha k)}\ (z \in \mathbb{C}, \mathrm{Re}(\alpha) > 0)$ is named as the one parameter Mittag-Leffler function. For α = 1 and 2 we get

$$E_1(z) \overset{\text{def}}{=} 1 + \frac{z}{\Gamma(1+1)} + \frac{z^2}{\Gamma(1+2)} + \frac{z^3}{\Gamma(1+3)} + \ldots\ldots\ldots\ldots\infty$$

$$= 1 + \frac{z}{1!} + \frac{z^2}{2!} + \frac{z^3}{3!} + \ldots\ldots\ldots\ldots\infty = e^z$$

$$E_2(z) \overset{\text{def}}{=} 1 + \frac{z}{\Gamma(1+2)} + \frac{z^2}{\Gamma(1+4)} + \frac{z^3}{\Gamma(1+6)} + \ldots\ldots\ldots\ldots\infty$$

$$= 1 + \frac{z}{2!} + \frac{z^2}{4!} + \frac{z^3}{6!} + \ldots\ldots\ldots\ldots\infty$$

$$= Cosh(\sqrt{z})$$

The integral representation of the Mittag-Leffler function is $E_{\alpha}(z) = \frac{1}{2\pi}\int_C \frac{t^{\alpha-1}e^t}{t^{\alpha}-z}dt, (z \in \mathbb{C}, \mathrm{Re}(\alpha) > 0)$ Here the path of the integral C is a loop which starts and ends at -∞ and encloses the circles of disk $\text{disk}\ |t| \le |z|^{1/\alpha}\ \text{in positive sense}: |\arg(t)| \le \pi\ \text{on}\ C$ The two parameter Mittag-Leffler function [21] was defined by

$$E_{\alpha,\beta}(z)=\sum_{k=0}^{\infty}\frac{z^k}{\Gamma(\beta+\alpha k)},(z,\beta\in\mathbb{C},\mathrm{Re}(\alpha)>0)$$

$$\text{Here, } \mathrm{E}_{\alpha,1}(z)=\mathrm{E}_{\alpha}(z)=\mathrm{e}^z\,(z\in\mathbb{C},\mathrm{Re}(\alpha)>0)$$

$$\mathrm{E}_{1,2}(z)=\frac{e^z-1}{z},\ \mathrm{E}_{2,2}(z)=\frac{Sinh(\sqrt{z})}{\sqrt{z}}.$$

and the corresponding integral representation of the two parameter Mittag-Leffler function is $E_{\alpha,\beta}(z)=\frac{1}{2\pi}\int_C\frac{t^{\alpha-\beta}e^t}{t^\alpha-z}dt,\quad(z\in\mathbb{C},\mathrm{Re}(\alpha)>0)$ where the contour C is already defined.

3.1. Some Properties Mittag-Leffler Function and Its Application

We rewrite the Mittag-Leffler [17, 18, 19] function in the following form by an infinite series

$$E_\alpha(at^\alpha)\overset{\text{def}}{=}1+\frac{at^\alpha}{\Gamma(1+\alpha)}+\frac{a^2t^{2\alpha}}{\Gamma(1+2\alpha)}+\frac{a^3t^{3\alpha}}{\Gamma(1+3\alpha)}+\ldots\ldots\ldots\ldots\ldots$$

is the one parameter Mittag-Leffler function.

Using Jumarie derivative of order α, with $0\le\alpha<1$ with start point as $a=0$ for $f(t)=t^{n\alpha}$, that is $(t)^{\alpha(n-1)}\Gamma(n\alpha+1)/\Gamma[\alpha(n-1)+1]$, for $n=1,2,3,\ldots$;, and also using Jumarie derivative of constant as zero, we get the following very useful identity. In all the subsequent sections we will say D^α is the Jumarie derivative with zero as start point

$$
\begin{aligned}
& D^{\alpha}(E_{\alpha}(at^{\alpha})) \\
& = D^{\alpha}\left(\begin{array}{l} 1+\frac{at^{\alpha}}{\Gamma(1+\alpha)}+\frac{a^2t^{2\alpha}}{\Gamma(1+2\alpha)} \\ +\frac{a^3t^{3\alpha}}{\Gamma(1+3\alpha)}+\ldots\ldots\ldots\ldots\ldots \end{array}\right) \\
& = 0+\frac{\Gamma(1+\alpha)a}{\Gamma(1)\Gamma(1+\alpha)}+\frac{\Gamma(1+2\alpha)a^2t^{\alpha}}{\Gamma(1+2\alpha)\Gamma(1+\alpha)} \\
& \quad +\frac{\Gamma(1+3\alpha)a^3t^{2\alpha}}{\Gamma(1+3\alpha)\Gamma(1+2\alpha)}+\ldots\ldots\ldots \\
& = a\left(\begin{array}{l} 1+\frac{at^{\alpha}}{\Gamma(1+\alpha)}+\frac{a^2t^{2\alpha}}{\Gamma(1+2\alpha)} \\ +\frac{a^3t^{3\alpha}}{\Gamma(1+3\alpha)}+\ldots\ldots\ldots\ldots \end{array}\right) \\
& = aE_{\alpha}(at^{\alpha})
\end{aligned}
$$

Thus

$$D^{\alpha}(E_{\alpha}(at^{\alpha})) = aE_{\alpha}(at^{\alpha}) \tag{1}$$

This shows that $AE_{\alpha}(at^{\alpha})$ is a solution is a solution of the fractional differential equation

$$D^{\alpha}y = ay \tag{2}$$

Where A is arbitrary constant.

Therefore

$$D^{\alpha}y = ay$$

with $y(0) = 1$ has solution

$$\mathrm{y} = E^{\alpha}(at^{\alpha})$$

Using this property of the Mittag-Leffler one can easily prove the following theorem.

Again integrating the Mittag-Leffler function in the interval $[0, x]$ we get

$$ {}_0D_a^{-\alpha}(E_\alpha(at^\alpha)) = D^{-\alpha}\left(1+\frac{at^\alpha}{\Gamma(1+\alpha)}+\frac{a^2t^{2\alpha}}{\Gamma(1+2\alpha)}+\frac{a^3t^{3\alpha}}{\Gamma(1+3\alpha)}+\ldots\ldots\ldots\ldots\right) $$

$$ = \frac{t^\alpha}{\Gamma(1+\alpha)}+\frac{\Gamma(1+\alpha)at^{2\alpha}}{\Gamma(1+2\alpha)\Gamma(1+\alpha)}+\frac{\Gamma(1+2\alpha)a^2t^{3\alpha}}{\Gamma(1+3\alpha)\Gamma(1+2\alpha)}+\ldots\ldots\ldots $$

$$ = \frac{1}{a}\left(1+\frac{at^\alpha}{\Gamma(1+\alpha)}+\frac{a^2t^{2\alpha}}{\Gamma(1+2\alpha)}+\frac{a^3t^{3\alpha}}{\Gamma(1+3\alpha)}+\ldots\ldots\ldots\infty-1\right) $$

$$ = \frac{1}{a}\{E_\alpha(at^\alpha)-1\} $$

Theorem 1: The Mittag-Leffler function $E_\alpha(at^\alpha)$ $E_\alpha(at^\alpha)$ satisfies the relation

$$ E_\alpha(at^\alpha)E_\alpha(bt^\alpha) = E_\alpha((a+b)t^\alpha) $$

Proof: Let $y = E_\alpha(at^\alpha)E_\alpha(bt^\alpha)$ then

$$ y = E_\alpha(at^\alpha)E_\alpha(bt^\alpha) $$

$$ D^\alpha y = E_\alpha(bt^\alpha)D^\alpha(at^\alpha)+E_\alpha(at^\alpha)D^\alpha(bt^\alpha) $$

$$ D^\alpha y = aE_\alpha(at^\alpha)E_\alpha(bt^\alpha)+bE_\alpha(at^\alpha)E_\alpha(bt^\alpha) $$

$$ = (a+b)E_\alpha(at^\alpha)E_\alpha(bt^\alpha) $$

$$ D^\alpha y = (a+b)y \tag{3} $$

Using the solution of the equation (2) we get the solution of the equation (3) in the following form

$$ y = AE_\alpha\left([a+b]t^\alpha\right) $$

From the definition of y we get $y(0) = 1$. Therefore we have $y = E_\alpha((a+b)t^\alpha)$.

Thus we get

$$E_\alpha((a+b)t^\alpha) = E_\alpha(at^\alpha)E_\alpha(bt^\alpha) \tag{4}$$

We get useful property of one parameter Mittag-Leffler function.

Using the above property of Mittag-Leffler function we get

$$E_\alpha(at^\alpha)E_\alpha(-at^\alpha) = 1 \text{ or } E_\alpha(-at^\alpha) = \frac{1}{E_\alpha(at^\alpha)} \tag{5}$$

$$E_\alpha(at^\alpha)E_\alpha(at^\alpha) = E_\alpha(2at^\alpha) \tag{6}$$

4. COMPLEX MITTAG-LEFFLER FUNCTION AND ITS PROPERTIES

Jumarie [22] defined the complex Mittag-Leffler in the following form

$$E_\alpha(it^\alpha) \overset{\text{def}}{=} \cos_\alpha(t^\alpha) + i\sin_\alpha(t^\alpha)$$

$$\cos_\alpha(t^\alpha) = \frac{E_\alpha(it^\alpha)+E_\alpha(-it^\alpha)}{2} = \sum_{k=1}^{\infty}(-1)^k \frac{t^{2k\alpha}}{(2k\alpha)!}$$

$$\sin_\alpha(t^\alpha) = \frac{E_\alpha(it^\alpha)-E_\alpha(-it^\alpha)}{2} = \sum_{k=1}^{\infty}(-1)^k \frac{t^{(2k+1)\alpha}}{(2k\alpha+\alpha)!}$$

On the other hand Jumarie [22] defined period ($M\ \alpha$) of the function $E_\alpha(it^\alpha)$ in the following form, taking $E_\alpha(i(M_\alpha)^\alpha)=1$ and therefore

$$\cos_\alpha(t+M_\alpha)^\alpha = \cos_\alpha(t^\alpha), \sin_\alpha(t+M_\alpha)^\alpha = \sin_\alpha(t^\alpha)$$

$$\cos_\alpha((-t)^\alpha) = \cos_\alpha(t^\alpha), \sin_\alpha((-t)^\alpha) = (-1)^\alpha \sin_\alpha(t^\alpha).$$

The series presentation of $\cos_\alpha(t^\alpha)$is

$$\cos_\alpha(t^\alpha) = 1 - \frac{t^{2\alpha}}{\Gamma(1+2\alpha)} + \frac{t^{4\alpha}}{\Gamma(1+4\alpha)} - \frac{t^{6\alpha}}{\Gamma(1+6\alpha)} + \ldots..$$

Taking term by term Jumarie derivative we get

$$
\begin{aligned}
&D^{\alpha}[\cos_{\alpha}(t^{\alpha})]\\
&=0-\frac{\Gamma(1+2\alpha)t^{2\alpha-\alpha}}{\Gamma(1+2\alpha)\Gamma(1+\alpha)}+\frac{\Gamma(1+4\alpha)t^{4\alpha-\alpha}}{\Gamma(1+4\alpha)\Gamma(1+3\alpha)}\\
&\quad-\frac{\Gamma(1+6\alpha)t^{6\alpha-\alpha}}{\Gamma(1+6\alpha)\Gamma(1+5\alpha)}+\dots\dots\\
&=-\left[\frac{t^{\alpha}}{\Gamma(1+\alpha)}-\frac{t^{3\alpha}}{\Gamma(1+3\alpha)}+\dots\dots\right]\\
&=-\sin_{\alpha}(t^{\alpha})
\end{aligned}
$$

5. SOLUTION OF LINEAR SECOND ORDER FRACTIONAL DIFFERENTIAL EQUATION

Let us consider the function

$$y = AE_{\alpha}(at^{\alpha}) + BE_{\alpha}(bt^{\alpha})$$

with A and B is constants. Differentiating α-times with respect to t, for $0 < \alpha < 1$, with Jumarie derivative we get

$$
\begin{aligned}
D^{\alpha}y &= AaE_{\alpha}(at^{\alpha}) + BbE_{\alpha}(bt^{\alpha})\\
D^{\alpha}y - ay &= AaE_{\alpha}(at^{\alpha}) + BbE_{\alpha}(bt^{\alpha}) - ay\\
&= AaE_{\alpha}(at^{\alpha}) + BbE_{\alpha}(bt^{\alpha})\\
&\quad - a\left(AE_{\alpha}(at^{\alpha}) + BE_{\alpha}(bt^{\alpha})\right)\\
&= B(b-a)E_{\alpha}(bt^{\alpha})\\
D^{\alpha}y - ay &= B(b-a)E_{\alpha}(bt^{\alpha})
\end{aligned}
$$

Differentiating above by Jumarrie derivative and re-arranging, we get

$$
\begin{aligned}
&D^{2\alpha}y - aD^{\alpha}y = Bb(b-a)E_{\alpha}(bt^{\alpha})\\
&D^{2\alpha}y - (a+b)D^{\alpha}y + aby = 0
\end{aligned}
$$

This shows that the fractional differential equation

$$D^{2\alpha}y-(a+b)D^{\alpha}y+aby=0$$

has solution in the form

$$y=AE_{\alpha}(at^{\alpha})+BE_{\alpha}(bt^{\alpha})$$

On the other hand consider the differential equation

$$D^{2\alpha}y-(a+b)D^{\alpha}y+aby=0$$

it can be express in the following form

$$(D^{\alpha}-a)(D^{\alpha}-b)y(t)=0 \tag{7}$$

Let, $(D^{\alpha}-b)y(t)=x(t)$ then equation (7) reduce to the form

$$(D^{\alpha}-a)x(t)=0 \quad \text{or} \quad D^{\alpha}x(t)=ax(t)$$

Solution of the above equation is same as the solution of the equation (2) which is

$$x(t)=A_1E_{\alpha}(at^{\alpha})$$
$$(D^{\alpha}-b)y(t)=A_1E_{\alpha}(at^{\alpha})$$
$$D^{\alpha}y-by=A_1E_{\alpha}(at^{\alpha})$$
$$E_{\alpha}(-bt^{\alpha})(D^{\alpha}y-by)=A_1E_{\alpha}(at^{\alpha})E_{\alpha}(-bt^{\alpha})$$
$$D^{\alpha}(yE_{\alpha}(-bt^{\alpha}))=\frac{A_1}{a-b}D^{\alpha}(E_{\alpha}(at^{\alpha})E_{\alpha}(-bt^{\alpha}))$$

On integrating both side we get that is applying $D^{-\alpha}$ on both sides of above, we get

$$yE_{\alpha}(-bt^{\alpha})=\frac{A_1}{a-b}(E_{\alpha}(at^{\alpha})E_{\alpha}(-bt^{\alpha}))+B$$
$$y=AE_{\alpha}(at^{\alpha})+BE_{\alpha}(bt^{\alpha}) \qquad \text{where} \qquad A=\frac{A_1}{a-b}$$

Therefore $y = AE_{\alpha}(at^{\alpha}) + BE_{\alpha}(bt^{\alpha})$ is a solution of the differential equation. Thus we can state the following theorem

Theorem 2: The fractional differential equation

$$(D^{\alpha} - a)(D^{\alpha} - b)y(t) = 0$$

has solution of the form

$$y = AE_{\alpha}(at^{\alpha}) + BE_{\alpha}(bt^{\alpha})$$

where A and B are constants.

Proof of the theorem is follows from the previous arguments.

Similarly one can generalized the solution of the differential equation in the following form

If

$$(D^{\alpha} - a_1)(D^{\alpha} - a_2)(D^{\alpha} - a_3)\ldots(D^{\alpha} - a_n)y(t) = 0$$

with all a_i's are distinct be a fractional differential equation with $0 \le \alpha < 1$ then solution of the differential equation will be

$$y = \sum_{i=1}^{n} A_i E_{\alpha}(a_i t^{\alpha})$$

where A_i are arbitrary constants and $E_{\alpha}(a_i t^{\alpha})$ is one parameter Mittag-Leffler function.

Let us consider the function

$$y = (At^{\alpha} + B)E_{\alpha}(at^{\alpha})$$

Where A and B are constants.

Then

$$D^{\alpha}y = \Gamma(1+\alpha)AE_{\alpha}(at^{\alpha}) + (At^{\alpha}+B)aE_{\alpha}(at^{\alpha})$$

$$\begin{aligned} D^{\alpha}y - ay &= \Gamma(1+\alpha)AE_{\alpha}(at^{\alpha}) \\ &\quad + (At^{\alpha}+B)aE_{\alpha}(at^{\alpha}) - a\{(At^{\alpha}+B)E_{\alpha}(at^{\alpha})\} \\ &= \Gamma(1+\alpha)AE_{\alpha}(at^{\alpha}) \end{aligned}$$

$$\begin{aligned} D^{2\alpha}y - aD^{\alpha}y &= a\Gamma(1+\alpha)AE_{\alpha}(at^{\alpha}) \\ &= a(D^{\alpha}y - ay) \end{aligned}$$

$$D^{2\alpha}y - 2aD^{\alpha}y + a^2y = 0$$

Thus solution of the differential equation

$$D^{2\alpha}y - 2aD^{\alpha}y + a^2y = 0$$

is

$$y = (At^{\alpha}+B)E_{\alpha}(at^{\alpha})$$

A and B are constants.

On the other hand consider the differential equation

$$(D^{\alpha}-a)^2y = 0 \quad \text{or} \quad (D^{2\alpha}-2aD^{\alpha}+a^2)y = 0$$

$$D^{2\alpha}y - 2aD^{\alpha}y + a^2y = 0 \tag{8}$$

Let $(D^{\alpha}-a)y = v$ then equation (8) reduce to the form

$$(D^{\alpha}-a)v = 0 \tag{9}$$

Solution of this differential equation is $v(t) = A_1E_{\alpha}(at^{\alpha})$

$$(D^{\alpha}-a)y = A_1 E_{\alpha}(at^{\alpha})$$
$$E_{\alpha}(-at^{\alpha})(D^{\alpha}y - ay) = A_1 E_{\alpha}(at^{\alpha})E_{\alpha}(-at^{\alpha})$$
$$D^{\alpha}[yE_{\alpha}(-at^{\alpha})] = A_1 = D^{\alpha}\left[\frac{A_1 t^{\alpha}}{\Gamma(1+\alpha)}\right]$$
$$yE_{\alpha}(-at^{\alpha}) = \frac{A_1}{\Gamma(1+\alpha)} + B$$
$$y = \left(At^{\alpha} + B\right)E_{\alpha}(at^{\alpha}) \qquad \text{where} \qquad A = \frac{A_1}{\Gamma(1+\alpha)}$$

A and B are constants.

Thus the following theorem can be stated

Theorem 3: The fractional differential equation

$$D^{2\alpha}y - 2aD^{\alpha}y + a^2 y = 0$$

has solution of the form

$$y = (At^{\alpha} + B)E_{\alpha}(at^{\alpha})$$

where A and B are constants.

The proof of the theorem is already explained in the previous arguments.

Theorem 4: Solution of the fractional differential equation

$$D^{2\alpha}y - 2aD^{\alpha}y + (a^2 + b^{\alpha})y = 0$$

is of the form

$$y = E_{\alpha}(at^{\alpha})[A\cos_{\alpha}(bt^{\alpha}) + B\sin_{\alpha}(bt^{\alpha})]$$

Proof: The given differential equation can be written in the following form

$$((D^{\alpha}-a)^2 + b^2)y = 0$$
$$\text{or } (D^{\alpha} - a + ib)(D^{\alpha} - a - ib)y = 0 \tag{10}$$

Using theorem 3 we get the solution of the fractional differential (10) can be written in the following form

$$y = A_1 E_\alpha((a+ib)t^\alpha) + B_1 E_\alpha((a-ib)t^\alpha)$$

$$\begin{aligned} y &= A_1 E_\alpha(at^\alpha) E_\alpha(ibt^\alpha) + B_1 E_\alpha(at^\alpha) E_\alpha(-ibt^\alpha) \\ &= E_\alpha(at^\alpha)[A_1\{\cos_\alpha(bt^\alpha) + i\sin_\alpha(bt^\alpha)\} \\ &\quad + B_1\{\cos_\alpha(bt^\alpha) - i\sin_\alpha(bt^\alpha)\}] \\ &= E_\alpha(at^\alpha)[A\cos_\alpha(bt^\alpha) + iB\sin_\alpha(bt^\alpha)] \end{aligned}$$

Where $A = A_1 + B_1$ and $B = A_1 - B_1$.

Thus we get useful results.

6. CONCLUSIONS

There are several methods to solve fractional differential equations, and the solution depends on the type of fractional derivative used. Here we develop an analytical method to find the solutions of linear fractional differential equation, composed by Jumarie fractional derivative in terms of one parameter Mittag-Leffler function. Some well known properties of Mittag-Leffler have been used to find solution of the fractional differential equations. The solutions obtained are similar as the solutions obtained usual calculus obtained in terms the exponential function. This conjugation with ordinary calculus when Jumarie type fractional derivative is used to compose the fractional differential equations is useful in several physical problems.

ACKNOWLEDGEMENT

Acknowledgments are to **Board of Research in Nuclear Science** (BRNS), Department of Atomic Energy Government of India for financial assistance received through BRNS research project no. 37(3)/14/46/2014-BRNS with BSC BRNS, title "Characterization of unreachable (Holderian) functions via Local Fractional Derivative and Deviation Function.

REFERENCES

1. de Oliveira, E. C., J. A. T. Machado. A review of definitions of fractional derivatives and Integral. Mathematical Problems in Engineering. Hindawi Publishing Corporation. 2014. 1-6.
2. Jumarie, G. Modified Riemann-Liouville derivative and fractional Taylor series of non-differentiable functions Further results, Computers and Mathematics with Applications, 2006. (51), 1367-1376.
3. G. Jumarie, "On the solution of the stochastic differential equation of exponential growth driven by fractional Brownian motion," *AppliedMathematics Letters*, vol. 18, no. 7, pp. 817-826, 2005.
4. G. Jumarie, "An approach to differential geometry of fractional order via modified Riemann-Liouville derivative," *Acta Mathematica Sinica*, vol. 28, no. 9, pp. 1741-1768, 2012.
5. G. Jumarie, "On the derivative chain-rules in fractional calculus via fractional difference and their application to systems modelling," *Central European Journal of Physics*, vol. 11, no. 6, pp. 617-633, 2013.
6. X. J. Yang, *Local Fractional Functional Analysis and Its Applications*, Asian Academic Publisher Limited, Hong Kong, 2011.
7. X. J. Yang, *Advanced Local Fractional Calculus and Its Applications*, World Science, New York, NY, USA, 2012.
8. Miller KS, Ross B. An Introduction to the Fractional Calculus and Fractional Differential Equations.John Wiley & Sons, New York, NY, USA; 1993.
9. Das. S. Functional Fractional Calculus 2nd Edition, Springer-Verlag 2011.
10. Podlubny I. Fractional Differential Equations, Mathematics in Science and Engineering, Academic Press, San Diego, Calif, USA. 1999; 198.
11. M. Caputo, "Linear models of dissipation whose q is almost frequency independent-ii," *GeophysicalJournal of the Royal Astronomical Society*, 1967. vol. 13, no. 5, pp. 529-539.
12. K M Kolwankar and A D. Gangal. Local fractional Fokker plank equation, Phys Rev Lett. 80 1998.
13. Abhay Parvate, A. D. Gangal. Calculus of fractals subset of real line: formulation-1; World Scientific, Fractals Vol. 17, 2009.
14. Abhay Parvate, Seema satin and A.D.Gangal. Calculus on a fractal curve in R^n arXiv:00906 oo76v1 3.6.2009; also in Pramana-J-Phys.
15. Abhay parvate, A. D. Gangal. Fractal differential equation and fractal time dynamic systems, Pramana-J-Phys, Vol 64, No. 3, 2005 pp 389-409.
16. E. Satin, Abhay Parvate, A. D. Gangal. Fokker-Plank Equation on Fractal Curves, Seema, Chaos Solitons & Fractals-52 2013, pp 30-35.
17. Erdelyi, A. Asymptotic expansions, Dover (1954).
18. Erdelyi, A. (Ed). Tables of Integral Transforms. vol. 1, McGraw-Hill, 1954.
19. Erdelyi, A. On some functional transformation Univ Potitec Torino 1950.
20. Mittag-Leffler, G. M.. Sur la nouvelle fonction *Eα* (*x*), *C. R. Acad. Sci. Paris*, (Ser. II) 137, 554-558 (1903).
21. Wiman, A. Uberden fundementalsatz in the theorie der funktionem Eα(z) Acta Math., 29. 1905. 191-201.
22. G. Jumarie. Fourier's transformation of fractional order via Mittag-Leffler function and modified Riemann-Liouville derivatives. J. Appl. Math. & informatics. 2008. 26. 1101-1121.

CHAPTER 16

Mathematical Analysis of a Large Scale Vector SIS Malaria Model in a Patchy Environment

Josephine Wairimu[1,2*], Sallet Gauthier[2], Wandera Ogana[1]

[1]*School of Mathematics, University of Nairobi, Nairobi, Kenya*

[2]*INRIA, Metz and University of Lorraine, Lorraine, France*

ABSTRACT

We answer the stability question of the large scale SIS model describing transmission of highland malaria in Western Kenya in a patchy environment, formulated in [1]. There are two equilibrium states and their stability depends on the basic reproduction number, R_0[2]. If $\mathcal{R}_0 \leq 1$, the disease-free steady solution is globally asymptotically stable and the disease always dies out. If $\mathcal{R}_0 > 1$, there exists a unique endemic equilibrium which is globally stable and the disease persists. Application is done on data from Western Kenya. The age structure reduces the level of infection and the populations settle to the equilibrium faster than in the model without age structure.

Keywords: Highland Malaria, Differentiated Susceptibility and Infectivity, Monotone Dynamical Systems

1. INTRODUCTION

We recall the large scale system developed in [1] reduced into a compact form as

$$\dot{X} = \mathcal{A}\mathrm{diag}(1 - X)\mathcal{B}X + (-\mathcal{C} + \tilde{\mathcal{M}})X \tag{1}$$

where

X = (x,y,z), is a vector representing; x is the proportion of infectious children, y is the proportion of infectious adults, and z is the proportion of infectious mosquitoes.

A, $\mathcal{B}$ and the matrix C are the matrices.

$$\mathcal{A} = \begin{pmatrix} \mathrm{diag}(\beta_1^C)\mathrm{diag}(\overline{N})^{-1}\mathrm{diag}(\overline{V}) & 0 & 0 \\ 0 & \mathrm{diag}(\beta_1^A)\mathrm{diag}(\overline{N})^{-1}\mathrm{diag}(\overline{V}) & 0 \\ 0 & 0 & \mathrm{diag}(\overline{N})^{-1} \end{pmatrix}$$

$$\mathcal{B} = \begin{pmatrix} 0 & 0 & I_n \\ 0 & 0 & I_n \\ \mathrm{diag}(\beta_2^C)\mathrm{diag}(\overline{N}^C) & \mathrm{diag}(\beta_2^A)\mathrm{diag}(\overline{N}^A) & 0 \end{pmatrix}$$

$$\mathcal{C} = \begin{pmatrix} \mathrm{diag}(\gamma^C + v.1 + \mu_h^C) & 0 & 0 \\ -vI_n & \mathrm{diag}(\gamma^A + \mu_h^A) & 0 \\ 0 & 0 & \mathrm{diag}(\mu_v) \end{pmatrix}$$

$$\tilde{\mathcal{M}} = \begin{pmatrix} \tilde{M}^C & 0 & 0 \\ 0 & \tilde{M}^A & 0 \\ 0 & 0 & \tilde{M}^v \end{pmatrix}$$

The authors used the preceding matrices and the vector X = (x,y,z) to rewrite Equation (10) in [1] in a compact form as

$$X' = \mathcal{A}\mathrm{diag}(1 - X)\mathcal{B}X + \left(-\mathcal{C} + \tilde{\mathcal{M}}\right)X \tag{2}$$

This system evolves on the unit cube of $\mathbb{R}^{3n}$.

1.1 Calculation of the Basic Reproduction Number

We use the classical framework defined in [3] [4].

The application $F(x) = A\mathrm{diag}(1\text{-}X)BX$ represents the rate of appearance of new infections in the compartments in the patches.

The function $\mathcal{V}(X) = \left(-\mathcal{C} + \tilde{\mathcal{M}}\right)X$ is the rate of transfer of individuals in compartments.

If $F(x)$ is set to zero, system becomes $\dot{X} = \left(-\mathcal{C} + \tilde{\mathcal{M}}\right)X$, which is a linear system, and we have already seen that $\left(-\mathcal{C} + \tilde{\mathcal{M}}\right)$ is a stable Metzler matrix.

Proposition 1.1

The basic reproduction number of system (1) is

$$\mathcal{R}_0 = \rho\left(-\left(-\mathcal{C} + \tilde{\mathcal{M}}\right)^{-1} \mathcal{A}\mathcal{B}\right)$$

spProof

This is straightforward since the Jacobian of F computed at the DFE $X = 0$ is

$$F = ABs$$

and the Jacobian of $\mathcal{V}(X)$ computed at the DFE is $\mathfrak{V} = \left(-\mathcal{C} + \tilde{\mathcal{M}}\right)$.

The next generation matrix is then $K = -\left(-\mathcal{C} + \tilde{\mathcal{M}}\right)^{-1} \mathcal{A}\mathcal{B}$

We can develop the expression of R_0 further:

$$F = \mathcal{A}\mathcal{B} = \begin{pmatrix} 0 & 0 & \mathrm{diag}(\beta_1^C)\mathrm{diag}(\bar{N})^{-1}\mathrm{diag}(\bar{V}) \\ 0 & 0 & \mathrm{diag}(\beta_1^A)\mathrm{diag}(\bar{N})^{-1}\mathrm{diag}(\bar{V}) \\ \mathrm{diag}(\mu_v)\mathrm{diag}(\beta_2^C)\mathrm{diag}(\bar{N}^C) & \mathrm{diag}(\mu_v)\mathrm{diag}(\beta_2^A)\mathrm{diag}(\bar{N}^A) & 0 \end{pmatrix} \tag{3}$$

$$\mathfrak{V} = \left(-\mathcal{C} + \tilde{\mathcal{M}}\right) = \begin{pmatrix} -\text{diag}\left(\gamma + \nu 1 + \mu_h^C\right) + \tilde{M}^C & 0 & 0 \\ \nu I_n & -\text{diag}\left(\gamma + \mu_h^A\right) + \tilde{M}^A & 0 \\ 0 & 0 & -\text{diag}\left(\mu_v\right) + \tilde{M}^v \end{pmatrix} \quad (4)$$

Then we can compute the nonnegative matrix $-\mathfrak{V}^{-1}$, which is a lower triangular matrix

$$-\mathfrak{V}^{-1} = \begin{pmatrix} V_{11}^{-} & 0 & 0 \\ V_{21}^{-} & V_{22}^{-} & 0 \\ 0 & 0 & V_{33}^{-} \end{pmatrix}$$

with

$$\begin{aligned}
V_{11}^{-} &= -\left(-\text{diag}\left(\gamma + \nu.1 + \mu_h^C\right) + \tilde{M}^C\right)^{-1} \\
V_{21}^{-} &= \nu\left(-\text{diag}\left(\gamma + \mu_h^A\right) + \tilde{M}^A\right)^{-1}\left(-\text{diag}\left(\gamma + \nu.1 + \mu_h^C\right) + \tilde{M}^C\right)^{-1} \\
V_{22}^{-} &= -\left(-\text{diag}\left(\gamma + \mu_h^A\right) + \tilde{M}^A\right)^{-1} \\
V_{33}^{-} &= -\left(-\text{diag}\left(\mu_v\right) + \tilde{M}^v\right)^{-1}
\end{aligned}$$

The next generation matrix is a block matrix

$$K = \begin{pmatrix} 0 & 0 & K_{13} \\ 0 & 0 & K_{23} \\ K_{31} & K_{32} & 0 \end{pmatrix}$$

with

$$\begin{aligned}
K_{13} &= -\text{diag}\left(\beta_1^C\right)\text{diag}\left(\bar{N}\right)^{-1}\text{diag}\left(\bar{V}\right)\left(-\text{diag}\left(\mu_v\right) + \tilde{M}^v\right)^{-1} \\
K_{23} &= -\text{diag}\left(\beta_1^A\right)\text{diag}\left(\bar{N}\right)^{-1}\text{diag}\left(\bar{V}\right)\left(-\text{diag}\left(\mu_v\right) + \tilde{M}^v\right)^{-1} \\
K_{31} &= -\text{diag}\left(\mu_v\right)\text{diag}\left(\beta_2^C\right)\text{diag}\left(\bar{N}^C\right)\left(-\text{diag}\left(\gamma + \nu.1 + \mu_h^C\right) + \tilde{M}^C\right)^{-1} \\
&\quad + \nu\,\text{diag}\left(\mu_v\right)\text{diag}\left(\beta_2^A\right)\text{diag}\left(\bar{N}^A\right)\left(-\text{diag}\left(\gamma + \mu_h^A\right) + \tilde{M}^A\right)^{-1}\left(-\text{diag}\left(\gamma + \nu.1 + \mu_h^C\right) + \tilde{M}^C\right)^{-1} \\
K_{32} &= -\text{diag}\left(\mu_v\right)\text{diag}\left(\beta_2^A\right)\text{diag}\left(\bar{N}^A\right)\left(-\text{diag}\left(\gamma + \mu_h^A\right) + \tilde{M}^A\right)^{-1}
\end{aligned}$$

The block structure of K implies (see [4]) that

$$\mathcal{R}_0^2 = \rho\left(K_{31}K_{13} + K_{32}K_{23}\right)$$

When $\mathcal{R}_0 < 1$, the DFE is locally asymptotically stable, and if $\mathcal{R}_0 > 1$ the DFE is unstable, see [3] [4].

2. MAIN RESULT

In this section we establish a global stability result for the DFE when $\mathcal{R}_0 \leq 1$ and a global stability result when $\mathcal{R}_0 \leq 1$. We have the following theorem

Theorem 2.1

We consider the system (1) with the matrix $M^C + M^A + M^v$ irreducible.

Then If $\mathcal{R}_0 \leq 1$, then the system (1) is globally asymptotically stable at the origin If $\mathcal{R}_0 \leq 1$, then there exists a unique endemic equilibrium X^*, which is globally asymptotically stable on $\mathfrak{K} \in [0,1]^{3n}$

spProof

We recall system (1),

$$\dot{X} = \mathcal{A}\text{diag}(1-X)\mathcal{B}X + \left[-\mathcal{C} + \tilde{\mathcal{M}}\right]X$$

The Jacobian at the origin will be given by

$$J(X) = \mathcal{A}\text{diag}(1-X)\mathcal{B} - \mathcal{A}\text{diag}(\mathcal{B}X) + \left[-\mathcal{C} + \tilde{\mathcal{M}}\right]$$

and

$$J(0) = \mathcal{A}\mathcal{B} + \left[-\mathcal{C} + \tilde{\mathcal{M}}\right]$$

To prove the first part of the proposition above we assume that $\mathcal{R}_0 \leq 1$ Following [5] , $A = B + C$ is a regular splitting of A if B is Metzler stable and $C \geq 0$. Thus in our case $\left[-\mathcal{C} + \tilde{\mathcal{M}}\right]$ has to be a stable Metzler matrix which is invertible and $\mathcal{A}\mathcal{B} \geq 0$, or equivalently, $\left[-\mathcal{C} + \tilde{\mathcal{M}}\right]$ has to be an M-matrix.

We know from Thieme [6] , Driessche [4] , and Varga [7] that

$$\mathcal{R}_0 \le 1 \Leftrightarrow s\left(-\mathcal{A}\mathcal{B}+\left[-\mathcal{C}+\tilde{\mathcal{M}}\right]^{-1}\right)\le 0$$

From the preceding section we know that the Jacobian J is an irreducible Metzler matrix. So, by PerronFrobenius, there exists a positive vector $\mathbf{c} \gg 0$, such that

$$\left(\mathcal{A}\mathcal{B}-\mathcal{C}+\tilde{\mathcal{M}}\right)^T \mathbf{c} = s\left(J(0)\right)\mathbf{c} \le 0$$

To prove the global stability of the DFE we consider the Lyapunov function

$$L(X)=\langle \mathbf{c} | X\rangle$$

where $\langle | \rangle$ denotes the inner product. From the definition of $\mathbf{c} \gg 0$, this function is actually positive definite in the nonnegative orthant.

We compute the derivative of L along the trajectories of (1) and find that it is equivalent to

$$\begin{aligned} L'(X) &= \left\langle \mathbf{c} \middle| \mathcal{A}\mathrm{diag}(1-X)\mathcal{B}+\left[-\mathcal{C}+\tilde{\mathcal{M}}\right]X\right\rangle \le \left\langle \mathbf{c} \middle| \left(\mathcal{A}\mathcal{B}+\left[-\mathcal{C}+\tilde{\mathcal{M}}\right]\right)X\right\rangle \\ &= \left\langle \left(\mathcal{A}\mathcal{B}+\left[-\mathcal{C}+\tilde{\mathcal{M}}\right]\right)^T \mathbf{c} \middle| X\right\rangle = \left\langle s(J(0))\mathbf{c} \middle| X\right\rangle = s(J(0))\left\langle \mathbf{c} \middle| X\right\rangle \le 0 \end{aligned} \tag{5}$$

We see that $0 \le X \le 1$, it is clear that $(1-X)\mathcal{B} \le X\mathcal{B}$, hence the above inequality.

Since $s\left(J(0)\right)\le 0$ the derivative is non positive. The DFE is stable.

We will prove the asymptotic stability when $\mathcal{R}_0 \le 1$

First we consider the case when $\mathcal{R}_0 \le 1$. Since we know that $\mathcal{R}_0 \le 1$ implies $s(J)<0$, L', $L'$is negative definite, since $\mathbf{c} \gg 0$. This proves the asymptotic stability of the DFE.

When $R_0 = 1$, we consider the largest invariant set contained in the set

$$\mathcal{L}=\left\{X \middle| L'(X)=0\right\}$$

For such an X we have

$$0=\left\langle \mathbf{c} \middle| \mathcal{A}\mathrm{diag}(1-X)\mathcal{B}X+\left[-\mathcal{C}+\tilde{\mathcal{M}}\right]X\right\rangle = \left\langle \mathbf{c} \middle| \left(\mathcal{A}\mathcal{B}+\left[-\mathcal{C}+\tilde{\mathcal{M}}\right]\right)X - \mathcal{A}\mathrm{diag}(X)\mathcal{B}X\right\rangle$$

but since $L'(x)=0$, we have by the inequality (5),

$$\left\langle \mathbf{c} \middle| \left(\mathcal{AB}+\left[-\mathcal{C}+\tilde{\mathcal{M}}\right]\right)X\right\rangle = 0$$

Hence

$$V'(X) = -\left\langle \mathbf{c} \middle| \mathcal{A}\mathrm{diag}(X)\mathcal{B}X\right\rangle$$

Since $\mathbf{c} \gg 0$ then $A\mathrm{diag}(X)BX = 0$ or equivalently $\mathrm{diag}(X)BX = 0$

Now we show that the largest invariant set in $\mathcal{L}$ is reduced to the origin.

$$\mathrm{diag}(X)\mathcal{B}X = \begin{pmatrix} \mathrm{diag}(x)z \\ \mathrm{diag}(y)z \\ \mathrm{diag}(z)\left(\mathrm{diag}(\beta_2^C)\mathrm{diag}(\overline{N}^C)x+\mathrm{diag}(\beta_2^A)\mathrm{diag}(\overline{N}^A)y\right) \end{pmatrix}$$

We must have, for any index i, $x_i z_i = 0$ and $y_i z_{i} = 0$

Suppose $x_{i} = 0$ since

$$x_i' = \beta_{1i}^C \frac{\overline{V}_i}{\overline{\overline{N}}_i} z_i(1-x_i) - \left(\gamma_i^C + \mu_{h,i}^C + \nu\right)x_i + \sum_{j=1,\, j\neq i}^{n} m_{ij}^C \frac{\overline{N}_j^C}{\overline{\overline{N}}_i^C} x_j - x_i \sum_{j=1,\, j\neq i}^{n} m_{ji}^C$$

we have

$$x_i' = \beta_{1i}^C \frac{\overline{V}_i}{\overline{\overline{N}}_i} z_i + \sum_{j=1,\, j\neq i}^{n} m_{ij}^C \frac{\overline{N}_j^C}{\overline{\overline{N}}_i^C} x_j = 0$$

Then $z_{i} = 0$ and $x_{j} = 0$ for any patch j, with a "children" arc leaving j and entering i.

Since $z_{i} = 0$ and $x_{i} = 0$ and since

$$z_i' = \beta_{2i}^C \frac{\overline{N}_i^C}{\overline{N}_i} x_i(1-z_i) + \beta_{2i}^A \frac{\overline{N}_i^A}{\overline{N}_i} y_i(1-z_i) - \mu_{v,i} z_i + \sum_{j=1,\, j\neq i}^{n} m_{ij}^v \frac{\overline{V}_j}{\overline{V}_i} z_j - z_i \sum_{j=1,\, j\neq i}^{n} m_{ji}^v$$

we have

$$z_i' = \beta_{2i}^A \frac{\overline{N}_i^A}{\overline{N}_i} y_i + \sum_{j=1, j\neq i}^{n} m_{ij}^v \frac{\overline{V}_j}{\overline{V}_i} z_j = 0$$

Again $y_i = 0$ and $z_j = 0$ for any patch j with a "mosquito" arc leaving j and entering i.

Now $x_i = y_i = z_i = 0$ implies

$$y_i' = \sum_{j=1, j\neq i}^{n} m_{ij} \frac{N_j^A}{N_i^A} y_j$$

which implies that $y_j = 0$ for any patch with a "adult" arc leaving j and entering i.

Now, since any patch can be reached by a path composed of "children", "adult" or "mosquito" arcs, this proves that $x_i = y_i = z_i = 0$ for any index.

This ends the proof for the global asymptotic stability of the DFE from LaSalle's Invariance Principle [8].

To prove the second part of our theorem, when $\mathcal{R}_0 > 1$, we need the following theorem from [9].

Theorem 2.2

Let F be a C^1 vector field in $\mathbb{R}^n$, whose flow θ preserves $\mathbb{R}_+^n$ for $t \geq 0$ and is strongly monotone in $\mathbb{R}_+^n$. Assume that the origin is an equilibrium and that all trajectories in $\mathbb{R}_+^n$ are bounded. Suppose the matrixvalued map $DF : \mathbb{R}_+^n \to \mathbb{R}_+^n \times \mathbb{R}_+^n$ is strictly anti monotone, in the sense that,

If $x < y$, then $DF(x) > DF(y)$

then either all the trajectories in $\mathbb{R}_+^n \setminus \{0\}$ tend to the origin, or else there is a unique equilibrium $p \in Int\, \mathbb{R}_+^n \ (p \gg 0)$ and all the trajectories in $\mathbb{R}_+^n$ tend to p.

For our case we shall consider the positively invariant set $[0,1]^{3n}$, which is diffeomorphic to the nonnegative orthant $\mathbb{R}_+^{3n}$. Since the faces of the cube of type $x_i = 1$ are repulsive for the vector field associated to (1), all the trajectories are bounded in $[0,1]^{3n}$.

We recall system 1.

$$X' = \mathcal{A}\text{diag}(1 - X)\mathcal{B}X + (-\mathcal{C} + \tilde{\mathcal{M}})X$$

If we take $X_1 < X_2 \in X$, then

$$F(X_1) = \mathcal{A}\mathrm{diag}(1 - X_1)(\mathcal{B}X_1) + \left[-\mathcal{C} + \tilde{\mathcal{M}}\right] X_1$$
$$F(X_2) = \mathcal{A}\mathrm{diag}(1 - X_2)(\mathcal{B}X_2) + \left[-\mathcal{C} + \tilde{\mathcal{M}}\right] X_2$$

Clearly

$$\left[-\mathcal{C} + \tilde{\mathcal{M}}\right] X_1 < \left[-\mathcal{C} + \tilde{\mathcal{M}}\right] X_2$$

since the quantities are positive.

Now we prove that

$$\mathcal{A}\mathrm{diag}(1 - X_1)(\mathcal{B}X_1) < \mathcal{A}\mathrm{diag}(1 - X_2)(\mathcal{B}X_2)$$

and hence show that the system is strongly monotone. That is

$$\mathcal{A}\mathrm{diag}(\mathcal{B}X_1) - \mathcal{A}\mathrm{diag}(X_1)\mathcal{B}X_1 < \mathcal{A}\mathrm{diag}(X_2) - \mathcal{A}\mathrm{diag}(X_2)\mathcal{B}X_2$$

or

$$-\mathcal{A}\mathrm{diag}(X_1)\mathcal{B}X_1 < -\mathcal{A}\mathrm{diag}(X_2)\mathcal{B}X_2$$

Considering the structure of A and $\mathcal{B}$ and having $X_1 < X_2$ in $[0,1]^{3n}$ and the fact that a sign change reverses the inequality, then

$$\mathcal{A}\mathrm{diag}(X_1)\mathcal{B}X_1 > \mathcal{A}\mathrm{diag}(X_2)\mathcal{B}X_2$$

hence

$$-\mathcal{A}\mathrm{diag}(X_1)\mathcal{B}X_1 < -\mathcal{A}\mathrm{diag}(X_2)\mathcal{B}X_2$$

To prove that theorem, we recall the Jacobian $J(X)$ of system (1)

$$J(X) = -\mathcal{A}\mathrm{diag}(\mathcal{B}X) + \mathcal{A}\mathrm{diag}(1 - X)\mathcal{B} + \left[-\mathcal{C} + \tilde{\mathcal{M}}\right]$$

Again for any $X_1 < X_2 \in X$, then Diag$(1-X_2) < \mathrm{diag}(1-X_1)$. Since the matrix $\mathcal{B}$ has on each row a positive term, since A is a diagonal matrix with positive terms, we deduce

$$\mathcal{A}\mathrm{diag}(1-X_2)\mathcal{B} < \mathcal{A}\mathrm{diag}(1-X_1)\mathcal{B}$$

Considering the structure of $\mathcal{B}$ we have, if $X_1 < X_2$, the relation $\mathcal{B}X_1 < \mathcal{B}X_2$ holds and consequently $-\mathcal{A}\mathrm{diag}(\mathcal{B}X_2) < -\mathcal{A}\mathrm{diag}(\mathcal{B}X_1)$.

Finally we have $J(X_2) < J(X_1)$, therefore the anti monotone criteria is met.

We will prove that no trajectory tends to the origin.

We have $\mathcal{R}_0 > 1$ which is equivalent to $s(J(0)) > 1$. Then there exists a positive vector $\mathbf{c} \gg 0$ such that

$$J(0)^T \mathbf{c} = \left(\mathcal{A}\mathcal{B} - \mathcal{C} + \tilde{\mathcal{M}}\right)^T \mathbf{c} = s(J(0))\mathbf{c}$$

We consider the Chetaev function on a neighborhood of the origin

$$V(X) = \langle \mathbf{c} | X \rangle$$

An simple computation gives

$$V'(X) = \left\langle \mathbf{c} \middle| \left[s(J(o)) I_n - \mathcal{A}\mathrm{diag}(X)\mathcal{B} \right] X \right\rangle$$

Then in a sufficiently small neighborhood of the origin, in $[0,1]^{3n}$, $V'(X) > 0$. This proves that for $\varepsilon > 0$ sufficiently small, the hyperplane $\{X | \langle \mathbf{c} | X \rangle = \varepsilon\}$ is a barrier for the vector field associated to (1). This proves that no trajectory tends to the origin. Then we conclude, by Hirsch theorem, the existence of an attractive endemic equilibrium X^* in the interior of the cube.

To prove stability, we shall compute $J(X^*)X^*$:

$$J(X^*)X^* = -\mathcal{A}\mathrm{diag}(\mathcal{B}X^*)X^* + \mathcal{A}\mathrm{diag}(1-X^*)\mathcal{B}X^* + \left[-\mathcal{C} + \tilde{\mathcal{M}}\right]X^*$$

Taking into account that X^* is an equilibrium gives

$$\mathcal{A}\mathrm{diag}(1-X^*)\mathcal{B}X^* + \left(-\mathcal{C} + \tilde{\mathcal{M}}\right)X^* = 0$$

therefore

$$J\left(X^{*}\right)X^{*} = -\mathcal{A}\mathrm{diag}\left(\mathcal{B}X^{*}\right)X^{*} \ll 0$$

since $X^{*} \gg 0$

We have proved that there exists a vector $X^{*} \gg 0$ such that for the Metzler matrix $J\left(X^{*}\right)$ we have. $J\left(X^{*}\right)X^{*} \ll 0$ This implies that $J\left(X^{*}\right)$ is Hurwitz [10] [11].

This completes the proof of the global asymptotic stability of the endemic equilibrium.

3. AN EXAMPLE IN TWO PATCHES

In this section we give a result to the case of two patches. We shall use the structure defined in Subsection 1.1.

$$\mathcal{F} = \begin{pmatrix} \beta_{11}^{C} I_{v.1} \dfrac{\left(N_1^{C} - I_{h.1}^{C}\right)}{N_1} \\ \beta_{12}^{C} I_{v.2} \dfrac{\left(N_2^{C} - I_{h.2}^{C}\right)}{N_2} \\ \beta_{11}^{A} I_{v.1} \dfrac{\left(N_1^{A} - I_{h.1}^{A}\right)}{N_1} \\ \beta_{12}^{A} I_{v.2} \dfrac{\left(N_2^{A} - I_{h.2}^{A}\right)}{N_2} \\ \dfrac{\left(V_1 - I_{v.1}\right)}{N_1}\left(\beta_{21}^{C} I_{h.1}^{C} + \beta_{21}^{A} I_{h.1}^{A}\right) \\ \dfrac{\left(V_2 - I_{v.2}\right)}{N_2}\left(\beta_{22}^{C} I_{h.2}^{C} + \beta_{22}^{A} I_{h.2}^{A}\right) \end{pmatrix}, \quad \mathcal{V} = \begin{pmatrix} -\left(\gamma_1^{C} + \mu_1^{C} + v\right) I_{h.1}^{C} + m_{12}^{c} I_{h.2}^{C} - m_{21}^{c} I_{h.1}^{C} \\ -\left(\gamma_2^{C} + \mu_2^{C} + v\right) I_{h.2}^{C} + m_{21}^{c} I_{h.1}^{C} - m_{12}^{c} I_{h.2}^{C} \\ -\left(\gamma_1^{A} + \mu_1^{A}\right) I_{h.1}^{A} + m_{12}^{A} I_{h.2}^{A} - m_{21}^{A} I_{h.1}^{A} \\ -\left(\gamma_2^{A} + \mu_2^{A}\right) I_{h.2}^{A} + m_{21}^{A} I_{h.2}^{A} - m_{12}^{A} I_{h.1}^{A} \\ -\mu_v I_{v.1} + m_{12}^{v} I_{v.2} - m_{21}^{v} I_{v.1} \\ -\mu_v I_{v.2} + m_{21}^{v} I_{v.1} - m_{12}^{v} I_{v.2} \end{pmatrix}.$$

The basic reproduction number is given by $\rho\left(-FV^{-1}\right)$ From our example and at the DFE, this matrix is defined by FV^{-1}

which has the values

$$\begin{pmatrix} 0 & 0 & 0 & 0 & \frac{\beta_{11}^C N_1^C}{\mu_v N_1} & 0 \\ 0 & 0 & 0 & 0 & 0 & \frac{\beta_{12}^C N_2^C}{\mu_v N_2} \\ 0 & 0 & 0 & 0 & \frac{\beta_{11}^A N_1^A}{\mu_v N_1} & 0 \\ 0 & 0 & 0 & 0 & 0 & \frac{\beta_{12}^A N_2^A}{\mu_v N_2} \\ \frac{\beta_{21}^C V_1 (\gamma_2^C + v + \mu_2^C + m_{12}^C)}{N_1 (D^C)} & \frac{\beta_{21}^C V_1 m_{12}^C}{N_1 (D^C)} & \frac{\beta_{21}^A V_1 (\gamma_2^A + \mu_2^A + m_{12}^A)}{N_1 (D^A)} & \frac{\beta_{21}^A V_1 m_{12}^A}{N_1 (D^A)} & 0 & 0 \\ \frac{\beta_{22}^C V_2 m_{21} C}{N_2 (D^C)} & \frac{\beta_{22}^C V_2 (\gamma_1^C + v + \mu_1^C + m_{21}^C)}{N_2 (D^C)} & \frac{\beta_{22}^A V_2 m_{21}^A}{N_2 (D^A)} & \frac{\beta_{22}^A V_2 (\gamma_1^A + \mu_1^A + m_{21}^A)}{N_2 (D^A)} & 0 & 0 \end{pmatrix}$$

where

$D^C = (\gamma_1^C + m_{21}^C)(\gamma_2^C + m_{12}^C) - m_{12}^C m_{21}^C$, $D^A = (\gamma_1^A + m_{21}^A)(\gamma_2^A + m_{12}^A) - m_{12}^A m_{21}^A$. To get the basic reproduction number we need to solve $\det|\lambda I - J|$ which is a 6 matrix. Rewriting the matrix in the form

$$\Theta = \begin{pmatrix} 0I_4 & B \\ A & 0I_2 \end{pmatrix}$$

The determinant of Θ is given by

$$\det(\Theta - \lambda I_6) = \begin{pmatrix} -\lambda I_4 & B \\ A & -\lambda I_2 \end{pmatrix} = \lambda^4 \begin{vmatrix} I_4 & -\frac{1}{\lambda} B \\ A & -\lambda I_2 \end{vmatrix} = \lambda^4 \left| \begin{pmatrix} I_4 & 0 \\ A & I_2 \end{pmatrix} \begin{pmatrix} I_4 & -\frac{1}{\lambda} B \\ 0 & -\lambda I_2 + \frac{1}{\lambda} AB \end{pmatrix} \right|$$

Using the properties of determi $\det(\Theta - \lambda I_6) = \lambda^4 \det\left(-\lambda I_2 + \frac{1}{\lambda} AB\right) = \lambda^2 \det(AB - \lambda^2 I_2) = 0$ nants we have

We see after some calculation, that

$$\mathcal{R}_0^2 = \frac{a_0 + \sqrt{a_0^2 - 4a_1}}{2}$$

where

$$a_0 = (A_{11}B_{11} + A_{13}B_{31} + A_{22}B_{22} + A_{24}B_{42})$$

$$a_1 = \left[\left(A_{11}B_{11} + A_{13}B_{31}\right)\left(A_{22}B_{22} + A_{24}B_{42}\right)\right] - \left[\left(A_{21}B_{11} + A_{23}B_{31}\right)\left(A_{12}B_{22} + A_{14}B_{42}\right)\right]$$

The expression for $\mathcal{R}_0^2$, is complex due to the large number of parameters involved, but from the expression of the matrices A and B above, we can gain some insight. For example if there is no human migration between the two patches, then $m_{12}^C = m_{21}^C = m_{12}^A = m_{21}^A = 0$ and

$$\mathcal{R}_0^2 = \frac{a_0 + \sqrt{a_0^2 - 4a_1^*}}{2}$$

where

$$a_1^* = \left(A_{11}B_{11} + A_{13}B_{31}\right)\left(A_{22}B_{22} + A_{24}B_{42}\right)$$

This is the product of the maximum basic reproduction number for patch 1 and patch 2.

If there is no infective vectors in patch 2, and no vector migration, then $\beta_{12}^C = \beta_{12}^A = 0$, and

$$\mathcal{R}_0^2 = \frac{\beta_{11}^C \beta_{21}^C \left(\gamma_2 + \tilde{\mu}_2^C + m_{12}^C\right) N_1^C}{N_1^2\left[\left(\gamma_1 + m_{21}^C\right)\left(\gamma_2 + m_{12}^C\right) - m_{21}^C m_{12}^C\right]} + \frac{\beta_{11}^A \beta_{21}^A \left(\gamma_2 + \mu_2^A\right) N_1^A V_1}{N_1^2\left[\left(\gamma_1^A + m_{21}^A\right)\left(\gamma_2^A + m_{12}^A\right) - m_{21}^A m_{12}^A\right]}$$

which is the total children and adult contribution to $\mathcal{R}_0^2$.

It is clear that this new value of the basic reproduction number highly depends on the migration rates of the two age groups. If we increase the migration rates then $\mathcal{R}_0^2$ increases.

Biologically, this implies that back and forth movement between the patches, would introduce malaria infection in an otherwise malaria free patch.

4. SIMULATION

In this section we obtain baseline values for two sites: the U-shaped valleys and V-shaped valleys. and use them to simulate equation 1, which is a compact form of equation 5 in [1]. For the human population in our model, we consider two patches, Umutete and Iguhu for the U-shaped valleys, and, Marani and Fort Tenan for the V-shaped valleys. From the study on the different ecosystems, the plateaus and the U-shaped valleys ecosystem have the characteristic, such that the results for the V-shaped valleys apply to the plateau ecosystem. This results from the fact that on the plateaus, the terraine is characterised by raised but flat topography with very little stagnant water as the water darains down the rivers, to support breeding places for mosquitoes. The only notable differences is where there are large water bodies like dams and reservoirs. In these cases high mosquito population is likely to survive and hence increase malaria transmission and infection. Some suitable references for our values are [12] -[16].

Data for N_1, N_2, and N_v was estimated to be 10000 people and we assume that the population is evenly distibuted to the two patches so that we have 5000 people on each patch (1500 children and 3500 adults for each ecosystem). The mosquito population likewise was estimated to be 80,000 mosquitoes in the U-shaped valleys and 10,000 mosquitoes in the V-shaped valleys.

The summary of the parameter values used is given in Table1

4.1. The U-Shaped Valley Sites: Iguhu and Umutete, When $R_0 = 5.36$

When the age structuring is considered, the dynamics of the host population in the U-shaped valleys is represented on Figure 1. When there is no age structuring, the dynamics for the U-shaped valleys are shown in Figure 2. If we consider the U-shaped and the V-shaped valleys as one epidemiological region representing Western Kenya, then the dynamics are represented by Figure 3. The disease in the age structured model fades out faster. The steady states also settle to the endemic equilibrium faster.

If there is no spatialization the values for the U-shaped valleys for both ecosystems has host population variation represented on Figure 3. The interaction between the patches raises infection rate, so that the disease persists in the total population, while it fades out fast when the patches are isolated.

4.2. The V-Shaped Valley Sites: Fort Tenan and Marani, $R_0 = 1.67$

The dynamics of the model in the V-shaped valleys sites with age structure is given by Figure 4. When the age structuring is ignored the variation of the host population in the V-shaped valleys is represented by Figure 5.

5. CONCLUSIONS

Highland malaria in Western Kenya remains a source of mortality and morbidity. Concerned efforts have been put in place by the stakeholders to bring the disease under control with less than expected results. This study captures important factors key to endemicity of of malaria in Western Kenya, which could direct control measure and targets effectively.

Table 1 . Parameter values and ranges for System 1 and Equation (5) in [1].

Parameter	U-Shaped valleys	V-shaped valleys
a^C	0.52	0.42
a^A	0.15	0.12
b_1^C	0.011	0.08
b_1^A	0.011	0.08
b_2^C	0.048	0.24
b_2^A	0.024	0.018
μ_h^C	0.079	0.059
μ_h^A	0.033	0.033
μ_v	0.033	0.033
ν	0.000283	0.000283
m_{12}^C	0.08	0.08
m_{21}^C	0.08	0.08
m_{12}^A	0.5	0.5
m_{21}^A	0.5	0.5
m_{12}^v	0.02	0.02
m_{21}^v	0.02	0.02
γ^C	0.0035	0.0035
γ^A	0.0035	0.0035
Λ_h	0.04	0.04
Λ_v	0.13	0.07
N^C	1500	1500
N^A	3500	3500

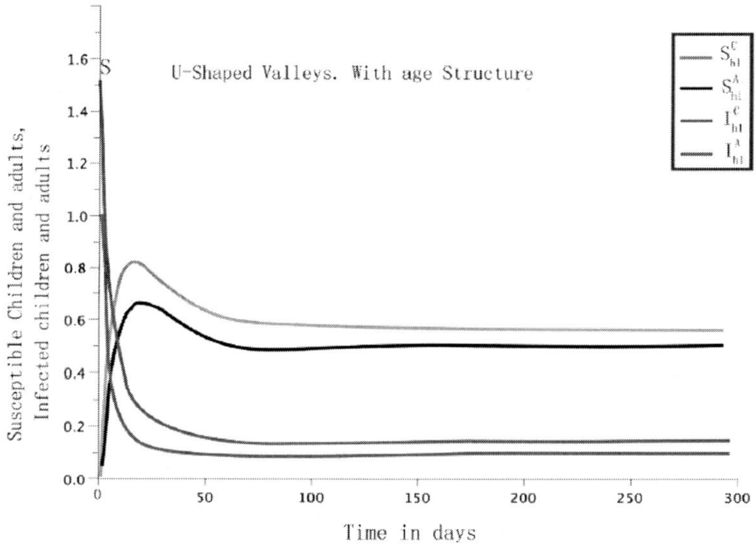

Figure 1. A numerical simulation for the variation of the two age classes population using Equation (1) and parameter values defined in Table 1 for the U-shaped valleys system with $R_0 = 5.36$.

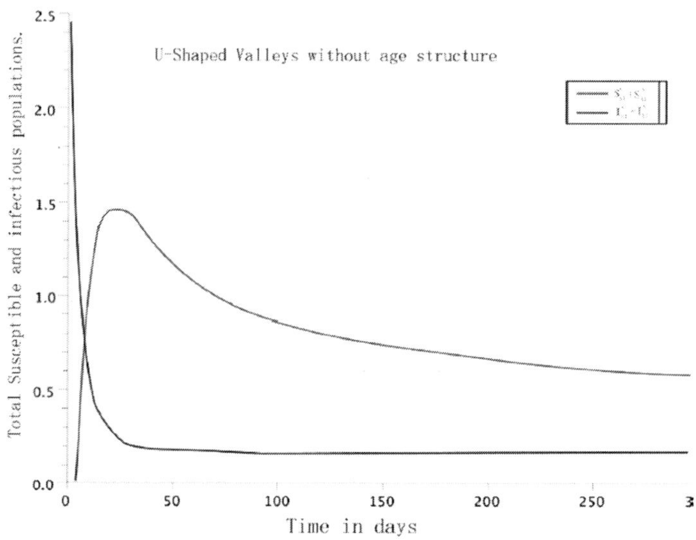

Figure 2. A numerical simulation of Equation (1) no age class in the population and $R_0 = 5.36$.

Age structure helps us differentiate between child's infectivity and susceptibility to malaria infection. It is clear from Balls [17] that children are a significant source of mosquito infection compared to adults. The biting rate for the two age groups differs [18] [19] as children are bitten more than adults did. The other difference captured in the model is death rates for the children (which may include malaria induced deaths). Most malaria deaths occur in children under the age of five years. While the adults also suffer morbidity due to severe infection of highland malaria, there are fewer deaths due to acquired immunity compared to children. We note that the populations settle to the endemic equilibrium faster than in the age-structured than in the unstructured system as shown in Figure 5, and the stable equilibrium is achieved faster in the structured than the unstructured system. Adding age structure allows age specific control strategies to reduce disease prevalence.

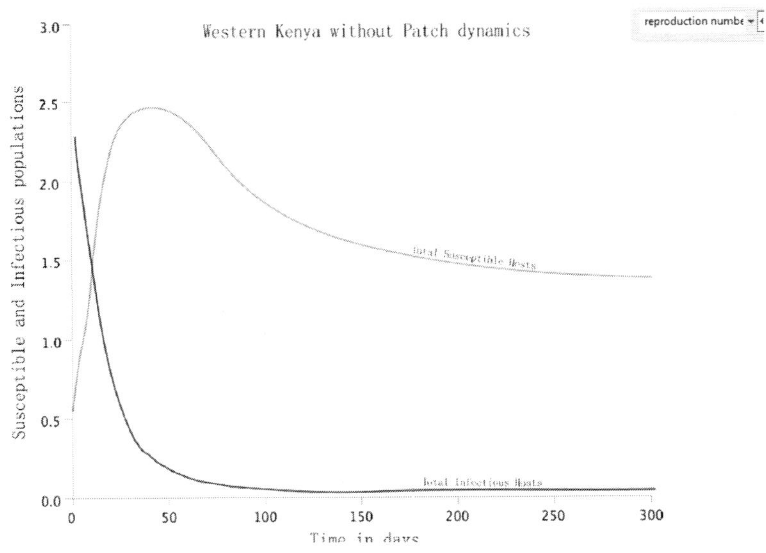

Figure 3. The variation of the total population in the region, no age class and the two ecosystems are treated as a single U-shaped valley ecosystem.

Our model suggests that a suitable model for malaria should be one that captures: age structure; differentiated patch or region susceptibility, which depends on the immunity of the inhabiting population; differentiated infectivity, which also depends on the immunity and age of the host population, and the mosquito population dynamics. Intervention then can be done with guidance from the model.

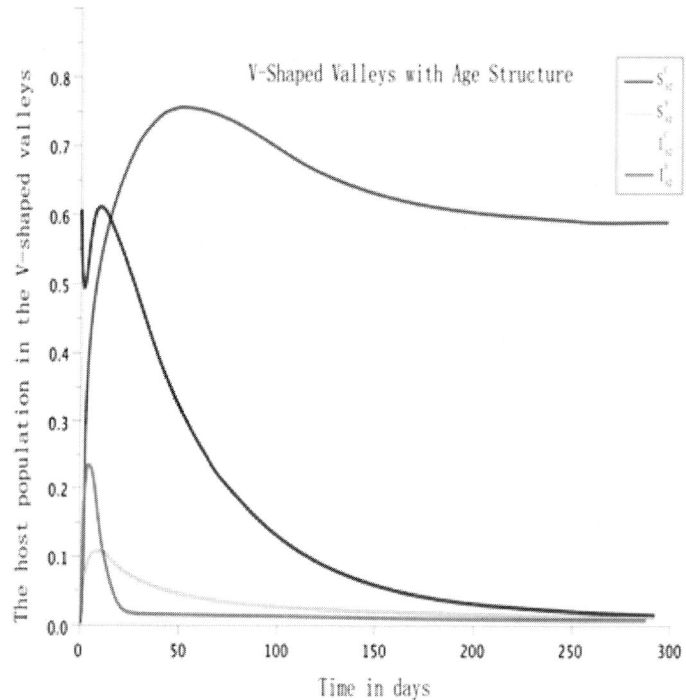

Figure 4. A numerical simulation of model 1 using parameter values defined in Table 1 for the V-shaped valleys ecosystem with age structure. In this case $R_0 = 0.96<1$.

A more comprehensive characterization of results would have to include other types of patches that may not be terraine related but have different epidemiological characteristics from the U-shaped and the V-shaped valleys. Such patches could take care of cities like Nairobi, where human migration has transferred malaria, and central Kenya where the cool highland ecosystem is disturbed by creation of dams for irrigation, rice cultivation, climate change and migration of population to the economically endowed part of the county. Adding age structure allows age specific control strategies to reduce disease prevalence.

We assumed that vectors migrate especially to nearby patches, and the migration parameters for hosts are constant, similar and independent of the compartment. For the compartments that are far apart, the migration of mosquitoes is negligible and is set to zero, since the mosquitoes are only able to fly about 2 kilometers away.

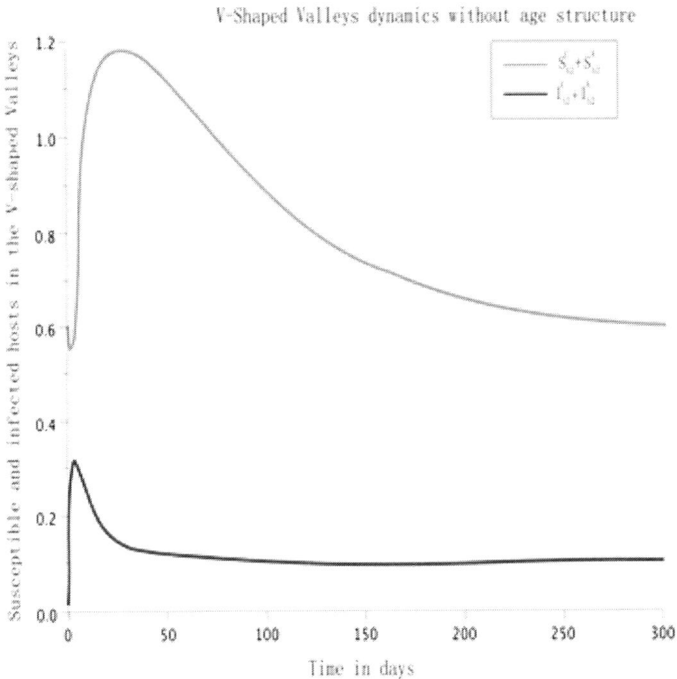

Figure 5. A numerical simulation of model 1 using parameter values defined in Table 1 for the V-shaped valleys ecosystem without age structure. In this case R_0 = 0.96<1, A numerical simulation of model 1 using parameter values defined in Table 1 for the V-shaped valleys ecosystem. The is no age structures in the populations. In this case R_0 = 0.96<1.

An explicit formula for R_0 is obtained, which although complex due to the infinite number of patches, can be used to explore the effects of the parameters of the model. This formula will allow theoretical exploration of the options and efficiency of targeted public health intervention policies. The example in the two ecosystems simplifies the expression for R_0, which we use to simulate our model with some realistic data from Western Kenya. This parameter is inversely related to migration of the hosts between the patches. This implies that to reduce $\mathcal{R}_0^2$, we have to i) administer effective treatment through provision of proper health care facilities in both patches, ii) promote drug adherence, iii) reduce malaria drugs abuse through self administration to shorten the infectious period and arrest human to mosquito infection, hence increase the rate of recovery represented by.

An example in two patches is given with an expression of $\mathcal{R}_0$ which is still complex; but insight is given in a case when no migration takes place. If the disease exists in one patch, with the back and forth movement, the disease in the otherwise free patch would be reintroduced. We want to mention that an example in three patches is also possible, but meaningful insight for the basic reproduction number may only be gained by simulation, with relevant data. This model can be extended to include intervention strategies by the Ministry of health in Kenya, through ITNs and IRS. Since the disease causes death especially in children, the model can also include disease related death rate in the human population. An important factor which is under investigation is the impact of climate related factors to the resurgent epidemics. Resistance of vectors to ITNs and IRS is also an important factor which may cause the disease to remain a menace in the region, not to mention the possibility of drugs resistance in human and possible emergence if new malaria strains.

So far, we have formulated an analytical and numerical analysis which is a foundation of more research and also applicable to other vector borne disease like chikungunya [20].

ACKNOWLEDGEMENTS

Josephine Wairimu, Sallet Gauthier, Wandera Ogana (IRD), the French Embassy in Nairobi and the university of Nairobi, Kenya, for their financial, logistic and moral, support during the writing of this article. We are very grateful to Dr. Githeko, KEMRI Kisumu for the great insight and literature he gave us during this study.

REFERENCES

1. Josephine, W.K., Gauthier, S. and Ogana, W. (2013) Formulation of a Vector SIS Malaria Model in a Patchy Environment with Two Age Classes. Applied Mathematics, 222, 4444.
2. Diekmann, O., Heesterbeek, J.A.P. and Metz, J.A.J. (1990) On the Definition and the Computation of the Basic Reproduction Ratio R0 in Models for Infectious Diseases in Heterogeneous Populations. Journal of Mathematical Biology, 28, 365-382. http://dx.doi.org/10.1007/BF00178324
3. Diekmann, O. and Heesterbeek, J.A.P. (2000) Mathematical Epidemiology of Infectious Diseases in Mathematical and Computational Biology. Wiley Series, Hoboken.
4. Van Den Driessche, P. and Watmough, J. (2002) Reproduction Numbers and Subthreshold Endemic Equilibria for Compartmental Models of Disease Transmission. Mathematical Biosciences, 180, 29-48. http://dx.doi.org/10.1016/S0025-5564(02)00108-6
5. Varga, R.S. (1962) Matrix Iterative Analysis. Prentice-Hall, Upper Saddle River.
6. Thieme, H.R. (2009) Spectral Bound and Reproduction Number for Infinite-Dimensional Population Structure and Time-Heterogeneity. SIAM Journal on Applied Mathematics, 70, 188-211. http://dx.doi.org/10.1137/080732870
7. Varga, R.S. (1960) Factorisation and Normalised Iterative Methods Boundary Problems in Differential Equation. University of Wisconsin Press, Madison.

8. La Salle, J.P. (1976) The Stability of Dynamical Systems. S0ociety for Industrial and Applied Mathematics. Regional Conference Series in Applied Mathematics.
9. Hirsch, M.W. (1982) Systems of Differential Equations that Are Competitive or Cooperative: I. Limit Sets. SIAM Journal on Mathematical Analysis, 13, 167-179. http://dx.doi.org/10.1137/0513013
10. Berman, A. and Plemmons, R.J. (1994) Nonnegative Matrices in the Mathematical Sciences, Volume 9 of Classics in Applied Mathematics. Society for Industrial and Applied Mathematics (SIAM), Philadelphia.
11. Hirsch, H.W. and Smith, H.L. (2005) Monotone Dynamical Systems. In: Handbook of Differential Equations: Ordinary Differential Equations, Vol. II, Elsevier B. V., Amsterdam, 239-357.
12. Chitnis, N., Hyman, J.M. and Cushing, J.M. (2008) Determining Important Parameters in the Spread of Malaria through the Sensitivity Analysis of a Mathematical Model. Bulletin of Mathematical Biology, 70, 1272-1296. http://dx.doi.org/10.1007/s11538-008-9299-0
13. Githeko, A.K., Branding-Bennet, D., Beier, M., Atieli, F., Owaga, M. and Collins, F.H. (1992) The Reservoir of Plasmodium Falciparum Malaria in a Holoendemic Area of Western Kenya. Transactions of the Royal Society of Tropical Medicine and Hygiene, 86, 335-358.
14. Ndenga, B., Githeko, A., Omukunda, E., Munyekenye, G., Atieli, H., Wamai, P., Mbogo, C., Minakawa, N., Zhou, G. and Yan, G. (2006) Population Dynamics of Malaria Vectors in Western Kenya Highlands. Journal of Medical Entomology, 43, 200-206. http://dx.doi.org/10.1603/0022-2585(2006)043[0200:PDOMVI]2.0.CO;2
15. UNICEF (2010) Kenya Statistics. Technical Report, United Nation, New York City.
16. Wanjala, C.L., Waitumbi, J., Zhou, G. and Githeko, A.K. (2011) Identification of Malaria Transmission and Epidemic Hotspots in the Western Kenya Highlands: Its Application to Malaria Epidemic Prediction. Parasites and Vectors, 4, 81. http://dx.doi.org/10. 1186/1756-3305-4-81
17. Balls, M.J., Bodker, R., Thomas, C.J., Kisinza, W., Msangeni, H.A. and Lindsay, S.W. (2004) Effect of Topography on the Risk of Malaria Infection in the Usambara Mountains, Tanzania. Transactions of the Royal Society of Tropical Medicine and Hygiene, 98, 400-408.
18. Mukabana, W.R., Takken, W., Richard, C. and Knols, B.G.J. (2002) Host-Specific Cues Cause Differential Attractiveness of Kenyan Men to the African Malaria Vector Anopheles Gambiae. Malaria Journal, 1, 17. http://dx.doi.org/10.1186/1475-2875-1-17
19. Smith, D.L., Guerra, C.A., Snow, R.W. and Simon, H.I. (2007) Standardizing Estimates of the Plasmodium falciparum Parasite Rate. Malaria Journal, 6, 131. http://dx.doi.org/10. 1186/1475-2875-6-131
20. Bowong, S., Dumont, Y. and Tewa, J.J. (2013) A Patchy Model for Chikungunya-Like Diseases. Biomath, 2, 1-19

CHAPTER 17

Realization of functions on the symmetrized bidisc ☆

Jim Agler[a], N.J. Young[b, c,*]

[a]*Department of Mathematics, University of California, San Diego, CA 92103, USA*

[b]*School of Mathematics and Statistics, Newcastle University, Newcastle upon Tyne NE17RU, UK*

[c]*School of Mathematics, Leeds University, Leeds LS29JT, UK*

ABSTRACT

We prove a realization formula and a model formula for analytic functions with modulus bounded by 1 on the symmetrized bidisc

$$G \stackrel{\text{def}}{=} \{(z+w, zw) : |z| < 1, \; |w| < 1\}$$

As an application we prove a Pick-type theorem giving a criterion for the existence of such a function satisfying a finite set of interpolation conditions.

Keywords: Analytic functions; Hilbert space model; Schur class; Pick theorem

1. INTRODUCTION

The fascination of the symmetrized bidisc G lies in the fact that much of the classical function theory of the disc D and bidisc D^2 generalizes in an explicit way to G, but with some surprising twists. The original motivation for the study of G was its connection with the spectral Nevanlinna–Pick problem [3] ; [5], wherefore the emphasis was on analytic maps from the unit disc D into G. However, in studying such maps one is inevitably drawn into studying maps from G to D; indeed, the duality between these two classes of maps is a central feature of the theory of hyperbolic complex spaces in the sense of Kobayashi [18].

The idea of a *realization formula* for a class of functions has proved potent in both engineering and operator theory. Out of hundreds of papers on this topic in the mathematical literature alone, we mention [20]; [15]; [16]; [1]; [9]; [10]; [11] ; [13]. The simplest realization formula provides an elegant connection between function theory (the Schur class of the disc) and contractive operators on Hilbert space. It is as follows.

Let f be an analytic function on D such that $|f(z)|\leq 1$ for all $z\in$ D. There exists a Hilbert space M, a scalar $A\in$ C, vectors $\beta,\gamma\in$ M and an operator D on M such that the operator

$$\begin{bmatrix} A & 1\otimes\beta \\ \gamma\otimes 1 & D \end{bmatrix} \quad \text{is a contraction on } \mathbb{C}\oplus\mathcal{M} \qquad (1.1)$$

and, for all z$\in$D,

$$f(z) = A + \left\langle z(1-Dz)^{-1}\gamma,\beta\right\rangle_{\mathcal{M}} \qquad (1.2)$$

Conversely, any function f on D expressible in the form (1.1), (1.2) is an analytic function in D satisfying $|f|\leq 1$ on D.

In an earlier paper [7] we gave a realization formula for analytic maps from D to the closure of G; in this paper we present the dual notion, a realization formula for analytic maps from G to D^-.

For any open set $U\subset C^d$ the set of analytic functions on U with values in the closed unit disc D^- is called the *Schur class of U* and is denoted by S (U).

We shall use superscripts to denote the components of points in C^d.

For any point $s=(s^1,s^2)\in G$ and any contractive linear operator T on a Hilbert space M, we define the operator

$$s_T = (2s^2T - s^1)(2 - s^1T)^{-1} \quad \text{on } \mathcal{M} \tag{1.3}$$

Note that $|s^1|<2$ for $s\in G$, and therefore the inverse in equation (1.3) exists.

We shall derive both 'model formulae' and a realization formula for functions in S (G). The latter is the following.

Theorem 1.1. *Let* $\varphi\in$ S (G). *There exist a Hilbert space* M *and unitary operators*

$$T \text{ on } \mathcal{M} \quad \text{and} \quad \begin{bmatrix} A & B \\ C & D \end{bmatrix} \text{ on } \mathbb{C}\oplus\mathcal{M} \tag{1.4}$$

such that, for all $s\in G$,

$$\varphi(s)=A+B_{sT}(1-D_{sT})^{-1}C. \tag{1.5}$$

Conversely, any function φ *on* G *expressible by the formula* (1.5), *where* $T,A,B,C,$D *are such that the operators in formula* (1.4)*are unitary, is an analytic function from* G *to* D⁻.

Both instances of the word 'unitary' in the above theorem can validly be replaced by 'contractive'.

The classical realization formula (1.2) is in terms of a single unitary operator (or contraction), whereas our formula for functions in S (G) requires the *pair* of unitaries (or contractions) (1.4); this is a consequence of the fact that our derivation invokes two separate lurking isometry arguments.

The model formula for functions in S (G) is derived in Section 2 from the known model formula for S $(\mathbb{D}^2)$ by a symmetrization argument. The realization formula is then deduced from the model formula in Section 3. A second model formula, involving an integral with respect to a spectral measure, is proved in Section 4. Finally a Pick-type interpolation theorem, giving a solvability criterion for interpolation problems in S $(\mathbb{D})$, is demonstrated in Section 5. We also give a realization formula for bounded analytic *operator*-valued functions on *G*. The proof requires only notational changes from that ofTheorem 1.1.

This paper is based on a short course of lectures [2] given by the first-named author at the International Centre for the Mathematical Sciences in Edinburgh in 2014.

Two sources for basic facts about the function theory and geometry of G are [17, Chapter 7] and [8, Appendix A].

Many authors have generalized the classical realization formula (1.1) to bounded functions on domains other than the disc. The paper [1] first made it clear that the appropriate class of holomorphic functions for realization theory on certain more general domains Ω is a subclass of $S(\Omega)$, which has become known as the *Schur–Agler class* of Ω. For the disc, the bidisc and the symmetrized bidisc the Schur–Agler class coincides with the Schur class, and so we have no need for its definition in this paper.

The fact that the Schur and Schur–Agler classes of G are equal was proved in [3] (see also [4]) with the aid of Ando's Theorem on commuting pairs of contractions and a symmetrization argument. In this paper we show that essentially the same argument, only with a different ending, yields a realization formula for functions in $S(G)$. We believe that the symmetrization argument is a significant item in the toolkit of realization theory.

Model formulae and realization formulae for the Schur–Agler class of Ω, for any domain Ω having a matrix-polynomial or even a holomorphic operator-valued defining function, are given in [9] ; [11], together with several applications. The question therefore arises as to whether G has a holomorphic operator-valued defining function, and accordingly whether a realization formula for the Schur–Agler class of G can be simply deduced from a known general result. More specifically, is there a continuous operator-valued function F on the closure G^- of G, holomorphic in G, which defines G in the following sense?

$$G = \{s \in G^- : \|F(s)\| < 1\}. \tag{1.6}$$

If so one immediately obtains a realization formula for the general function φ in the Schur–Agler class of G of the form

$$\varphi(s) = A + BF(s)(1 - DF(s))^{-1}C$$

for some contractive (or unitary) operator colligation $ABCD$. It is therefore significant for this paper that the symmetrized bidisc cannot be defined by a *matrix*-valued holomorphic function [19], nor is it known to be defined by an operator-valued holomorphic function. We say a little more about this question at the end of the paper.

A generalization of the realization theory of the polydisc to much more general domains, based on *test functions*, has been developed by Dritschel, McCullough and others [13]; [14] ; [10]. We thank a referee for the observation that a realization formula for functions in the Schur–Agler class of G can be derived from the 'abstract realization theorem' [13, Theorem 2.2] by the choice of the functions

$$s \mapsto \frac{2\lambda s^2 - s^1}{2 - \lambda s^1}$$

(for $|\lambda| < 1$) as the test functions on G. This procedure is essentially carried out in [12], where a realization formula somewhat similar to ours is given [12, Realization theorem, page 5]. However, this approach only yields a realization formula for the Schur–Agler class, not the Schur class, and so to prove Theorem 1.1 in this way one must invoke [3], implicitly utilizing the symmetrization argument we use in this paper.

We are grateful to an anonymous referee for some very helpful remarks which enabled us to improve the presentation of this paper.

2. A MODEL FORMULA FOR *G*

The notion of a Hilbert space model for a function on the polydisc was introduced in [1]. A *model on* $\mathbb{D}^2$ is a pair $(\mathcal{M}, u)$ where $\mathcal{M} = (\mathcal{M}^1, \mathcal{M}^2)$ is a pair of Hilbert spaces and $u = (u^1, u^2)$ is a pair of analytic maps from $\mathbb{D}^2$ to $\mathcal{M}^1, \mathcal{M}^2$ respectively. If φ is a function on $\mathbb{D}^2$ then $(\mathcal{M}, u)$ is a *model of* φ if, for all $\lambda, \mu \in \mathbb{D}^2$,

$$1 - \overline{\varphi(t)}\varphi(s) = \langle (1 - t_T^* s_T) u(s), u(t) \rangle_{\mathcal{M}} \tag{2.1}$$

It is shown in [1] that a function φ on $\mathbb{D}^2$ belongs to the Schur class $\mathcal{S}(\mathbb{D}^2)$ if and only if φ has a model. In this section we shall adapt the notion of model to G and prove an analogous result by means of a symmetrization argument.

Definition 2.1. A *G-model* for a function φ on G is a triple $(\mathcal{M}, T, u)$ where $\mathcal{M}$ is a Hilbert space, T is a contraction acting on $\mathcal{M}$ and $u: G \to \mathcal{M}$ is an analytic function such that, for all $s, t \in G$,

$$1 - \overline{\varphi(t)}\varphi(s) = \langle (1 - t_T^* s_T) u(s), u(t) \rangle_{\mathcal{M}}. \tag{2.2}$$

The following is the main result of this section.

Theorem 2.2. *Let φ be a function on G. The following three statements are equivalent.*

(1) $\varphi \in S(G)$;

(2) φ *has a G-model;*

(3) φ *has a G-model* (M, T, u) *in which T is a unitary operator on* M.

Proof. (2)⇒(1). Suppose φ has a G-model (M, T, u). By holding t fixed in equation (2.2) one can deduce that φ is analytic on G, and on choosing $t=s$ one has

$$1-|\varphi(s)|^2=\langle(1-s_T^*s_T)u(s),u(s)\rangle. \tag{2.3}$$

Now for $s\in G$ we have $|s^1|<2$ and so the function

$$f_s(\lambda)=\frac{2\lambda s^2-s^1}{2-\lambda s^1}$$

is analytic for λ in a neighborhood of D^-. Moreover $|f_s|$ is bounded by 1 on D [6, Theorem 2.1, (1)⇒(4)]. By von Neumann's inequality $f_s(T)$ is a contraction, that is, $\|s_T\|\le 1$. Hence, by equation (2.3), $|\varphi(s)|\le 1$.

(3)⇒(2) is trivial. To prove that (1)⇒(3) we first symmetrize the model (2.1) for the Schur class of the bidisc. Denote by superscript σ the transposition of co-ordinates in C^2, so that

$$(\lambda^1,\lambda^2)^\sigma=(\lambda^2,\lambda^1).$$

Say that a function h on $D^2\times D^2$ is *doubly symmetric* if it is symmetric with respect to σ in each variable separately, that is, if

$$h(\lambda,\mu)=h(\lambda^\sigma,\mu)=h(\lambda,\mu^\sigma)$$

for all $\lambda,\mu\in D^2$.

A doubly symmetric function $h(\lambda,\mu)$ on $D^2\times D^2$ that is analytic in λ and $\bar{\mu}$ can be written in terms of the elementary symmetric functions $\lambda^1+\lambda^2$, $\lambda^1\lambda^2$, $\mu^1+\mu^2$ and $\mu^1\mu^2$. Specifically, if h has a Hilbert space model on the bidisc in the sense of the next proposition, then it induces a function on G having a Hilbert space model of the following form.

Lemma 2.3. *Let h be a doubly symmetric function on* $\mathbb{D}^2 \times \mathbb{D}^2$ *such that there exists a model* $(\mathcal{H}, u)$ *on the bidisc satisfying, for all* $\lambda, \mu \in \mathbb{D}^2$,

$$h(\lambda,\mu) = (1-\overline{\mu^1}\lambda^1)\left\langle u^1(\lambda), u^1(\mu)\right\rangle_{\mathcal{H}^1} + (1-\overline{\mu^2}\lambda^2)\left\langle u^2(\lambda), u^2(\mu)\right\rangle_{\mathcal{H}^2} \quad (2.4)$$

Then there exist a Hilbert space $\mathcal{M}$, *a unitary operator U on* $\mathcal{M}$ *and an analytic function* $x: G \to \mathcal{M}$ *satisfying*

$$h(\lambda,\mu) = \left\langle \left[1 - \bar{t}^2 s^2 - \tfrac{1}{2}(\bar{t}^1 - s^1\bar{t}^2)U - \tfrac{1}{2}(s^1 - \bar{t}^1 s^2)U^*\right] x(s), x(t)\right\rangle_{\mathcal{M}} \quad (2.5)$$

for all $\lambda, \mu \in \mathbb{D}^2$, *where* $s = (\lambda^1+\lambda^2, \lambda^1\lambda^2)$ *and* $t = (\mu^1+\mu^2, \mu^1\mu^2)$.

Proof. We shall write u_λ in place of $u(\lambda)$ throughout the proof.

Replace λ with λ^σ and μ with μ^σ in equation (2.4) to deduce that

$$h(\lambda,\mu) = (1-\bar{\mu}^2\lambda^2)\left\langle u^1_{\lambda^\sigma}, u^1_{\mu^\sigma}\right\rangle + (1-\bar{\mu}^1\lambda^1)\left\langle u^2_{\lambda^\sigma}, u^2_{\mu^\sigma}\right\rangle \quad (2.6)$$

for all $\lambda, \mu \in \mathbb{D}^2$. On averaging equations (2.4) and (2.6) we obtain

$$h(\lambda,\mu) = \tfrac{1}{2}(1-\bar{\mu}^1\lambda^1)\left\langle \begin{bmatrix} u^1_\lambda \\ u^2_{\lambda^\sigma}\end{bmatrix}, \begin{bmatrix} u^1_\mu \\ u^2_{\mu^\sigma}\end{bmatrix}\right\rangle + \tfrac{1}{2}(1-\bar{\mu}^2\lambda^2)\left\langle \begin{bmatrix} u^1_{\lambda^\sigma} \\ u^2_{\lambda}\end{bmatrix}, \begin{bmatrix} u^1_{\mu^\sigma} \\ u^2_{\mu}\end{bmatrix}\right\rangle \quad (2.7)$$

For every $\lambda \in \mathbb{D}^2$ define $v_\lambda \in \mathcal{H}^1 \oplus \mathcal{H}^2$ by

$$v_\lambda = \begin{bmatrix} u^1_\lambda \\ u^2_{\lambda^\sigma}\end{bmatrix}$$

for $\lambda \in \mathbb{D}^2$. Equation (2.7) becomes

$$h(\lambda,\mu) = \tfrac{1}{2}(1-\bar{\mu}^1\lambda^1)\left\langle v_\lambda, v_\mu\right\rangle + \tfrac{1}{2}(1-\bar{\mu}^2\lambda^2)\left\langle v_{\lambda^\sigma}, v_{\mu^\sigma}\right\rangle \quad (2.8)$$

So far we have only used the 'weak symmetry' $h(\lambda^\sigma, \mu^\sigma) = h(\lambda, \mu)$. Now use the hypothesis $h(\lambda^\sigma, \mu) = h(\lambda, \mu)$. On substituting into equation (2.8) we deduce that

$$(1-\bar{\mu}^1\lambda^1)\left\langle v_\lambda, v_\mu\right\rangle + (1-\bar{\mu}^2\lambda^2)\left\langle v_{\lambda^\sigma}, v_{\mu^\sigma}\right\rangle = (1-\bar{\mu}^1\lambda^2)\left\langle v_{\lambda^\sigma}, v_\mu\right\rangle + (1-\bar{\mu}^2\lambda^1)\left\langle v_\lambda, v_{\mu^\sigma}\right\rangle$$

Rearrange the terms in this formula to obtain

$$\langle v_\lambda, v_\mu\rangle + \langle v_{\lambda^\sigma}, v_{\mu^\sigma}\rangle - \langle v_{\lambda^\sigma}, v_\mu\rangle - \langle v_\lambda, v_{\mu^\sigma}\rangle$$
$$= \bar{\mu}^1\lambda^1 \langle v_\lambda, v_\mu\rangle + \bar{\mu}^2\lambda^2 \langle v_{\lambda^\sigma}, v_{\mu^\sigma}\rangle - \bar{\mu}^1\lambda^2 \langle v_{\lambda^\sigma}, v_\mu\rangle - \bar{\mu}^2\lambda^1 \langle v_\lambda, v_{\mu^\sigma}\rangle.$$

Both sides of this equation factor, to yield

$$\langle v_\lambda - v_{\lambda^\sigma}, v_\mu - v_{\mu^\sigma}\rangle = \langle \lambda^1 v_\lambda - \lambda^2 v_{\lambda^\sigma}, \mu^1 v_\mu - \mu^2 v_{\mu^\sigma}\rangle \tag{2.9}$$

In other words, the Gramian in $H^1 \oplus H^2$ of the family of vectors $\{v_\lambda - v_{\lambda^\sigma} : \lambda \in \mathbb{D}^2\}$ is equal to the Gramian of the family $\{\lambda^1 v_\lambda - \lambda^2 v_{\lambda^\sigma} : \lambda \in \mathbb{D}^2\}$. Hence there exists a linear isometry $L : \overline{\text{span}}\{v_\lambda - v_{\lambda^\sigma} : \lambda \in \mathbb{D}^2\} \to \overline{\text{span}}\{\lambda^1 v_\lambda - \lambda^2 v_{\lambda^\sigma} : \lambda \in \mathbb{D}^2\}$ such that

$$L(v_\lambda - v_{\lambda^\sigma}) = \lambda^1 v_\lambda - \lambda^2 v_{\lambda^\sigma} \tag{2.10}$$

for all $\lambda \in \mathbb{D}^2$. Extend L to a unitary operator U on a Hilbert space $\mathcal{M} \supseteq H^1 \oplus H^2$.

Rearrange equation (2.10) (with L replaced by U) to obtain

$$(U - \lambda^1) v_\lambda = (U - \lambda^2) v_{\lambda^\sigma}$$

or equivalently,

$$(U - \lambda^2)^{-1} v_\lambda = (U - \lambda^1)^{-1} v_{\lambda^\sigma}. \tag{2.11}$$

Therefore, if we define w_λ by the formula

$$w_\lambda = (U - \lambda^2)^{-1} v_\lambda \qquad 2.12)$$

then

$$v_\lambda = (U - \lambda^2) w_\lambda \quad \text{and} \quad v_{\lambda^\sigma} = (U - \lambda^1) w_\lambda \tag{2.13}$$

If we substitute these formulae into equation (2.8) we obtain

$$\begin{aligned} h(\lambda,\mu) &= \tfrac{1}{2}(1-\bar{\mu}^1\lambda^1)\left\langle (U-\lambda^2)w_\lambda,(U-\mu^2)w_\mu\right\rangle + \tfrac{1}{2}(1-\bar{\mu}^2\lambda^2)\left\langle (U-\lambda^1)w_\lambda,(U-\mu^1)w_\mu\right\rangle \\ &= \tfrac{1}{2}(1-\bar{\mu}^1\lambda^1)\left\langle (U-\mu^2)^*(U-\lambda^2)w_\lambda,w_\mu\right\rangle + \tfrac{1}{2}(1-\bar{\mu}^2\lambda^2)\left\langle (U-\mu^1)^*(U-\lambda^1)w_\lambda,w_\mu\right\rangle \\ &= \left\langle Zw_\lambda,w_\mu\right\rangle \end{aligned} \tag{2.14}$$

where

$$Z = \tfrac{1}{2}\Big[(1-\bar{\mu}^1\lambda^1)(U-\mu^2)^*(U-\lambda^2) + (1-\bar{\mu}^2\lambda^2)(U-\mu^1)^*(U-\lambda^1)\Big] \tag{2.15}$$

Gathering terms in equation (2.15) we find that

$$Z = (1-\bar{\mu}^1\bar{\mu}^2\lambda^1\lambda^2) - \tfrac{1}{2}\left(\overline{\mu^1+\mu^2} - (\lambda^1+\lambda^2)\overline{\mu^1\mu^2}\right)U - \tfrac{1}{2}\left(\lambda^1+\lambda^2-\overline{\mu^1+\mu^2}\lambda^1\lambda^2\right)U^*$$

which, in the symmetric variables

$$s^1 = \lambda^1+\lambda^2, \qquad s^2 = \lambda^1\lambda^2 \tag{2.16}$$

and

$$t^1 = \mu^1+\mu^2, \qquad t^2 = \mu^1\mu^2 \tag{2.17}$$

Becomes

$$Z = 1-\bar{t}^2 s^2 - \tfrac{1}{2}(\bar{t}^1 - s^1\bar{t}^2)\,U - \tfrac{1}{2}(s^1-\bar{t}^1 s^2)\,U^* \tag{2.18}$$

Hence

$$h(\lambda,\mu) = \left\langle (1-\bar{t}^2 s^2 - \tfrac{1}{2}(\bar{t}^1 - s^1\bar{t}^2)\,U - \tfrac{1}{2}(s^1-\bar{t}^1 s^2)\,U^*)w_\lambda, w_\mu\right\rangle \tag{2.19}$$

for all $\lambda,\mu\in D^2$.

From the definition (2.12) of w_λ it is clear that $w: D^2 \to M$ is analytic, and from equation (2.11) we have

$$w_{\lambda^\sigma} = (U-\lambda^1)^{-1}v_{\lambda^\sigma} = (U-\lambda^2)^{-1}v_\lambda = w_\lambda.$$

Thus w, being symmetric, factors through G : there exists an analytic function $x: G \to \mathcal{M}$ such that, for all $\lambda \in \mathbb{D}^2$,

$$w_\lambda = x(\lambda^1 + \lambda^2, \lambda^1\lambda^2) = x(s^1, s^2).$$

On combining this equation with equation (2.19) we obtain the desired model formula (2.5) for h□ .

We resume the proof of (1)⇒(3) in Theorem 2.2. Let $\varphi \in \mathcal{S}(G)$. The function

$$\tilde{\varphi}(\lambda) = \varphi(\lambda^1 + \lambda^2, \lambda^1\lambda^2)$$

belongs to $\mathcal{S}(\mathbb{D}^2)$, and therefore, by [1, Theorem 1.12], has a model $(\mathcal{H}, v)$ on $\mathbb{D}^2$, which is to say that

$$1 - \overline{\tilde{\varphi}(\mu)}\tilde{\varphi}(\lambda) = (1 - \overline{\mu^1}\lambda^1)\left\langle v^1(\lambda), v^1(\mu)\right\rangle_{\mathcal{H}^1} + (1 - \overline{\mu^2}\lambda^2)\left\langle v^2(\lambda), v^2(\mu)\right\rangle_{\mathcal{H}^2}$$

for all $\lambda, \mu \in \mathbb{D}^2$. The left hand side of this equation is clearly a doubly symmetric function of (λ, μ), and so, by Lemma 2.3, there exist a Hilbert space $\mathcal{M}$, a unitary operator U on $\mathcal{M}$ and an analytic function $x: G \to \mathcal{M}$ satisfying (in terms of the variables s, t defined in equations (2.16) and (2.17))

$$1 - \overline{\varphi(t)}\varphi(s) = \left\langle \left[1 - \bar{t^2}s^2 - \tfrac{1}{2}(\bar{t^1} - s^1\bar{t^2})U - \tfrac{1}{2}(s^1 - \bar{t^1}s^2)U^*\right] x(s), x(t)\right\rangle_{\mathcal{M}} \tag{2.20}$$

By inspection,

$$1 - \bar{t^2}s^2 - \tfrac{1}{2}(\bar{t^1} - s^1\bar{t^2})U - \tfrac{1}{2}(s^1 - \bar{t^1}s^2)U^* = (1 - \tfrac{1}{2}t^1U)^*(1 - \tfrac{1}{2}s^1U) - (t^2U - \tfrac{1}{2}t^1)^*(s^2U - \tfrac{1}{2}s^1)$$

In the notation s_U introduced in Definition 1.3,

$$s_U = (s^2U - \tfrac{1}{2}s^1)(1 - \tfrac{1}{2}s^1U)^{-1}$$

and we have

$$1 - \bar{t^2}s^2 - \tfrac{1}{2}(\bar{t^1} - s^1\bar{t^2})U - \tfrac{1}{2}(s^1 - \bar{t^1}s^2)U^* = (1 - \tfrac{1}{2}t^1U)^*(1 - t_U^*s_U)(1 - \tfrac{1}{2}s^1U)$$

For $s \in G$ let

$$u(s) = (1 - \tfrac{1}{2}s^1 U)x(s)$$

Then $u : G \to \mathrm{M}$ is analytic, and equation (2.20) can be written

$$\begin{aligned} 1 - \overline{\varphi(t)}\varphi(s) &= \left\langle (1 - \tfrac{1}{2}t^1 U)^* (1 - t_U^* s_U)(1 - \tfrac{1}{2}s^1 U)x(s), x(t) \right\rangle \\ &= \left\langle (1 - t_U^* s_U)u(s), u(t) \right\rangle . \end{aligned}$$

Thus (M, U, u) is a G-model for φ. Therefore (1)⇒(3).

There is an analogue of Theorem 2.2 for operator-valued functions. It is proved by making only notational changes in the above proof. If H, K are Hilbert spaces, L (H, K) is the Banach space of bounded linear operators from H to K in the operator norm, then we define the corresponding Schur class S $(\mathrm{G}; \mathrm{H}, \mathrm{K})$ to be the set of analytic maps $\varphi : G \to \mathrm{L}(\mathrm{H}, \mathrm{K})$ such that $\varphi(\lambda)$ is a contraction for all $\lambda \in G$. The notion of G-model is extended as follows.

Definition 2.4. A *G-model* for an operator-valued function $\varphi : G \to \mathrm{L}(\mathrm{H}, \mathrm{K})$ is a triple (M, T, u) where M is a Hilbert space, T is a contraction acting on M and $u : G \to \mathrm{L}(\mathrm{H}, \mathrm{M})$ is an analytic function such that, for all $s, t \in G$,

$$1 - \varphi(t)^* \varphi(s) = u(t)^* (1 - t_T^* s_T) u(s) \qquad (2.21)$$

The generalization of Theorem 2.2 is then:

Theorem 2.5. *Let φ be a function from G to (H,K). The following three statements are equivalent.*

(1) $\varphi \in \mathrm{S}(\mathrm{G}; \mathrm{H}, \mathrm{K})$;

(2) *φ has a G-model;*

(3) *φ has a G-model (M, T, u) in which T is a unitary operator on* M.

Details of the proof of this theorem can be found in [3, Lemmas 3.3 and 3.4], where the result was used to derive a certain integral representation formula [3, Theorem 3.5]and thereafter to show that if G^- is a spectral set for a commuting pair of operators then G^- is a *complete* spectral set. The fact that the Schur and Schur–Agler classes of G coincide then follows by standard manoeuvres based on the Arveson Extension and Stinespring Representation Theorems. In this paper we use Theorem 2.2 and its analogue for operator-valued functions to take a more direct route to realization formulae for S (G) and S $(G;\mathrm{H},\mathrm{K})$ (Theorem 3.1 ; Theorem 3.2 in the next section).

3. THE REALIZATION FORMULA

There is a standard way to deduce a realization formula from a model formula with the aid of a 'lurking isometry' argument. We shall apply such an argument to derive the following slight strengthening of Theorem 1.1 in the introduction.

Theorem 3.1. *Let $\varphi \in$ S (G). There exist a scalar A, a Hilbert space* M, *vectors $\beta,\gamma \in$ M and operators D, U on* M *such that U is unitary, the operator*

$$\begin{bmatrix} A & 1\otimes\beta \\ \gamma\otimes 1 & D \end{bmatrix} \text{ is unitary on } \mathbb{C}\oplus\mathcal{M} \tag{3.1}$$

and, for all $s \in G$,

$$\varphi(s) = A + \langle s_U(1 - Ds_U)^{-1}\gamma, \beta\rangle_{\mathcal{M}} \tag{3.2}$$

Conversely, if a scalar A, a Hilbert space M, vectors $\beta,\gamma \in$ M *and operators T, D on M are given such that T is a contraction and*

$$\begin{bmatrix} A & 1\otimes\beta \\ \gamma\otimes 1 & D \end{bmatrix} \text{ is a contraction on } \mathbb{C}\oplus\mathcal{M} \tag{3.3}$$

then the function φ on G defined by

$$\varphi(s) = A + \langle s_T(1 - D_{sT})^{-1}\gamma, \beta\rangle \tag{3.4}$$

belongs to S (G).

Proof. Let $\varphi \in$ S (G). By Theorem 2.2, φ has a G -model (M,U,u) where U is a unitary operator on M. By the definition of a G-model we have

$$1 - \overline{\varphi(t)}\varphi(s) = \langle (1 - t_U^* s_U)u(s), u(t)\rangle$$

for all $s,t \in G$. Rearrange to obtain

$$1 + \langle s_U u(s), t_U u(t)\rangle = \overline{\varphi(t)}\varphi(s) + \langle u(s), u(t)\rangle$$,

which is to say that the two families of vectors

$$\begin{pmatrix} 1 \\ s_U u(s) \end{pmatrix}_{s\in G} \quad \text{and} \quad \begin{pmatrix} \varphi(s) \\ u(s) \end{pmatrix}_{s\in G}$$

in $\mathbb{C} \oplus M$ have the same Gramians. Hence there exists an isometry

$$L : \overline{\text{span}}\left\{\begin{pmatrix} 1 \\ s_U u(s) \end{pmatrix}_{s\in G}\right\} \to \overline{\text{span}}\left\{\begin{pmatrix} \varphi(s) \\ u(s) \end{pmatrix}_{s\in G}\right\} \tag{3.5}$$

such that

$$L\begin{pmatrix} 1 \\ s_U u(s) \end{pmatrix} = \begin{pmatrix} \varphi(s) \\ u(s) \end{pmatrix}$$

for every $s \in G$. If necessary enlarge the Hilbert space M (and simultaneously the unitary operator U on M) so that the isometry L extends to a unitary operator

$$L^\sharp \sim \begin{bmatrix} A & 1\otimes\beta \\ \gamma\otimes 1 & D \end{bmatrix} \text{ on } \mathbb{C}\oplus\mathcal{M}$$

for some vectors $\beta, \gamma \in M$. By equation (3.5), for any $s \in G$,

$$\begin{aligned} A + \langle s_U u(s), \beta\rangle &= \varphi(s), \\ \gamma + D s_U u(s) &= u(s). \end{aligned} \tag{3.6}$$

Now $s_U = f_s(U)$ where

$$f_s(\lambda) = \frac{2\lambda s^2 - s^1}{2 - \lambda s^1}$$

for λ in a neighborhood of $\mathbb{D}^-$. The linear fractional map f_s maps $\mathbb{D}$ onto the open disc with centre and radius

$$2\frac{\overline{s^1}s^2 - s^1}{4 - |s^1|^2} \quad \text{and} \quad \frac{|(s^1)^2 - 4s^2|}{4 - |s^1|^2}.$$

Therefore, by von Neumann's inequality,

$$\|s_U\| \leq \sup_{\mathbb{D}} |f_s| = \frac{2|s^1 - \overline{s^1}s^2| + |(s^1)^2 - 4s^2|}{4 - |s^1|^2}$$

But, by [6, Theorem 2.1], the right hand side of this equation is less than one for $s \in G$. Hence $1 - Ds_U$ is invertible for any $s \in G$, and we may eliminate $u(s)$ from equations (3.6) to obtain the realization formula (3.2) for $\varphi(s)$.

The converse statement is easy since equation (3.4) expresses $\varphi(s)$ as a linear fractional transform of the contraction s_T with a contractive coefficient matrix. □

Again, there is an analogue for operator-valued functions. The proof above requires only minimal changes.

Theorem 3.2. *Let* H, K *be Hilbert spaces.*

If $\varphi \in S(G; H, K)$ *then there exist a Hilbert space* M, *a unitary operator* U *on* M *and a unitary operator*

$$\begin{bmatrix} A & B \\ C & D \end{bmatrix} : \mathcal{H} \oplus \mathcal{M} \to \mathcal{K} \oplus \mathcal{M} \tag{3.7}$$

such that, for all $s \in G$,

$$\varphi(s) = A + Bs_U(1 - Ds_U)^{-1}C \tag{3.8}$$

Conversely, if a Hilbert space M, *a contraction* T *on* M *and a contraction*

$$\begin{bmatrix} A & B \\ C & D \end{bmatrix} : \mathcal{H} \oplus \mathcal{M} \to \mathcal{K} \oplus \mathcal{M} \tag{3.9}$$

are given, then the function $\varphi : G \to L(H, K)$ *defined by*

$$\varphi(s) = A + Bs_T(1 - Ds_T)^{-1}C \tag{3.10}$$

belongs to $S(G; H, K)$.

4. A SECOND MODEL FORMULA FOR G AND SPECTRAL DOMAINS

The model formula in Section 2 has an alternative expression as an integral formula.

We shall need the rational functions

$$\Phi_\omega(s) = \frac{2\omega s^2 - s^1}{2 - \omega s^1}, \qquad s \in G$$

for $\omega \in \mathbb{T}$ (in the notation of the proof of Theorem 2.2, $\Phi_\omega(s) = f_s(\omega)$). These functions have been used in many papers on G. By [6, Theorem 2.1], each Φ_ω maps G into $\mathbb{D}$.

Now invoke the spectral theorem to rewrite the model formula (2.2). Consider a function $\varphi \in \mathcal{S}(G)$. By Theorem 2.2, φ has a G-model $(\mathcal{M}, T, u)$ in which T is a unitary operator on $\mathcal{M}$. By the spectral theorem,

$$T = \int_{\mathbb{T}} \omega \, \mathrm{d}E(\omega),$$

for some $\mathcal{L}(\mathcal{M}, \mathcal{M})$-valued spectral measure E on $\mathbb{T}$. Thus, for $s \in G$,

$$\begin{aligned} s_T &= (2s^2 T - s^1)(2 - s^1 T)^{-1} \\ &= \int_{\mathbb{T}} \Phi_\omega(s) \, \mathrm{d}E(\omega), \end{aligned}$$

and therefore

$$1 - t_T^* s_T = \int_{\mathbb{T}} 1 - \overline{\Phi_\omega(t)} \Phi_\omega(s) \, \mathrm{d}E(\omega).$$

On combining this formula with Theorem 2.2 we obtain the following statement.

Theorem 4.1. *Let $\varphi : G \to \mathbb{C}$ be a function. Then $\varphi \in \mathcal{S}(G)$ if and only if there exist a Hilbert space $\mathcal{M}$, an $\mathcal{L}(\mathcal{M}, \mathcal{M})$-valued spectral measure E on $\mathbb{T}$ and an analytic map $u : G \to \mathcal{M}$ such that, for all $s, t \in G$,*

$$1-\overline{\varphi(t)}\varphi(s)=\int_{\mathbb{T}}\left(1-\overline{\Phi_\omega(t)}\Phi_\omega(s)\right)\langle \mathrm{d}E(\omega)u(s),u(t)\rangle$$

One advantage of the integral form of the model formula is that it instantly yields a criterion for G to be a spectral domain of a commuting pair of operators. We recall the meaning of this notion.

Definition 4.2. If T is a d-tuple of pairwise commuting operators and U is an open set in $\mathbb{C}^d$ we say that U is a *spectral domain for T* if $\sigma(T)\subset U$ and

$$\varphi\in\mathcal{S}(U)\implies\|\varphi(T)\|\le 1$$

The following statement is contained in [3, Theorem 1.2].

Theorem 4.3.

Let $S=(S^1,S^2)$ be a commuting pair of operators acting on a Hilbert space with $\sigma(S)\subset G$. Then G is a spectral domain for S if and only if

$$\|\Phi_\omega(S)\|\le 1\qquad \textit{for all } \omega\in\mathbb{T}.$$

Proof. Since $\Phi_\omega\in\mathcal{S}(G)$, the condition is obviously necessary.

Conversely, assume that $\|\Phi_\omega(S)\|\le 1$ for all $\omega\in T$. We need to show that G is a spectral domain for S, i.e., that

$$\|\varphi(S)\|\le 1$$

whenever $\varphi\in\mathcal{S}(G)$.

But if $\varphi\in\mathcal{S}(G)$, it follows from Theorem 4.1 that $1-\overline{\varphi(t)}\varphi(s)$ can be uniformly approximated by convex combinations of functions of the form

$$\overline{f(t)}\left(1-\overline{\Phi_\omega(t)}\Phi_\omega(s)\right)f(s)$$

where $\omega\in\mathbb{T}$ and f is holomorphic on G. It follows that $1-\varphi(S)^*\varphi(S)$ can be approximated in the operator norm by operators of the form

$$f(S)^*\left(1-\Phi_\omega(S)^*\Phi_\omega(S)\right)f(S)$$

Since these operators are positive, it follows that $1-\varphi(S^*\varphi(S)$ is positive, that is, $\|\varphi(S)\|\leq 1$. □

5. A PICK THEOREM FOR G

A standard application of realization formulae is to prove Pick-type theorems, which provide necessary and sufficient conditions for the solvability of interpolation problems. For example, the realization formula for the Schur class of the bidisc in [1] yields the following criterion for analytic interpolation from D^2 to D^-.

Let $\lambda_1,\dots,\lambda_n$ be distinct points in D^2 and let $w_1,\dots,w_n$ belong to D^-. There exists a function φ in $S\ (D^2)$ such that $\varphi(\lambda_j)=w_j$ for $\mathbf{j}=1,\dots,n$ if and only if there exist positive semidefinite $n\times n$ matrices $a^1=[a^1_{ij}]^n_{i,j=1}$ and $a^2=[a^2_{ij}]^n_{i,j=1}$ such that

$$1-\overline{w_i}w_j=a^1_{ij}(1-\overline{\lambda^1_i}\lambda^1_j)+a^2_{ij}(1-\overline{\lambda^2_i}\lambda^2_j) \qquad (5.1)$$

for $i,j=1,\dots,n$.

This result reduces the interpolation problem to the feasibility of a linear matrix equality for $n\times n$ matrices, a task which can be efficiently solved by standard engineering packages such as Matlab.

Consider the analogous problem in which the bidisc is replaced by the symmetrized bidisc. Given distinct points $s_1,\dots,s_n$ in G and target points $w_1,\dots,w_n$ in D^-, we wish to determine whether there exists an analytic function $\varphi:G\to D^-$ such that $\varphi(s_j)=w_j$ for $j=1,\dots,n$. One way to solve such an interpolation problem is to lift it to the bidisc. Let $\mu_1,\dots,\mu_m$ be the preimages in D^2 of the points $s_1,\dots,s_n$ under the natural map $\pi:D^2\to G$ given by

$$\pi(\mu)=(\mu^1+\mu^2,\mu^1\mu^2).$$

For any $s\in G$, the set $\pi^{-1}\{s\}$ comprises either one or two points, and therefore $n\leq m\leq 2n$. It is easily seen that our interpolation problem $s_j\mapsto w_j$ for G is equivalent to the lifted problem $\mu_j\mapsto w_{j'}$ on D^2, where j' is chosen so that $1\leq j'\leq n$ and $\pi(\mu_j)=s_{j'}$. The Pick criterion (5.1) applies to the lifted problem; since this criterion is necessarily symmetric with respect to the transposition map σ , it can be rewritten in terms of the symmetrized variables s_j (and w_j).

However, the model in Theorem 2.2 permits us to obtain directly a criterion for interpolation from G to D^- in terms of the symmetrized variables.

Theorem 5.1. *Let $s_1,\ldots,s_n$ be distinct points in G and let $w_1,\ldots,w_n \in D^-$. There exists an analytic function $\varphi: G \to D^-$ such that $\varphi(s_j)=w_j$ for $j=1,\ldots,n$ if and only if there exist a Hilbert space M, a contraction T on M and vectors $v_1,\ldots,v_n \in M$ such that*

$$1 - \overline{w_i}w_j = \langle (1-(s_i)_T^*(s_j)_T)\, v_j, v_i\rangle_{\mathcal{M}} \tag{5.2}$$

for $i,j=1,\ldots,n$.

Proof. *Necessity* . Suppose an interpolating function $\varphi \in S\ (G)$ exists. By Theorem 2.2, φ has a G -model, that is, there exist a Hilbert space M, a contraction T on M and an analytic map $u: G \to M$ such that, for all $s,t \in G$,

$$1 - \overline{\varphi(t)}\varphi(s) = \langle (1 - t_T^* s_T) u(s), u(t)\rangle_{\mathcal{M}} \tag{5.3}$$

On choosing $s=s_j$, $t=s_i$ and $v_i=u(s_i)$ for $i=1,\ldots,n$ we deduce that equation (5.2)holds for all i,j.

Sufficiency . Suppose that $M, T, v_1,\ldots,v_n$ exist such that equation (5.2) holds for each i,j, as in the statement of the theorem. Rearrange the equation to obtain

$$1 + \langle (s_j)_T v_j, (s_i)_T v_i\rangle_{\mathcal{M}} = \overline{w_i}w_j + \langle v_j, v_i\rangle_{\mathcal{M}}$$

for all i,j. This means that the family of vectors $(1,(s_j)_T v_j),\ j=1,\ldots,n$, in $C \oplus M$ has the same Gramian as the family $(w_j, v_j),\ j=1,\ldots,n$, also in $C \oplus M$. Hence there exists an isometry

$$L: \mathrm{span}\{(1,(s_j)_T v_j) \mid j=1,\ldots,n\} \to \mathrm{span}\{(w_j,v_j) \mid j=1,\ldots,n\}$$

such that $L(1,(s_j)_T v_j)=(w_j,v_j)$ for each j. Extend L to a contraction $L^\#$ mapping $C \oplus M$ to itself. Then $L^\#$ is expressible as a block operator matrix of the form

$$L^\sharp \sim \begin{bmatrix} A & 1\otimes\beta \\ \gamma\otimes 1 & D \end{bmatrix}$$

for some $A \in C$, vectors $\beta, \gamma \in M$ and operator D on M. Since $L^{\sharp}(1, (s_j)_T v_j) = (w_j, v_j)$ for each j,

$$A + \langle (s_j)_T v_j, \beta \rangle = w_j,$$

$$\gamma + D(s_j)_T v_j = v_j.$$

Thus

$$v_j = (1 - D(s_j)_T)^{-1} \gamma$$

and

$$A + \langle (s_j)_T (1 - D(s_j)_T)^{-1} \gamma, \beta \rangle = w_j \tag{5.4}$$

for each j.

Define a function $\varphi : G \to C$ by

$$\varphi(s) = A + \langle s_T (1 - D s_T)^{-1} \gamma, \beta \rangle .$$

By Theorem 2.2, $\varphi \in S(G)$, and by equation (5.4),

$$\varphi(s_j) = w_j \quad \text{for } j = 1, \ldots, n. \square$$

Remark 5.2. One can replace 'there exists a contraction T' in the statement of Theorem 5.1 by 'there exists a unitary operator T'.

Another criterion for the solvability of a finite interpolation problem in S (G) is given in [12, Theorem 6.1]. It is shown that, in the situation of Theorem 5.1, a desired interpolating function $\varphi \in S(G)$ exists if and only if there exists a $C(D^-)^*$-valued positive semidefinite kernel on $\{s_1, \ldots, s_n\}$ such that an analogue of equation (5.2) holds.

We conclude with an observation about the question raised in the introduction: is there a continuous *operator*-valued function F on the closure G^- of G, holomorphic in G, such that

$$G = \{s \in G^- : \| F(s) \| < 1\}? \tag{5.5}$$

Since a point $s \in C^2$ belongs to G if and only if $|f_s(\lambda)| < 1$ for all $\lambda \in D$, one could try

$$F(s) = \mathrm{diag}_{n \geq 1}[f_s(\lambda_n)],$$

where (λ_n) is a dense sequence in D. It is then true that F is well defined on G^- and $G = \{s \in G^- : \| F(s) \| < 1\}$, but F is discontinuous as a map from G^- to the space of bounded linear operators on ℓ^2 with the operator norm. Indeed, for any $\omega \in \mathrm{T}$ and $0 < r < 1$,

$$\begin{aligned}\|F(2\bar{\omega}, \bar{\omega}^2) - F(2r\bar{\omega}, r\bar{\omega}^2)\| &= (1-r) \sup_n \left| \frac{\lambda_n}{1 - r\lambda_n \bar{\omega}} \right| \\ &= 1.\end{aligned}$$

Thus F is discontinuous at every point $(2\bar{\omega}, \bar{\omega}^2) \in G^-$ for $\omega \in \mathrm{T}$. Indeed F is even discontinuous at these points with respect to the weak operator topology on the space of bounded linear operators on ℓ^2. We leave open the question of whether there exists a continuous holomorphic operator-valued defining function for G.

REFERENCES

1. J. Agler **On the Representation of Certain Holomorphic Functions Defined on a Polydisc** Operator Theory: Advances and Applications, vol. OT 48, Birkhäuser Verlag, Basel (1990)
2. J. Agler, Operator Theory and Function Theory on the Symmetrized Bidisc, unpublished lecture notes for a short course at the International Centre for the Mathematical Sciences, Edinburgh, July 2014.
3. J. Agler, N.J. Young **A commutant lifting theorem for a domain in $\mathbf{C}^2$ and spectral interpolation** J. Funct. Anal., 161 (1999), pp. 452–477
4. J. Agler, N.J. Young **Operators having the symmetrized bidisc as a spectral set** Proc. Edinb. Math. Soc., 43 (2000), pp. 195–210
5. J. Agler, N.J. Young **A Schwarz lemma for the symmetrized bidisc** Bull. Lond. Math. Soc., 33 (2001), pp. 175–186
6. J. Agler, N.J. Young **The hyperbolic geometry of the symmetrised bidisc** J. Geom. Anal., 14 (2004), pp. 375–403
7. J. Agler, F.B. Yeh, N.J. Young **Realization of functions into the symmetrised bidisc** ,in: D. Alpay (Ed.), Reproducing Kernel Spaces and Applications, Operator Theory: Advances and Applications, vol. 143, , Birkhaüser Verlag (2003), pp. 1–37
8. J. Agler, Z.A. Lykova, N.J. Young **Geodesics, retracts, and the norm-preserving extension property in the symmetrized bidisc** Mem. Amer. Math. Soc. (2017) in press arXiv:1603.04030
9. J.A. Ball, V. Bolotnikov **Realization and interpolation for Schur–Agler class functions on domains with matrix polynomial defining functions in $\mathbf{C}^n$** J. Funct. Anal., 213 (2004), pp. 45–87
10. J.A. Ball, M.D. Guerra Huamán **Test functions, Schur–Agler classes and transfer-function realizations: the matrix-valued setting** Complex Anal. Oper. Theory, 7 (2003), pp. 529–575
11. J.A. Ball, G. Marx, V. Vinnikov **Interpolation and transfer-function realization for the noncommutative Schur–Agler class** arXiv:1602.00762

12. T. Bhattacharyya, H. Sau **Holomorphic functions on the symmetrized bidisk – realization, interpolation and extension** arXiv:1511.08962v5
13. M.A. Dritschel, S. McCullough **Test functions, kernels, realizations and interpolation** ,in: M. Bakonyi, A. Gheondea, M. Putinar, J. Rovnyak (Eds.), Operator Theory, Structured Matrices and Dilations: T. Constantinescu Memorial Volume, Theta Series in Advanced Mathematics, Bucharest (2007), pp. 153–179
14. M.A. Dritschel, S. Marcantognini, S. McCullough **Interpolation in semigroupoid algebras** J. Reine Angew. Math., 606 (2007), pp. 1–40
15. J.W. Helton **Discrete time systems, operator models and scattering theory** J. Funct. Anal., 16 (1974), pp. 15–38
16. J.W. Helton **Operator Theory, Analytic Functions, Matrices and Electrical Engineering** CBMS Regional Conference Series in Mathematics, vol. 68, AMS (1987)
17. M. Jarnicki, P. Pflug **Invariant Distances and Metrics in Complex Analysis** (2nd extended edition)De Gruyter, Berlin (2013)
18. S. Kobayashi **Hyperbolic Complex Spaces** Grundlehren der mathematischen Wissenschaften, vol. 318, Springer Verlag (1998)
19. L. Kosinski, N.J. Young, The symmetrized bidisc cannot be defined by a matrix polynomial describing function, in preparation.
20. B. Sz.-Nagy, C. Foias **Harmonic Analysis of Operators on Hilbert Space** Akadémiai Kiadó, Budapest (1970)

INDEX

S

U

V

W

Z